COMPUTER AIDS FOR VLSI DESIGN

COMPUTER AIDS FOR VLSI DESIGN

Steven M. Rubin
Schlumberger Palo Alto Research

ADDISON-WESLEY PUBLISHING COMPANY

Reading, Massachusetts • Menlo Park, California • Don Mills, Ontario
Wokingham, England • Amsterdam • Sydney • Singapore • Tokyo
Madrid • Bogotá • Santiago • San Juan

This book is in the **Addison-Wesley VLSI Systems Series**
Lynn Conway and Charles Seitz, **Consulting Editors**

Peter S. Gordon, **Sponsoring Editor**
Bette J. Aaronson, **Production Supervisor**
Maureen Langer, **Text Designer**
Lyn Dupré, **Copy Editor**
Marshall Henrichs, **Cover Designer**
Hugh Crawford, **Manufacturing Supervisor**

Library of Congress Cataloging-in-Publication Data

Rubin, Steven M.
  Computer Aids for VLSI Design.

  Bibliography: p.
  Includes index.
  1. Integrated circuits—Very large scale integration—
Design and construction—Data processing. 2. Computer-
aided design. I. Title.
TK7874.R83   1987   621.395′0285   86-26571
ISBN 0-201-05824-3

Cover photograph courtesy of U.S. Geological Survey (photo, Ed. Garrigues).

ABCDEFGHIJ-DO-8987

The VLSI Systems Series
Lynn Conway   *Consulting*
Charles Seitz   *Editors*

FOREWORD

The subject of VLSI systems spans a broad range of disciplines, including semiconductor devices and processing, integrated electronic circuits, digital logic, design disciplines and tools for creating complex systems, and the architecture, algorithms, and applications of complete VLSI systems. The Addison-Wesley VLSI Systems Series is organized as a set of textbooks and research references that present the best current work spanning this exciting and diverse field, with each book providing for its subject a perspective that ties it to related disciplines.

Computer Aids for VLSI Design by Steven Rubin presents a broad and coherent view of the computational tools available to the VLSI designer. This book contains insights and information that will be valuable both to chip designers and to tool builders. Modern VLSI computer aided design (CAD) systems allow the chip designer to access in a consistent and convenient way a variety of synthesis and analysis tools. Such tools have advanced considerably in the past several years, both in their scope and in their ability to handle large designs.

Part of what distinguishes the expert chip designer from the novice is an understanding of the entire suite of tools available and how they work together to support the design flow of a project. One can come to understand the capabilities and limitations of individual tools in some cases from their external interfaces and roles in the design process. In other cases it is useful to appreciate how the tools work internally. Rubin presents both perspectives for all of the important categories of synthesis and analysis tools. The exposition is readily understandable to anyone familiar with chip design and computer programming.

The starting point for this book was a design system called "Electric," which Rubin designed and programmed almost single-handedly over the past four years. Electric is a complete and elegant system that has been used for the design and verification of many chips. The book is *not* a "manual" for using Electric. Rather, Electric is used as a vehicle for exposing design choices and internal interfaces that are common to many of the advanced design systems currently available. Thus this book allows one to learn in a realistic setting something of the art practiced by VLSI design system developers.

Lynn Conway
Ann Arbor, Michigan

Chuck Seitz
Pasadena, California

PREFACE

This book describes how computers can be programmed to help in the design of very-large-scale integrated (VLSI) circuits. Such circuits are becoming increasingly common due to their ease of manufacture, low cost, and simplified design methodologies. No longer must the designer study electronics and physics to build an integrated circuit. Digital electronic design is taught widely and is accessible to people with any scientific background.

As the complexity of these electronic circuits increases, the need to use computers for their design becomes more important. Although computer-aided design (CAD) systems have existed for quite some time, many of them are inadequate for current tasks, and a continuous flow of new tools is being developed. These tools perform more and more of the detailed and repetitive work involved in VLSI system design, thus reducing the time it takes to produce a chip.

The need for better VLSI design systems has fostered a need to study these systems more carefully. The book, *Introduction to VLSI Systems* by Mead and Conway, opened up this field of study to a wider audience than ever before. A follow-on book, *Principles of CMOS VLSI Design* by Weste and Eshraghian, combined with the first text to present a broad foundation for the construction of VLSI circuits. This book continues where the others have left off: It explores the computer-aided design systems that enable such construction.

Although students of this textbook need not have done so, they are advised to have studied one of the mentioned VLSI books and to have done some chip design. This will provide the proper background to distinguish VLSI design from more traditional circuit layout. Students should also be familiar with basic computer-programming issues. The reader of this text is therefore presumed to be a VLSI designer who would like to develop better design tools. In addition to descriptions of techniques, much reference information is provided in this book.

Another potential reader of this book is the VLSI designer who, although not interested in programming new tools, would like to develop a finer understanding of the tools that are already in use. This book illustrates the operation and interrelation of the parts of VLSI design systems. Designers who understand these concepts will be better able to work with their systems and to help specify future CAD directions.

Also, a person who has never designed chips can still benefit from this book. The subject matter is large system design, which includes many considerations that are independent of electronics. However, many examples do refer to transistors or logic gates as they appear in a circuit. If the reader is not familiar with these notions, then he or she is likely to find the discussions confusing. The reader needs to understand these concepts only abstractly, however; the book does not rely on circuit analysis or electrical-engineering concepts.

This book is geared for advanced undergraduate or graduate students. In a course taught to designers and tool builders at Schlumberger, the book was covered with one lecture per chapter. Although some chapters could easily stretch into two lectures, the pace was not too fast. The course provided a good exposition of CAD for VLSI systems. Questions at the end of each chapter range from simple exercises to unsolved problems that can be used as discussion points.

Previous texts on CAD systems have focused on specific aspects of the design problem, such as synthesis algorithms or hardware-description languages. This book takes a much broader view. It starts with basics, such as design environments and machine representations; covers the fundamental subjects of synthesis and analysis tools; and also discusses important peripheral notions such as output formats, programmability, graphics, and human engineering.

The last chapter of the book ties together the lessons by describing a working VLSI design system called "Electric." This is an instructive example because it embodies many of the features described in the text and because the source code is available to universities for experimentation. Interested schools can contact the author of this book to obtain the system. A more extensive version of Electric, called "Bravo3 VLSI," is sold commercially by the Applicon, a Schlumberger company.

Like any reference book, this one is certain to be inaccurate. Readers are cautioned against blindly accepting details, and should test all techniques thoroughly. The only way to ensure that an algorithm is correct is to test it in a system.

Acknowledgments

Thanks are due to a number of people who helped me with this book. The series editors, Lynn Conway and Chuck Seitz, provided good feedback on early drafts of the manuscript, as did the reviewers Bryan Ackland, AT&T Bell Laboratories; John P. Hayes, University of Michigan; Richard P. Lyon, Schlumberger Palo Alto Research; Neil Weste, Symbolics, Inc.; and Telle Whitney, Schlumberger Palo Alto Research. Thanks are also due to the two guest authors: Bob Hon and Sundar Iyengar. A broad note of appreciation goes to the employees of Schlumberger who attended the initial course and waded through the first draft.

On a personal level, I am indebted to my wife, Amy Lansky, and my family who have been supportive during this ordeal. Amy provided perspective by reminding me how similar the process of writing a book is to being in graduate school and struggling with a dissertation. I would also like to thank the punk-rock band, S.D.

Palo Alto, California   S.M.R.

CONTENTS

4
SYNTHESIS TOOLS

5

STATIC ANALYSIS TOOLS

6

DYNAMIC ANALYSIS TOOLS (BY ROBERT W. HON)

7

THE OUTPUT OF DESIGN AIDS

8

PROGRAMMABILITY

9

GRAPHICS   255

10

HUMAN ENGINEERING   283

11

ELECTRIC   305

CHARACTERISTICS OF DIGITAL ELECTRONIC DESIGN

1.1   Design

Design is the most significant human endeavor: It is the channel through which creativity is realized. Design determines our every activity as well as the results of those activities; thus it includes planning, problem solving, and producing. Typically, the term "design" is applied to the planning and production of artifacts such as jewelry, houses, cars, and cities. Design is also found in problem-solving tasks such as mathematical proofs and games. Finally, design is found in pure planning activities such as making a law or throwing a party.

More specific to the matter at hand is the design of manufacturable artifacts. This activity uses all facets of design because, in addition to the specification of a producible object, it requires the planning of that object's manufacture, and much problem solving along the way. Design of objects usually begins with a rough sketch that is refined by adding precise dimensions. The final plan must not only specify exact sizes, but also include a scheme for ordering the steps of production. Additional considerations depend on the production environment; for example, whether one or ten million will be made, and how precisely the manufacturing environment can be controlled.

This book is particularly concerned with the design of highly complex electronic circuits, referred to as **VLSI** (very-large-scale integrated) circuits. When doing design of VLSI systems, the same steps must be taken (see Fig. 1.1). An initial sketch shows the "black box" characteristics of a circuit: which wires will bring what information and which wires will get results. This is then refined into an architectural design that shows the major functional units of the circuit and how they interact. Each unit is then designed at a more detailed but still abstract level, typically using logic gates that perform inversion, conjunction, and disjunction. The final refinement converts this schematic specification into an integrated-circuit layout in whatever semiconductor technology will be used to build the chip. It is also possible to produce working circuits without the need for custom chip fabrication, simply by assembling standard chips to perform the specified function. Although design rarely proceeds strictly from the abstract to the specific, the refinement notion is useful in describing any design process. In actuality, design proceeds at many levels, moving back and forth to refine details and concepts, and their interfaces.

A **semiconductor process technology** is a method by which working circuits can be manufactured from designed specifications. There are many such technologies, each of which creates a different **environment** or style of design. In **integrated-circuit (IC)** design, which is the primary

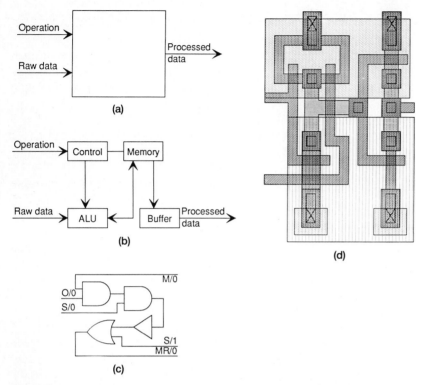

FIGURE 1.1 Refinements of electronic design: (a) Black-box (b) Architecture (c) Logic (d) Integrated-circuit layout.

focus of this book, the specification consists of polygons of conducting and semiconducting material that will be layered on top of each other to produce a working chip (see Fig. 1.2). **Printed-circuit (PC)** design also results in precise positions of conducting materials as they will appear on a circuit board; in addition, PC design aggregates the bulk of the electronic activity into standard IC packages the position and interconnection of which are essential to the final circuit (see Fig. 1.3). Printed circuitry may be easier to debug than integrated circuitry is, but it is slower, less compact, more expensive, and unable to take advantage of specialized silicon layout structures that make VLSI systems so attractive [Mead and Conway]. **Wire-wrap** boards are like printed-circuit boards in that they use packages that must be precisely positioned. However, they allow the wire locations to fall anywhere as long as they connect to the posts of the IC packages. Such boards are typically manufactured as prototypes for less expensive PC boards.

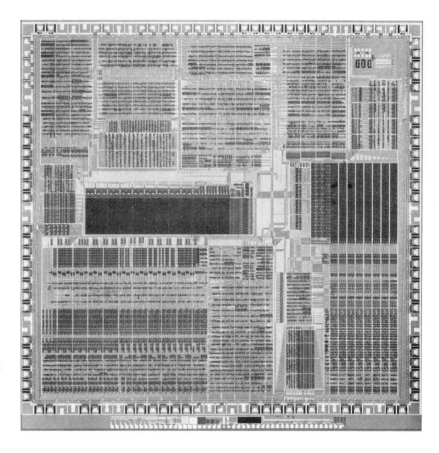

FIGURE 1.2 Photomicrograph of an integrated circuit (IC). The Fairchild Clipper CPU, courtesy Fairchild Semiconductor Corporation. Photo: Ed Garrigues.

The design of these electronic circuits can be achieved at many different refinement levels from the most detailed layout to the most abstract architectures. Given the complexity that is demanded at all levels, computers are increasingly used to aid this design at each step. It is no longer reasonable to use manual design techniques, in which each layer is hand etched or composed by laying tape on film. Thus the term **computer-aided design** or **CAD** is a most accurate description of this modern way and seems more broad in its scope than the recently popular term **computer-aided engineering (CAE)**. Although CAE implies a greater emphasis on circuit analysis, the term has instead been used to describe the recent spate of design systems that are attractively packaged on workstations. These systems all seek to aid in the design of circuitry.

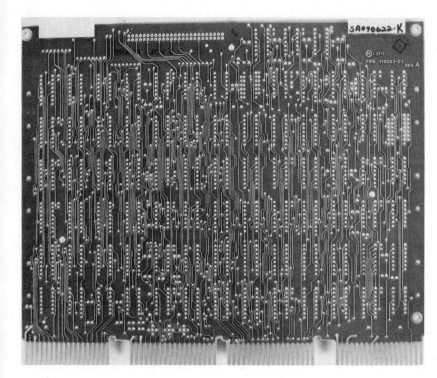

FIGURE 1.3 A printed-circuit (PC) board. The AED DMA graphics interface, courtesy of Advanced Electronics Design, Incorporated. Photo: Ed Garrigues.

There are many steps that a computer can take in the design of electronic circuits. Graphical drawing programs are used to sketch the circuits on a display screen, whereas hardware-description languages achieve the same result textually, employing the expressive power of programming languages (see Fig. 1.4). Special **synthesis** programs convert among the different refinement levels of a design, and **analysis** programs help to check circuit correctness. A final step converts the design to a manufacturing specification so that the circuit can be fabricated. Thus computer programming is used throughout the circuit design process both as an aid to, and as a metaphor of, the design activity. In fact, the parallels between programming and VLSI design are very compelling and will be seen frequently in this book.

1.1.1 The VLSI System Design Process

A fundamental assumption about VLSI circuits is that they are designed by humans and built by machines. Thus all CAD systems act as translators between the two. On one end of a CAD system is the

```
SYM mem
{
   DEF load %y load1, load2;
   TRAN (3,2) %270 t3, %270 t4, %270 t5, %270 t6;
   TRAN (12,2) t1, t2;
   DEF metaldiffusion md1, md2, md3, md4;
   DEF polydiffusion pd1, pd2, pd3;
   DEF metalpoly mp1, mp2, mp3, mp4;
   IDEF DIFF d1, d2, d3, d4, d5;
   IDEF METAL vi, vo S=vdd, gi, go S=gnd;

   # run power and ground
   W=4 vo RIGHT TO md1.m RIGHT 10 TO md2.m RIGHT TO vi;
   W=4 go RIGHT TO md3.m RIGHT TO md4.m RIGHT TO gi;

   #run pull-up 1
   md1.d DOWN 1 TO load1.v;
   load1.pp DOWN 8 TO pd1.p DOWN 7 LEFT 5 DOWN 3 LEFT 6 DOWN 9 RIGHT 0 TO t1.gw;
   pd1.d RIGHT 4 DOWN 8 RIGHT 4 DOWN 5 RIGHT 5 TO d4;
   d4 BELOW BY 1 t2.dn;
   t1.ds DOWN 2 TO md3.d;

   # run pull-up 2
   md2.d DOWN 1 TO load2.v;
   load2.pp DOWN 8 TO pd2.p DOWN 4 RIGHT 2 DOWN 6 RIGHT 9 DOWN 5 LEFT 0 TO t2.ge;
   t2.gw LEFT 0 UP 1 LEFT 4 TO pd3.p;
   pd3.d DOWN 2 LEFT 4 TO t1.dn;
   t2.ds DOWN 6 TO md4.d;

   # connect data lines
   t3.ds LEFT 4 TO d1 DOWN 9 TO d5 RIGHT 1 TO t5.ds;
   t4.dn RIGHT 4 TO d2 DOWN 9 TO d3 LEFT 1 TO t6.dn;
   t3.dn DOWN 0.5 RIGHT 4 TO load1.d;
   t5.dn UP 0.5 RIGHT 7 TO pd1.d;
   t4.ds DOWN 0.5 LEFT 4 TO load2.d;
   t6.ds UP 0.5 LEFT 7 TO pd2.d;

   # connect the select lines
   mp1.m RIGHT TO mp2.m;
   mp3.m RIGHT TO mp4.m;
   mp1.p DOWN 1 TO t3.gw;
   mp2.p DOWN 1 TO t4.gw;
   mp3.p UP 1 TO t5.ge;
   mp4.p UP 1 TO t6.ge;
}
```

FIGURE 1.4   Hardware-description language for an nMOS memory cell. Initial code defines symbols (loads, transistors, contacts, points). Bulk of the code runs wires between components, implicitly placing everything.

human interface that must be intelligent enough to communicate in a manner that is intuitive to the designer. On the other end is a generator of specifications that can be used to manufacture a circuit. In between are the many programming and design tools that are necessary in the production of a VLSI system.

The front end of a CAD system is the human interface and there are two basic ways that it can operate: graphically or textually. Graphic design allows the display of a circuit to be manipulated interactively, usually with a pointing device. Textual design allows a textual description, written in a **hardware-description language**, to be manipulated with a keyboard and a text editor. For example, suppose a designer wants to specify the layout of a transistor that is coupled to a terminal. This can be done graphically by first pointing on the display to the desired location for the transistor and issuing a "create" command. A similar operation will create the terminal. Finally, the connecting wire can be placed by tracing its intended path on the display. To do this same operation textually, the following might be typed:

```
transistor at (53,100).
  terminal below transistor by 30, left by 3 or more.
  wire from transistor left then down to terminal.
```

Notice that the textual description need not be completely specific ("left by 3 or more"). This is one of the advantages of textual descriptions: the ability to underspecify and let the computer fill in the detail. In this example, other parts of the circuit and other spacing rules will help to determine the exact location of these components. Additional advantages of text are the ease of verbal documentation, ease of parameterization, ease of moving the CAD system between computers, and a somewhat lower cost of a design workstation because of the reduced need for graphics display.

The disadvantage of text, however, is immediately clear: It is not as good a representation of the final circuit, because it does not visually capture the spatial organization. Text is one-dimensional and graphics is two-dimensional. Also, graphics provides faster and clearer feedback during design, so it is easier to learn and to use, which results in more productivity. Although graphics cannot handle verbal documentation as well, it does provide instant visual documentation, which can be more valuable. Even underspecified spacing can be achieved graphically by creating an abstract design that is subsequently fleshed out.

A number of design styles exist to bridge the gap between text and graphics. These attempt to be less demanding than are precise polygon drawing systems while still capturing the graphical flavor. In

sticks design [Williams], the circuit is drawn on a display, but the components have no true dimensions and their spacing is similarly inaccurate. **Virtual grid** design [Weste] also abstracts the graphics of a circuit, but it uses quasi-real component sizes to give something of the feel for the final layout. Closer to text is the **SLIC** design style [Gibson and Nance], which uses special characters in a text file to specify layout. For example, an "X" indicates a transistor and a " | " is used for metal wires, so the adjacency of these two characters indicates a connection.

At the back end of a design system is a facility for writing manufacturing specifications. Complex circuits cannot be built by hand, so these specifications are generally used as input to other programs in machines that control the fabrication process. There are many manufacturing devices (photoplotters, wafer etchers, and so on) and each has its own format. Although standardization is constantly proposed, there continue to be many output formats that a CAD system must provide. Figure 1.5 shows the structure of a simple CAD system with its front end and back end in place

Between the front-end user interface and the back-end manufacturing specification are the analysis and synthesis tools that help reduce the tedium of creating correct layout. **Analysis tools** are able to detect

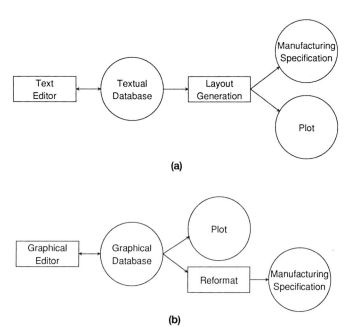

FIGURE 1.5 Simple CAD system structures: (a) Textual (b) Graphical.

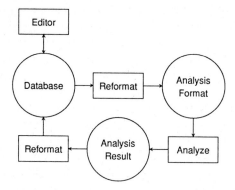

FIGURE 1.6 Steps typically required to analyze circuitry.

local layout errors such as design-rule violations, and more global design errors such as logical failures, short-circuits, and power inadequacies. Figure 1.6 illustrates the sequence of analysis steps in a typical design system. Analysis tools can also be used to compare different versions or different views of the same circuit that have been independently designed (see Fig. 1.7).

Synthesis tools help perform repetitious layout such as programmable logic array (PLA) generation, complex layout such as routing, and even layout manipulation such as compaction. Figure 1.8 illustrates the sequence of steps for a typical synthesis tool. As circuit design becomes more complex, these tools become more valuable and numerous.

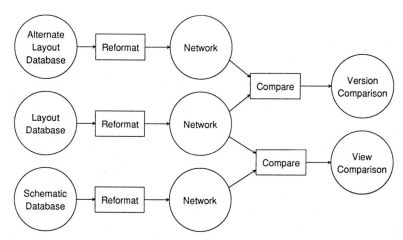

FIGURE 1.7 Comparing different versions of a circuit.

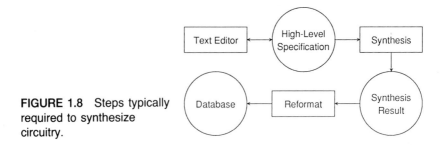

FIGURE 1.8 Steps typically required to synthesize circuitry.

Today's VLSI designers guide their circuit through the many different phases of the process outlined here. They must correctly control the initial creation of a design, the synthesis of additional detail, the analysis of the entire circuit, and the circuit's preparation for manufacturing. As synthesis tools become more reliable and complete, the need for analysis tools will lessen. Ultimately, the entire process will be automated, so that a system can translate directly from behavioral requirements to manufacturing specifications. This is the goal of **silicon compilers**, which can currently do such translation only in limited contexts by automatically invoking the necessary tools (see Fig. 1.9).

Of course, totally automated design has been sought for as long as there have been circuits to fabricate. Although today's systems can easily produce yesterday's circuits, new fabrication possibilities always seem to keep design capability behind production capability. Therefore it is unreasonable to expect complete automation, and a more realistic approach to CAD acknowledges the need for human guidance of the design process.

1.1.2   The Design System Computer

The number of different computers continues to grow with each passing year. Although it would be foolish to recommend any particular machines for design use, certain requirements are clear and should be mentioned. In particular, large address space, good graphics, and effective communication are desirable features.

Large designs consume large amounts of memory. Whether that memory is primary or on a disk, it must still be addressed. Therefore design workstations should have the capability for addressing many millions of bytes at once. This means that 16-bit computers are inadequate and even 24 bits of addressing will not be sufficient. Many modern machines have 32 bits of addressing, which should satisfy designers for a few more years. Nevertheless, a truly forward-looking machine must have 36 or 40 bits of addressing to prevent the agony of inadequacy.

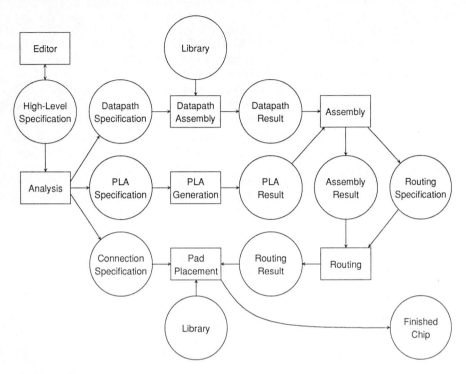

FIGURE 1.9 Structure of a typical silicon compiler.

Graphic displays are becoming an integral part of many computer workstations. Although not every designer wants such a facility, it must be available at some point in the manufacturing of circuits. Color graphics is also important in VLSI design. Chapter 9, Graphics, discusses the many different display types and their relative merits.

As computers get smaller and less expensive, people buy more of them than ever before. Thus the modern design computer does not stand alone in its laboratory and must be able to communicate with the other workstations. Designers need to exchange circuit data and even need to switch computers when one fails or is busy. Thus sophisticated communication facilities are demanded to create a feeling that everyone is using the same computer. Also, human-level communications such as electronic mail must be supported.

1.1.3 Design as Constraint Optimization

One step toward a unified solution to CAD is a look at the decision processes required of a designer. The goal of design is to produce precise specifications that will result in a product that fulfills its intended purpose. To achieve this goal, the designer must weave through the myriad alternatives at each step. Should the object be taller or wider? Need the corners be rounded? Should this wire run under or over the board? These sorts of decisions must be made constantly and their answers depend on the particular requirements of the design. The object should be wider *because* that gives it more stability. The corners should be rounded *because* they can be fabricated with fewer errors. The wire should run under the board *because* it will obstruct components if it runs on top. Such collected information forms a set of **constraints** on the design, examples of which are shown in Fig. 1.10.

Some constraints are inflexible and known to the designer before design begins. These constraints are the generic rules for design of a particular kind of object. For example, a logic designer may begin with the constraint of using strictly AND, OR, and NOT gates to build a circuit. More complex objects must all reduce to these at some point. Other prior knowledge is more subtle and is sometimes taken for granted, such as the constraint of having to do logic design with only two-valued logic (true or false) rather than multivalued logic or continuous (analog) logic. All of these are guiding factors to a designer.

Other design constraints are specific to the individual circuit being designed. Examples of these constraints are the particular components selected for the design, the location of power and ground paths, and

Constraint Class	Sample Constraint
Spacing	Polysilicon must remain 1 micron from diffusion
Topology	Transistor with depletion and source connected to gate is a pullup
Power	Metal migrates if too much power is on a narrow wire
Timing	Two-phase clock drops Φ_1 before raising Φ_2
Logic	$F \wedge F = F$ $F \wedge T = F$ $T \wedge F = F$ $T \wedge T = T$
Packaging	IC must be 300 mils square to fit in 16 pin DIP
Manufacturing	Fabrication bloats contacts, so mask must be shrunk
Bonding	Pads must be near outside to keep wires short
Hierarchy	All instances of a cell are identical in contents
Layout	Minimum metal wire width is 8 microns
Routing	Power and ground wires may not cross
Graphics	Top layer is visible and should contain messages
Testing	Only top and bottom layers of PC board can be probed
Defects	Chip yield is proportional to chip size

FIGURE 1.10   Design constraints.

the timing specifications. As design proceeds, the number of constraints increases. A good design leaves sufficient options available in the final stages so that corrections and improvements can be made at that point. A poor design can leave the designer "painted into a corner" such that the final stages become more and more difficult, even impossible to accomplish.

The hardest design constraints to satisfy are those that are continuous in their tradeoff and, because they compete with others, have no optimum solutions. An example of such constraints is the simultaneous desire to keep space low, power consumption low, and performance high. In general, these three criteria cannot all be optimized at once because they are at cross-purposes to each other. The designer must choose a scheme that meets specified needs in some areas and is not too wasteful in the others. The rules for this are complex, intertwined, and imprecise.

Given that design can be viewed as constraint optimization, one might expect that a circuit could be specified as a set of constraint rules that a computer could automatically solve. The result would be the best possible design, given the constraints. Although this can be done for very simple circuits, modern VLSI systems have so many constraints that no existing automatic technique can optimize them all. Also, many of these constraints are difficult for a computer to represent and for a human to specify. Simple design constraints that can be managed automatically have been the primary function of CAD tools, as the name suggests. These simple checks are very useful, because humans constantly lose track of the details and make flagrant design-constraint violations.

Computers do not lose track of detail, but they do have a more limited capacity than do humans to perform complex reasoning about nonspecific constraints. Recently, there have been attempts to mimic human design methods with the use of **heuristics**. These are imprecise rules that guide complex processes in ways that model human thought. Heuristics do not guarantee an optimal design; rather, they acknowledge the impossibility of considering every alternative and instead do the best they can. These rules plan the direction of design, eliminate many constraints as too insignificant to consider, and prioritize the remaining constraints to obtain a reasonably good result. This modeling of the human design process is part of the realm of **artificial intelligence**.

Regardless of whether a human or a machine does design, the endeavor is certainly one of constraint satisfaction. A good design system should therefore allow constraints to be specified, both as an aid to the human and as a way of directly controlling the design process. This enables CAD systems to evolve beyond the level of dumb sketchpads and fulfill their name by *aiding* in the design activity.

1.1.4   Four Characteristics of Digital Electronic Design

To understand VLSI CAD properly it is first necessary to discuss the characteristics of design in general and those of digital electronic design in particular. Two characteristics are universal to all design: the use of structural hierarchy to control detail and the use of differing views to abstract a design usefully. Two other characteristics are more specific to electronics: the emphasis on connectivity and the use of "flat" geometry in circuit layout.

Structural hierarchy views an object as parts composed of subparts in a recursive manner. For example, a radio may have hundreds of parts in it, but they are more easily viewed when divided into groups such as the tuner, amplifier, power supply, and speaker. Each group can then be viewed in subgroups; for example, by dividing the amplifier into its first stage and its second stage. The bottom of the structural hierarchy is reached when all the parts are basic physical components such as transistors, resistors, and capacitors. This hierarchical composition enables a designer to visualize an entire related aspect of some object without the confusing detail of subparts and without the unrelated generality of superparts. For example, the hierarchical view of an automobile fuel-supply system is best visualized without the distracting detail of the inner workings of the fuel pump and also without the unrelated generality of the car's acceleration requirements and maximum speed. Hierarchical organization is the most important characteristic in the design of any moderately complex object.

Another common technique is the use of multiple **views** to provide differing perspectives. Each view contains an **abstraction** of the essential artifact, which is useful in aggregating only the information relevant to a particular facet of the design. A VLSI circuit can be viewed physically as a collection of polygons on different layers of a chip, structurally as a collection of logic gates, or behaviorally as a set of operational restrictions in a hardware-description language. It is useful to be able to flip among these views when building a circuit because each has its own merit in aiding design. Designers of mechanical objects must also make use of multiple views so that the object can be understood completely. It is not sufficient to design a house solely with a set of floor-plans because that is only one way to express the information. Different views include the plumbing plan, perspective sketches showing solar exposure, and even parts lists that describe the house as comprising various quantities of wood, metal, and glass. Although different views can often be derived from each other in straightforward ways, they are useful to consider separately because they each cater to a different way of thinking and thus provide a "double check" for the designer.

More specific to VLSI design is the notion of **connectivity**. All electronic components have wires coming out of them to connect to other components. Every component is therefore interconnected with all other components through some path. The collection of paths through a circuit is its **topology**. This use of connectivity is not always present in mechanical artifacts such as planes or houses, which have optional connectivity and often contain unrelated components. However, there are other design disciplines that do use connectivity, in particular those that relate to flow between components. Thus plumbing, meteorology, and anatomy do make use of connectivity for their design.

The final characteristic of design is its particular geometric nature. All design disciplines have a spatial **dimensionality** in which they are significant. Automobiles have three dimensions; atomic structures and choreography have four because they are studied over time; music has one dimension to its presentation with an additional pseudodimension of the different notes; and physical VLSI design consists of two-dimensional layers that are placed on top of each other. An understanding of this dimensionality can be used to advantage when building design tools.

The rest of this chapter discusses these four aspects of design and introduces the necessary concepts for the constructions of VLSI CAD systems.

1.2   Hierarchy

1.2.1   Terms and Issues

The first significant characteristic of VLSI and all other design is a heavy reliance on hierarchical description. The major reason for using hierarchical description is to hide the vast amount of detail in a design. By reducing the distracting detail to a single object that is lower in the hierarchy, one can greatly simplify many CAD operations. For example, simulation, verification, design-rule checking, and layout constraints can all benefit from hierarchical representation, which makes them much more computationally tractable.

Since many circuits are too complicated to be easily considered in their totality, a complete design is often viewed as a collection of component aggregates that are further divided into subaggregates in a recursive and hierarchical manner (see Fig. 1.11). In VLSI design, these aggregates are commonly referred to as **cells** (or **modules**, **blocks**, **macros**, and so on); the use of a cell at a given level of hierarchy is called an **instance**. In schematic design, hierarchical units are called **function blocks** or **sections**; for consistency, however, the term "cell"

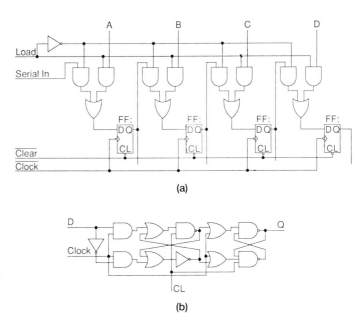

FIGURE 1.11 Hierarchy:
(a) Top-level of parallel-access
shift register (b) Flip-flop
subcomponent (FF).

will be used here. The use of a cell at some point in a circuit implies
that the entire contents of the cell's definition is present at that point
in the final circuit. Multiple uses of a cell indicate that the cell contents
are to be repeated at each use. Graphically, an instance can be seen
as an outline (called a **footprint** or a **bounding box**) that displays only
the boundary of the cell definition, or it can be displayed more fully
by showing the cell's contents.

Certain terms are used commonly in describing hierarchy. A cell
that does not contain any instances of other cells is at the bottom of
the hierarchy and is called a **leaf cell**. The one cell that is not contained
as an instance in any other cell is at the top of the hierarchy and is
called the **root cell**. All others are **composition cells** because they compose
the body of the hierarchy. A **separated hierarchy** is one that distinguishes
between leaf cells and the higher composition cells [Rowson].

By convention, hierarchy is discussed in terms of depth, with leaf
cells at the bottom considered deepest. A circuit with no hierarchy
has a depth of one and is said to be **flat**. Some aspects of CAD require
that hierarchy be eliminated because particular tools cannot handle
hierarchical descriptions. For example, many simulators are not hierarchical and insist that the circuit be represented as a simple collection
of components and connections. Also, many manufacturing-output
descriptions typically permit no hierarchy. A circuit that has all its
cell instances recursively replaced with their contents is thus reduced
to a hierarchical depth of one and is said to be **flattened**, or **fully
instantiated**.

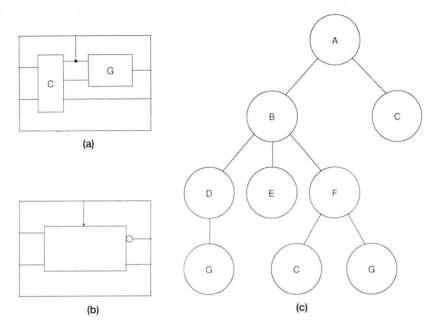

FIGURE 1.12 Cell definitions may be separate from hierarchy: (a) Hierarchical definition of cell "F" (b) Cell definition of cell "F" (c) Hierarchy.

Although cells and their definitions implement an elegant hierarchy, they do not always provide for the needs of the designer who may want an alternate view of a cell instance. For example, in Fig. 1.12, the contents of cell "F" can be described in pure hierarchical terms as consisting of cells "C" and "G." Alternately, the cell definition can break from the hierarchy and be described with a totally different image. In most systems, cell instances and definitions are the method used to implement hierarchy, because the definitions describe the contents of a hierarchical level and the instances link two levels by appearing in some other definition. In this situation, cell definitions are said to be unified with hierarchy. However, as the figure illustrates, it is possible to allow a cell definition that is separate from the hierarchy, so that a definition is not indicative of the overall structure of the circuit, but rather is an alternate description to be used when displaying instances. This adds extra complexity to a design system because each cell must be described twice: once in terms of its definition in the hierarchy, and once in terms of its actual appearance.

1.2.2 Hierarchical Organization

The most difficult problem that designers face is that of selecting a hierarchical organization for their circuit. This organization defines the way that the designer will think about the circuit, since layout is

typically examined one hierarchical level at a time. If the wrong organization is chosen, it may confuse the designer and obscure simple
solutions to the design problem. The circuit may get so convoluted
that an awkward hierarchy is worse than no hierarchy at all.

On the other hand, a clean hierarchical organization makes all
phases of design easier. If each level of the hierarchy has obvious
functionality and aggregates only those components that pertain to
that hierarchical level, then the circuit is easier to understand. For
example, with a good hierarchy, simulation can be done effectively
by completely testing each level of the hierarchy starting at the bottom.
Good circuit hierarchy is similar to good subroutine organization in
programming, which can lead to code that is more self-documenting.
For both structured programming and structured circuit design, a one-
to-one mapping of function to structure provides a clean view of the
final object. Nevertheless, a proper hierarchical organization is more
difficult to obtain than is a leaf-cell layout, because it embodies the
essence of the overall circuit and captures the intentions of the designer.

Unfortunately, there is no way to describe precisely how to choose
a good hierarchical organization. The proper planning of a circuit is
as much an art as is the design of the actual components. For this
reason, the techniques described here are only guidelines; they should
be followed not blindly, but rather with an understanding of the particular
design task at hand.

The first issue to consider is that of **top-down** versus **bottom-up**
design. In the former, a circuit is designed with successively more
detail; in the latter, a circuit is viewed with successively less detail.
The choice of direction depends on the nature of the task, and the
two techniques are almost always mixed in real design [Ackland *et
al.*]. The distinction is further clouded by the fact that the hierarchical
organization may be created with one method (top-down) and then
actually implemented in the opposite direction [Losleben]. Thus the
following examples are idealistic pictures of hierarchical design activity.

As an example of top-down design, consider a circuit that computes
the absolute difference between two numbers (see Fig. 1.13). This
problem starts with the most abstract specification: a description solely
in terms of inputs and outputs. In order to arrive at a hierarchical
organization that shows the complete circuit, a top-down design should
be done. So, starting at the top, the circuit is decomposed into the
subcircuits of the absolute value operation: subtraction and conditional
negation. Subtraction can be further decomposed into negation followed
by addition. Negation (in twos-complement number systems) can be
decomposed into inversion followed by an increment. The entire set
of blocks can be hierarchically divided into one-bit slices that do
adding, inverting, and incrementing on a bit-by-bit basis. The lowest

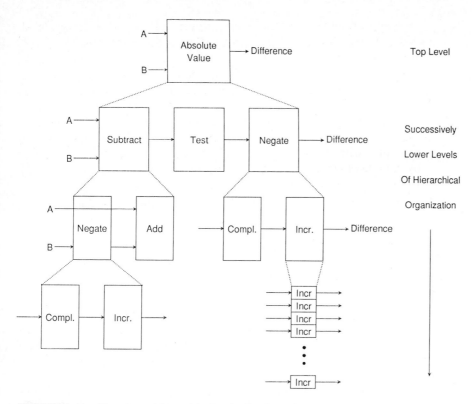

FIGURE 1.13 Top-down hierarchical organization of an absolute-value engine.

level of the design shows the layout for implementing these functions.

The opposite of top-down hierarchical organization is, of course, bottom-up. This style of design can be used when the details of a circuit are already known and must be properly implemented. For example, suppose that a 4K memory chip is to be designed, and further suppose that the design for a single bit of that memory is already done. Since the size of this bit of memory is the most important factor in the chip, all other circuitry must be designed to accommodate this cell. Therefore the design composition must proceed in a bottom-up manner, starting with the single bit of memory (see Fig. 1.14). In this example there are six levels of hierarchy starting at the single bit, aggregating a row of eight bits; stacking four of those vertically; stacking eight at the next higher level; and so on. The highest level of the hierarchy shows four arrays, each containing 32×32 bits. Memory-driving circuitry with the correct spacing is then placed around the bits.

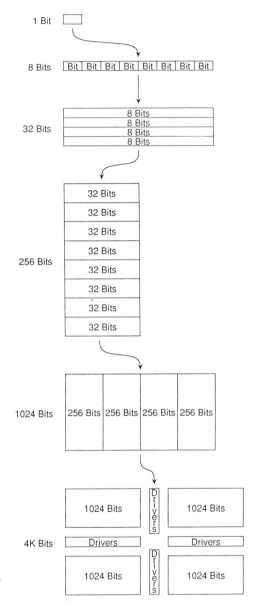

FIGURE 1.14 Bottom-up hierarchical composition of a 4K memory array.

1.2.3 Branching Factor

The previous two examples illustrate an important attribute of hierarchical organization: **branching factor**. When hierarchical organization is viewed as a tree, the branching factor is the average number of splits that are made at any point. Figure 1.15 shows the tree repre-

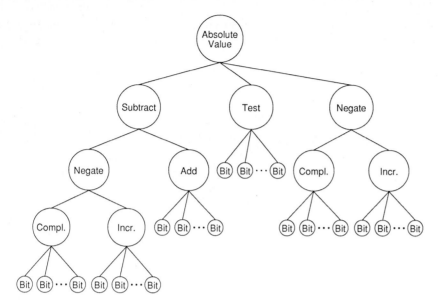

Branching Factor = (6n + 9) / 10

For n = 8, Factor = 5.7

For n = 16, Factor = 10.5

FIGURE 1.15 Branching factor for the circuit of Fig. 1.13.

sentation for the circuit in Fig. 1.13. The overall branching factor is the average number of branches made at any level of the hierarchy. In this example, the factor is a function of the word size. The branching factor for the example in Fig. 1.14 is 6.4 (presuming that the drivers are not further decomposed), which is low. Reasonably low branching factors are important in good hierarchical organization because, if they grow too large, the designer will no longer be able to consider a single hierarchical level at one time. Given that human working memory is said to contain from five to nine objects [Miller], this range also is good for the branching factors for circuit hierarchy [Mosteller]. Of course, the ability of a human to ''chunk'' similar objects means that there should be from five to nine different circuit structures at a particular level, even if one of those structures is an array of identical subcells. Branching factors that are smaller than five will be excessively simple and probably will have too much depth of hierarchy. Conversely, branching factors that are larger than nine probably will yield designs with too much detail at any level of hierarchy. Either situation will cause designers to be unable to perceive functionality effectively.

In addition to being concerned about the branching factor, the designer should ensure that the hierarchy does not ''bottom-out'' too soon. If the leaf cells of the hierarchy are excessively complex, then the organization has probably not gone deep enough. Leaf cells with hundreds of gates are likely to be too complex.

1.2.4   Spatial versus Functional Organization

Another factor in hierarchical organization is the nature of the branching. Hierarchical subdivision is done frequently along functional boundaries and occasionally along spatial boundaries. Within each category are other considerations for organization, such as the choice between structuring by flow of data or by flow of control [Sequin]. A purely **functional** organization is one that aggregates components according to the degree to which they interact, and therefore counts the number of wires used as interconnect. Such an organization is useful for circuits that have no layout associated with them, because it allows the function to be considered without any confusing physical considerations. However, layout must always enter into a design at some point, so it cannot be ignored in the hierarchy planning stage.

Purely **spatial** organization is a much more foolish thing to do. It presumes that the circuit is being hierarchically organized solely so that less of it can be displayed at a time, and therefore aggregates according to distance in the layout. This sort of organization is sometimes done retroactively to completed designs that need to be manipulated hierarchically but have no hierarchical structure. If the functional interactions are not considered, the resulting organization is sure to be more obtuse than the fully instantiated circuit.

The ideal design has a hierarchical organization that divides the circuit along both spatial and functional boundaries. Figure 1.16 illustrates an example of this. The fully instantiated circuit has 12 components: six NOR gates and six NAND gates. A purely spatial organization (Fig. 1.16b) ignores the connectivity and aggregates components by type. This is the sort of view that integrated-circuit placement and routing systems take when they try to make assignments of gates to multigate packages. However, it is not the best hierarchical organization because, among other faults, it leaves global wires running across the top of the cell instances [Mosteller]. The purely functional view (Fig. 1.16c) considers each functional aggregate of two gates to be worthy of composition into a hierarchical level because it has minimal interwiring needs and has small cells. This is a much better solution, but is still not optimal given the nature of the design. Perhaps the best organization is that shown in Fig. 1.16(d), which makes more use of spatial considerations by combining four gates in a cell along with the feedback wiring. This makes the cell connections easier to visualize without sacrificing functional organization. Of course, all of these subdivisions

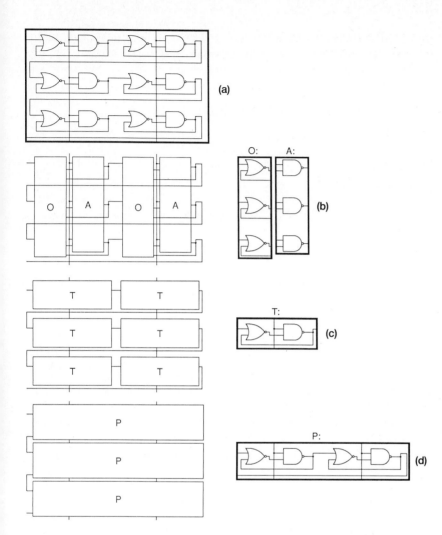

FIGURE 1.16 Hierarchical organization: (a) Complete circuit (b) Spatial organization (c) Functional organization (d) Functional and spatial organization.

have merit and present a different point of view to the designer. Good CAD systems allow ease of hierarchical reorganization so that the designer can find the best structure for the task at hand.

1.2.5   Layout Overlap

For circuit layout, there is an additional factor to consider when selecting a hierarchical organization: the extent to which objects at a given level of hierarchy will overlap spatially. There are times when the functions of two components are different enough to place them in different subcells of the hierarchy, yet their relative placement in the total design is very close. In more extreme circumstances, components and wires are shared by duplicating them in different subcells and having them overlap to combine (see Fig. 1.17). In cases such as these, the bounding boxes of the cell instances overlap and connecting wires become tiny or nonexistent. This is not a problem to the hierarchy but can cause difficulty to the analysis tools that examine the circuit.

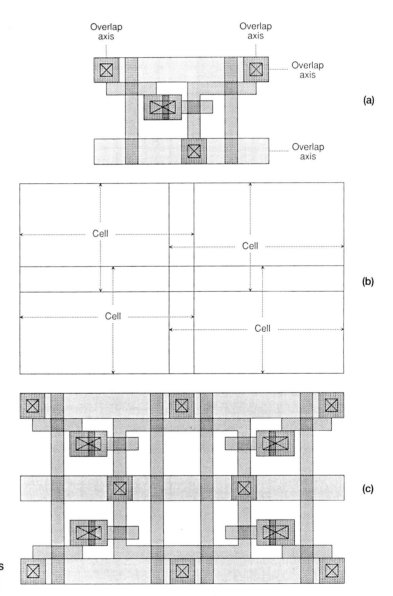

FIGURE 1.17 Component sharing causes cell overlap: (a) A cell (b) Four cell instances (c) Four cell instances showing component sharing.

Some design systems enforce methodologies that do not allow overlap. One such system allows irregular cell boundaries so that components can be efficiently placed [Scheffer]. Severe overlap should be avoided because it can become visually distracting to the designer, especially when wiring runs across cells to connect them. A small amount of overlap is acceptable provided that the design system allows it and that it does not cause clarity problems.

1.2.6   Parameterized Cells

An important issue in hierarchical design is the relationship between cell definitions and cell instances. How will changes to a definition affect the instances and how will changes to the instances affect the definition? For example, can the designer make a change to just one instance and not any others? The amount of **differentiation** allowed between multiple instances of a cell depends heavily on the design environment. If no differentiation is allowed, then every cell instance will be exactly the same as its cell definition and a change to the definition will require that all instances change too. If cell instances are allowed to be different, changes to one will not affect the others. Such cells, which can be instantiated differently according to their environment, are called **parameterized cells**.

When a cell definition is changed in a system that allows no differentiation, the change propagates upward from lower levels of the hierarchy. This is because each change to a cell definition affects all instances of the cell definition that reside higher up the hierarchy. The opposite situation is one in which parameterized cells are allowed and a change to one instance affects neither the other instances nor the original cell definition.

In systems that force all cell instances to be the same, there is still some differentiation allowed. After all, the cell instances are at different physical locations. Some of the instances may be rotated, indicating that their contents are altered in the overall circuit. So there are some attributes of cell instances that do differentiate them. A truly powerful design system allows even more parameterization so that instances can be completely differentiated. For example, a parameterized cell might change its contents as more input lines are added, so that it always has the correct circuitry to accommodate that particular instance.

The typical parameterized cell is described textually such that a change to an instance is actually a modification of the parameters in a line of code. These cell parameters are used like subroutine parameters to invoke the cell definition code and produce a particular circuit. Systems that function in this way propagate change information downward, from higher levels of the hierarchy. A significant aspect of a design system is the flexibility of its implementation of cell differentiation.

1.2.7   Hierarchy in Multidesigner Circuits

A final consideration in hierarchical organization is the proper partitioning of a circuit that is being built by more than one designer. Since such a partition is usually done on a functional basis, it lends itself well to hierarchical organization. Each designer can work on a single cell, and the collected set of cells can be put together to form the final circuit. Much planning must go into the specification of these cells to ensure that they will connect properly and that they can be designed independently. In addition, the use of libraries of common subcells should be specified in advance to make the final design be as uniform as possible. A CAD system must be able to mix previously completed designs into a single circuit. This is just one of the many hierarchical facilities that make a CAD system usable.

1.3   Views

The next characteristic of design is the use of different views to abstract an object. This certainly applies to electronic design; circuit designers view their work in many different ways. For example, the transistor can be drawn either as a schematic symbol, as a fully fleshed-out piece of geometry, or as a behavioral description in a textual language (see Fig. 1.18). Similarly, an inverter can be drawn as a

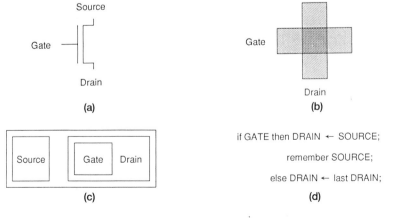

FIGURE 1.18 Multiple views for a transistor: (a) Schematic (b) MOS circuit (c) Bipolar circuit (d) Behavior.

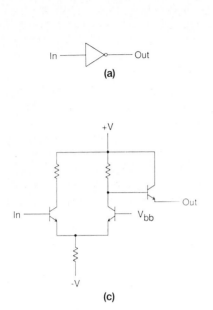

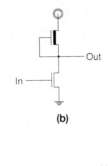

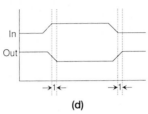

FIGURE 1.19 Multiple views for an inverter: (a) Schematic (b) nMOS (c) Bipolar (d) Temporal.

single symbol with one input and one (inverted) output, as a temporal diagram showing transitions, or as a complete gate diagram with all the transistors and connections that perform the inversion (see Fig. 1.19). Circuitry is often specified with differing views, and these representations combine to form a complete design. This section will discuss three view classes for electronic circuitry: topology, time, and behavior. Also of importance is the maintenance of correspondence among different views, which will be discussed at the end of the section.

A distinction now must be made between the terms "hierarchy" and "view." These terms blur because, in one sense, a hierarchical organization is a view of an object, and, in another sense, a collection of different views forms a hierarchy. Thus there is the **physical hierarchy** that was discussed in the previous section, which shows how the parts of a circuit are composed; there are many different views of a circuit, each one potentially hierarchical; and there is an organization of these views that is a hierarchy. The first hierarchy is a **physical hierarchy** and the last is a **view hierarchy**.

1.3.1 Topological Views

The **topology** of a circuit is its structure, or connectivity that describes the physical parts. Figures 1.18 and 1.19 illustrate a few topological views, the exceptions being Figs. 1.18(d) and 1.19(d), which are behavioral and temporal descriptions. Each view captures some essential abstraction of the circuit. The most abstract form is the architecture view, in which machine components such as processors and memories interconnect. Lower than that is the component view, in which datapaths and registers connect. This is often used to describe a complete hardware design without giving trivial detail. Below this, the gate view shows all necessary components without precise placement or size. At the bottom, there are layout or mask views that show all the geometry that will be manufactured. Chapter 2 describes these topological views and shows that their ordering is not able to be precisely specified. In fact, there are many other ways to consider the ordering of circuit views [Brown, Tong, and Foyster]. Nevertheless, all these views form a hierarchy in which the more abstract representations are higher than are the more detailed descriptions.

1.3.2 Temporal Views

Temporal views capture the activity of a circuit over time. The most precise view of a circuit's temporal activity is an analog plot of voltage over time (see Fig. 1.20a). Digital logic designers prefer to see their circuit as a more discrete process, so they use a threshold to separate true (or "on" or "1") from false (or "off" or "0"). In addition to forcing discrete logic values, they also view time discretely by presuming transitions to occur instantly (see Fig. 1.20b).

When multiple signals are described temporally, there is often a need to see the transitions and compare their start and end times (see Fig. 1.19d). When the switching time is all that matters and the direction of the switch is unimportant, timing diagrams are drawn with both transitions overlapped (see Fig. 1.21).

To relate circuit activity temporally, the designer must understand how each change affects subsequent activity in the design. A circuit that allows each value to flow directly into the next with no synchronization is an **asynchronous circuit**. Such circuits can use special structures to let each part of the design inform the next part when it is done [Seitz].

More common, however, is the use of **clocks**, which broadcast values at regular intervals. These values force the circuitry to function in lock-step fashion, making it **synchronous**. A popular clocking scheme is the **two-phase clock**, which consists of two signals called Φ_1 and Φ_2 that alternate being on and off (see Fig. 1.22). These values are generally **nonoverlapping** such that they are never both on at the same time (one must go off before the other goes on). Other clock schemes include single clocks and even complex combinations of six clocks.

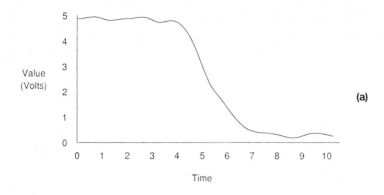

(a)

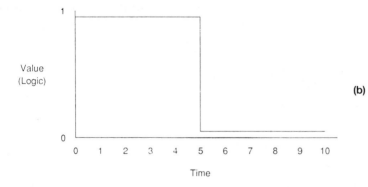

(b)

FIGURE 1.20 Continuous and discrete temporal views: (a) Analog (b) Digital.

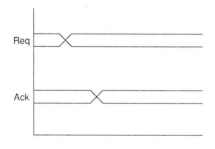

FIGURE 1.21 Temporal relationships: The Ack signal changes after the Req signal does.

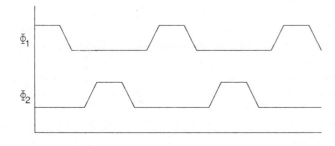

FIGURE 1.22 Two-phase nonoverlapping clocks: The two clock phases are never high at the same time.

There can also be major and minor cycles, with one clock signal changing many times during one pulse of another clock signal.

One of the most difficult problems for designers is the visualization of temporal relationships. The many synchronous and asynchronous methods of design help to alleviate the problem somewhat by allowing temporal assumptions to be made. A common assumption is that all activity triggered on one phase of a clock will complete during that phase. However, designers are still responsible for ensuring that the clock phases are of reasonable length and that there are no pathological pieces of circuitry that demand excessive clock time. When circuit activity in one clock phase extends into the next phase, **clock skew** problems arise. Although skew can be used to advantage when properly designed, it can also cause **race conditions** in which supposedly stable values are still in transition, causing unpredictable results.

Another source of trouble is the delay of clock signals as they propagate through the chip. This may cause the phases to overlap and the expected clock windows to be wrong. Designers must run the clock lines in parallel to ensure identical delays. Today's CAD systems do not help much with the problems of time and often leave the final analysis to the designer.

1.3.3   Behavioral Views

In addition to showing how a circuit actually acts, there must be ways of describing its expected behavior. A **behavioral** view of a circuit helps designers to visualize abstractly the interactions among elements of the circuit. Behavior shows both the topology and the temporal activity of a circuit, and is the highest level of description available.

Behavioral descriptions express topology in many ways, such as the algorithmic description of Fig. 1.18(d). Hardware-description languages provide convenient shorthand tools, such as ways of aggregating individual signals into integers and buses. Also common are iteration constructs and macros that allow similar behavior patterns to be expressed concisely. Of course, the behavioral description should be able to handle the dynamic nature of a circuit.

Behavioral descriptions can express time in absolute or relative ways. Absolute event specification is measured from the start of an activity. An example is the claim that at time zero, the circuit is in a given state and at time t, the circuit is in some other state. More common are relative event specifications because they combine easily. An example is a shift cell the output of which at the present time cycle is defined in terms of the input value during some previous clock cycle.

Temporal relationships can be expressed behaviorally with logic operators that have formal properties. By expressing events and state in terms of these operators, the designer can specify precise temporal order. For example, the unary operator *eventually* implies that something will take place in the future and the binary operator *prerequisite* implies that one event must occur before, or as a prerequisite to, another event. Combining these operators in proven ways allows an overall circuit analysis to be derived from the specification of sub-components. This static analysis is the basis of some simulators and timing verifiers.

Behavioral descriptions are often specified textually because they are algorithmic in nature. The use of such descriptive languages becomes a task in programming, with all of the ensuing problems of debugging, documentation, and understanding. However, since designers usually prefer graphical layout methods, there is no reason why behavior should not be specified in the same way. Thus the graphical representations in Figs. 1.18 and 1.19 are popular ways of describing behavior. For the same reasons that it is good to have graphic layout, it is also good to have graphic behavioral descriptions. Furthermore, since simulation is such a common way of testing behavioral specifications, the graphic display is best suited for monitoring that process. This is because many parts of a circuit may run in parallel, so only a spatial display can effectively capture the activity.

The drawbacks of graphical behavioral specification are the same as those of all graphic design. The lack of portability, verbal documentation, and parameterization can all be overcome with more expressive graphics languages. The expense of a graphics display, which is no longer as great as it used to be, is simply worthwhile given the advantages of graphic editing and can be defended only by saying that you get what you pay for. However, there is an overall problem concerning graphic programming: the lack of much theory or experience. Most programming languages and their studies have been textually based. Any reasonable graphical behavior language must have all the facilities that programmers have come to expect. Until such a language is available, textual hardware-description languages will be the most common.

1.3.4   View Correspondence

A fundamental problem in VLSI design systems is the maintenance of correspondence among different views. Designers rarely want to work at the lowest layout levels, so they often begin by specifying their circuit at the architectural or behavioral view. Conversion from more abstract descriptions to less abstract descriptions can be done in a number of ways; for example, by using silicon compilers to convert from more abstract behavior to less abstract layout. However, once converted, there is rarely any connection between the views.

The example in Fig. 1.19 illustrates this problem. It is not very hard to convert between a behavioral description and the schematic description of an inverter, maintaining a correspondence between the components. However, the transistor descriptions have more components so there arises a complication in maintaining a correspondence: the problem of maintaining a many-to-one mapping of component functionality. The transistors, their power and ground connections, and all the internal wiring are equivalent to a single component in the schematic. A change to that schematic component must result in a change to all the corresponding transistor components. But what can be done when a change is made to a single transistor component? There may be no equivalent schematic for the resulting circuit.

Many CAD systems ignore the problem of view correspondence. Either they provide no conversion between views (because they work in only one) or they provide conversion but no maintenance or enforcement. Once the circuit has been translated to a different view, it is effectively a new circuit with no relationship to the former view. This is a problem for designers who wish to adjust their converted circuit, but need to ensure that the original specification is not being violated.

One solution to the problem of view correspondence is to convert all views to a common one, usually a low-level gate description, and compare the results. This has a number of problems, not least of which is the amount of time it takes to perform the conversion and comparison. Because of the time, this process must be done separately from the actual design and so is less convenient, since the interactive design process must be frequently paused and checked. Also, feedback of errors is difficult because of the same correspondence problem between the original views and the comparison view. Finally, this method of view association relies on the ability to convert views effectively. In some cases, there may be conversion problems that ruin everything.

Another solution to the problem of view correspondence is to maintain a mapping between the views and to complain about or prevent changes that violate the mapping. Unfortunately, this approach is too restrictive because designers will want to take advantage of the features of a given view by rearranging components. For example, the power and ground wires that are shared in the transistor view are not connected in schematics. Rearrangement of these connections may cause spurious correspondence errors from the design system.

No general solution to the problem of view correspondence has been found. However, if silicon compilation improves to the point at which designers no longer manipulate the low-level layout, then correspondence with that view will be unimportant. This is effectively the state of software compilation today, which functions well enough that most programmers do not need to examine the resulting machine code.

1.4 Connectivity

1.4.1 Networks

The third distinguishing characteristic of design, and the first one specific to electronics, is connectivity. Since electronics involves the movement of charge, components are all viewed in terms of how they move signals back and forth across wires. All components have locations that attach to wires that make a connection to other such locations on other components. Thus an implicit constraint of VLSI design is that everything is connected and the connections carry information about the relationship of the components (see Fig. 1.23). This suggests that **networks** are well suited to the representation of circuits. Networks are merely collections of nodes that are connected with arcs. For many VLSI design systems, the nodes of the network are components and the arcs of the network are connecting wires. Some systems reverse this to view signal values on wires as network nodes and the signal-changing components as network arcs. Yet another variation is to view both wires and components as nodes, with network arcs used to indicate connectivity [Ebeling and Zajicek]. Regardless of the way in which it is implemented, the network is a useful representation of circuitry.

In general, connectivity finds less use outside of circuit design. The representation of most physical objects is done with a combination of volumes, polygons, and other spatial objects. These have no inputs or outputs and therefore do not connect. But if the representation of physical structure has no need for connectivity, physical behavior often does. Fluid-flow design is a classic example of mechanical connectivity. In fact, fluid models are often used to explain circuit connectivity and behavior (see Fig. 1.24).

Many IC design systems work without any explicit mention of connectivity. The layout is created in a **painting** style that describes a circuit only with polygons that have no electrical connections. When two polygons adjoin or overlap, they are presumed to connect (see

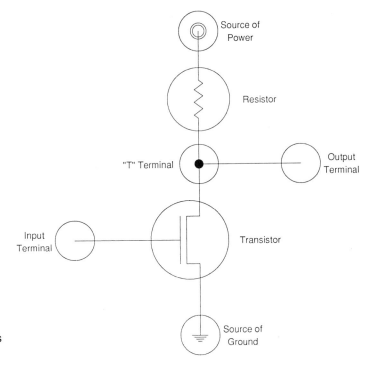

FIGURE 1.23 A circuit as connected components.

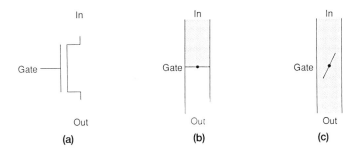

FIGURE 1.24 The fluid model of a transistor: (a) Electrical form (b) Fluid model with gate off (closed) (c) Fluid model with gate on (open).

Fig. 1.25). In these design environments, the system must derive the connections from the nature of the polygon overlap. This difficult and time-consuming operation is called **node extraction**. The step is necessary in order to simulate or to statically analyze an integrated circuit. It can also be used to ensure that the layout geometry does indeed connect where it should and is not short-circuited. Given the fact that geometry-based IC design systems are ignorant about connectivity, the node-extraction step is quite important.

In most schematic design systems and some IC design systems, connectivity is essential to layout. Components must be wired together

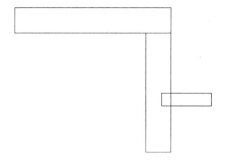

FIGURE 1.25 Three polygons forming a single net: Overlap and adjacency indicate electrical connectivity.

to form a completely connected circuit. An important concept is that of a **net**, which is a single electrical path in a circuit that has the same value at all of its points. Any collection of wires that carries the same signal between components is a net. If the components allow the signal to pass through unaltered (as in the case of a terminal), then the net continues on subsequently connected wires. If, however, the component modifies the signal (as in the case of a transistor or a logic gate), then the net terminates at that component and a new net begins on the other side.

Figure 1.26 illustrates the notion of nets. There are three components shown in the figure: two **active** ones, which modify signals, and one **passive** one, which does not. The active components are the NAND gate on the left, and the NOT gate in the middle. The passive component is the terminal connection on the right (the blob). The figure also contains three nets. Two of them have been given labels: In and Out. The third net is an internal net that has no name. This unnamed net connects the output of the NAND to the input of the NOT gate. Notice that the Out net extends from the output of the NOT gate all the way back to the input of the NAND gate.

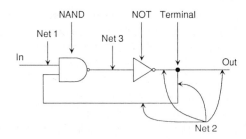

FIGURE 1.26 Three components and three nets.

1.4.2   The Importance of Nets

The notion of a net is important in many steps of design, analysis, and fabrication. Designers often want to see an entire net so that they can see the path of a given signal. This path will then identify the origin of the signal and those components that use the signal as an input. Although some design styles mandate that only one component may output to a net, there are many in which multiple components may output to a single net. In some cases, the multiple outputs combine logically to produce a composite net signal (called a **wired-or** or a **wired-and**, depending on how the sources combine). In other cases, the circuit timing is planned so that only one component will output to the net at any given time, and the other components will have inactive outputs. The ability to visualize a circuit by nets is therefore important in understanding the paths between components.

Nets are also used in simulation because, viewed abstractly, a circuit is merely a collection of gating components and their connections. A list of nets (called a **netlist**) ignores the passive components and the actual geometry of the circuit layout. Therefore, if the simulator is concerned not with exact timing but only with the general functionality, then the collection of nets and active components is the only information that the simulator needs. Such simulators run faster than do those that consider timing and layout.

1.4.3   Hierarchical Connectivity

Circuit connectivity must be handled properly in the presence of hierarchy. When a cell instance is placed in a circuit, its contents must be able to connect to other components surrounding that instance. There are two ways to connect cell instances, depending on how they are placed. When the instances abut directly, they are linked by **composition rules** determined entirely by the location of wires inside of the cells. Conversely, when instances are separated by a channel that contains interconnecting wires, they are linked by **connection rules** specified by the layout outside of the instances.

There are two ways to implement composition rules: implicitly and explicitly. In those IC design environments that support only geometry and no explicit connectivity, hierarchical connections are made implicitly by ensuring that the layout in the cell definition connects spatially with objects at higher levels of the hierarchy (see Fig. 1.27). This raises the problem of accurate connection alignment, which is difficult to check automatically when there are no declared connection sites. However, most VLSI design systems understand the notion of

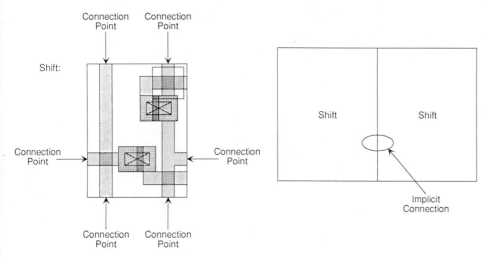

FIGURE 1.27   Cell instance connections implied by polygon adjacency.

an explicit connection point that is declared in a cell definition and is then available for use on instances of the cell at higher levels of the hierarchy (as in Fig. 1.11). This style of connection is similar to the notion of subroutine parameters in programming languages because the scope applies to the immediate higher level of hierarchy only. As mentioned before, the parallels between VLSI design and programming are significant and are seen frequently.

1.4.4   Connectivity for Constraint Specification

Another important use of connectivity is in the programming of design constraints. Many constraints specify relationships among objects, so it is convenient to implement these constraints as special components that are connected into the circuit. As an example, suppose that a designer wishes to connect two transistors in a layout and to ensure that these transistors always stay on the same horizontal level. Figure 1.28(a) illustrates this layout using two MOS transistors and a connecting wire. In addition to the layout components and their connectivity, this figure shows an additional component: a constraint called "horizontal." It is connected to the transistors in much the same way that the transistors are connected to each other, the only difference being that the transistors are connected with a diffusion wire and the constraint is connected with a "constraining" wire. Thus the figure shows three

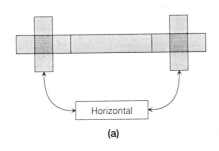

(a)

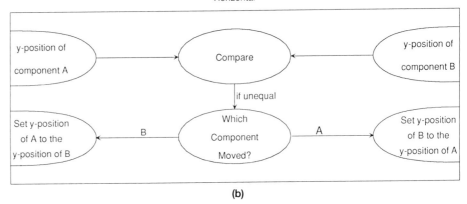

(b)

FIGURE 1.28   Graphical layout constraint: (a) Layout (b) Constraint contents.

components (two transistors and a constraint) and three connections (two constraints and a wire).

The contents of the constraint are shown in Fig. 1.28(b). The elements of this "layout" access data values, make decisions, and take actions concerning the updating of data values. It is the connection of this constraint to the layout in Fig. 1.28(a) that enables its contents to access and manipulate that design. More detail of this style of constraint specification will be given in Chapter 8, Programmability.

1.5   Spatial Dimensionality

The last important characteristic of VLSI design is that it has no depth in its spatial geometry. Circuits are always specified in two dimensions and are not made into solid models. Two possible contradictory views to this depth-free characteristic are the issues of time and multiple circuit layers. Time is the extra dimension that must be considered by some circuit-analysis tools. However, time is not represented as a full extra dimension because that would be very memory-inefficient. Also, circuit structure does not change with time so there is no need to show it as a physical relation. Rather, time is an abstract view of a circuit, as was discussed previously.

Multiple circuit layers add the only true extra spatial dimension to circuitry. IC design must accommodate the multiple layers of the fabrication process, and printed circuitry often uses multilayer boards to allow wires to cross. However, circuitry does not extend in the third dimension to the extent that it exists in the first two. Even process designers, who explore new layering combinations for integrated circuits, must limit the space above the wafer that a design may use. Some people have concerned themselves with the possibility of truly three-dimensional integrated circuits in which signals extend arbitrarily far in three-space, forming a design volume rather than a design plane [Koppelman and Wesley]. The ability to manufacture such designs is not a near prospect, so full three-dimensional design methods are not considered here.

The ability of a human to do three-dimensional design is additionally limited by both the depth-free nature of most human interface devices and the inability of a human to keep track of complex three-dimensional interactions. Notice that even in the design of buildings, architects use only two-dimensional sketches from above (called the plan view), from the side (called the elevation view), and from fixed angles (called perspective views). Although the limitation of graphic displays, plots, and input devices seems to be more serious than that of the ability of human perception, it is still a human limit that keeps us from being able to use three-dimensional devices effectively. Therefore, given the ability to fabricate true three-dimensional circuits, one would still find CAD systems using two-dimensional interaction and designing in slices that could be interconnected. For the present, it can be assumed that the third dimension—that of multiple layers—is a pseudodimension of multiple parallel planes. This is more than two-dimensional, but has many simplifications that make it less complicated than three-dimensional.

This same environment of design (using a limited set of two-dimensional planes) can be found in a very different field: that of cartoon animation. The backdrop is the deepest plane and each animated character has a separate plane in front of the backdrop. There are no complex interactions between these planes: Objects in the closest plane always obscure those in deeper planes. In the animation business, this style of graphics is referred to as **two-and-one-half-dimensional** and the term applies well to VLSI design.

The advantages of two-and-one-half-dimensional graphics over full three-dimensional graphics are that many display algorithms are simpler and faster. In fact, many color displays are built for two-and-one-half-dimensional graphics and efficiently implement nonintersecting planes. Chapter 9, Graphics, discusses algorithms that can be used on such hardware.

1.6   Summary

This chapter has described the VLSI design process and the characteristics of such design as a foundation for understanding CAD systems. Hierarchy is a major consideration in design because it is difficult to specify properly. In addition, there are many aspects of hierarchical design that place heavy demands on CAD systems. The use of differing views is important in all design. These views should be handled uniformly so that structure can be related to behavior. Coupled with this are the problems of conversion and correspondence between views. An important characteristic, specific to circuit design, is connectivity, which is related to the notion of nets. Not only are connected nets important in circuitry, but connected constraints help make the design process more flexible. Finally, the basically two-dimensional nature of circuits is another characteristic that should be recognized as important to VLSI design. An understanding of these issues sets the stage for the computer-aided design tools described in this book.

Questions

1 Name some two-and-one-half-dimensional design environments other than VLSI.

2 What is the most important reason for the use of hierarchy: (a) to save space in memory, (b) to allow humans to design circuits more effectively, (c) to enable multiperson projects, (d) to speed display by drawing bounding boxes.

3 Design a scheme for view correspondence when the two corresponding hierarchies are different.

4 How does a net differ from a wire?

5 Describe a situation in which a cell definition would be different from the cell's hierarchical contents.

6 What is the meaning of the term "computer-aided engineering"? How did it arise?

7 List some attributes of mechanical design that are not found in VLSI design.

References

Ackland, Bryan; Dickenson, Alex; Ensor, Robert; Gabbe, John; Kollaritsch, Paul; London, Tom; Poirier, Charles; Subrahmanyam, P.; and Watanabe, Hiroyuki, "CADRE—A System of Cooperating VLSI Design Experts," Proceedings IEEE International Conference on Computer Design, 99-104, October 1985.

Brown, Harold; Tong, Christofer; and Foyster, Gordon, "Palladio: An Exploratory Environment for Circuit Design," *IEEE Computer*, 16:12, 41-56, December 1983.

Ebeling, Carl and Zajicek, Ofer, "Validating VLSI Circuit Layout by Wirelist Comparison," *ICCAD '83*, 172-173, September 1983.

Gibson, Dave and Nance, Scott, "SLIC—Symbolic Layout of Integrated Circuits," Proceedings 13th Design Automation Conference, 434-440, June 1976.

Koppelman, George M. and Wesley, Michael A., "OYSTER: A Study of Integrated Circuits as Three-Dimensional Structures," *IBM Journal of Research and Development*, 27:2, 149-163, March 1983.

Losleben, P., "Computer Aided Design for VLSI," *Very Large Scale Integration (VLSI)* 5 (Barbe, ed), Springer-Verlag, Berlin, 89-127, 1980.

Mead, C. and Conway, L., *Introduction to VLSI Systems*, Addison-Wesley, Reading, Massachusetts, 1980.

Miller, George A., "The Magical Number Seven, Plus or Minus Two: Some Limits on Our Capacity for Processing Information," *Psychological Review*, 63:2, 81-97, March 1956.

Mosteller, R. C., "REST—A Leaf Cell Design System," *VLSI '81* (Gray, ed), Academic Press, London, 163-172, August 1981.

Rowson, James A., *Understanding Hierarchical Design*, PhD dissertation, California Institute of Technology, TR 3710, April 1980.

Scheffer, Louis K., "A Methodology for Improved Verification of VLSI Designs Without Loss of Area," Proceedings 2nd Caltech Conference on VLSI (Seitz, ed), 299-309, January 1981.

Seitz, Charles L., "System Timing," *Introduction to VLSI Systems* (Mead and Conway), Addison-Wesley, Reading, Massachusetts, 1980.

Sequin, Carlo H., "Managing VLSI Complexity: An Outlook," *Proceedings IEEE*, 71:1, 149-166, January 1983.

Weste, Neil, "Virtual Grid Symbolic Layout," Proceedings 18th Design Automation Conference, 225-233, June 1981.

Williams, John D., "STICKS—A graphical compiler for high level LSI design," Proceedings AFIPS Conference 47, 289-295, June 1978.

DESIGN
ENVIRONMENTS

2.1   Introduction

2.1.1   Primitives and Environments

Design has two fundamental aspects: its methods and its materials. Methods are the techniques that when used creatively, achieve a result, and materials are the medium on which the methods are used. For example, writers design literature by applying the methods of syntax, grammar, and semantics to the materials called words, letters, and punctuation. Sculptors apply the methods of marker, hammer, and chisel to the materials wood, marble, and ice. Closer to home, engineers design circuits by applying the methods of synthesis and analysis to the materials called gates, wires, and silicon layers.

Although most of this book is concerned with the automation of VLSI design methods, this chapter focuses on the materials. There are many different ways to construct a circuit, and each has a somewhat different collection of materials. The collection, for a particular type of circuit, forms a **semiconductor process technology**. In addition, there are many styles of circuit design and many variations of fabrication techniques, all within a given technology. These variations of style and manufacturing methodology give rise to a large number of **environments** for VLSI design. Each environment has its own set of materials, called **primitive components**. This chapter describes a number of environments for VLSI design and illustrates the primitives in each.

One might think that the two-dimensional nature of circuits allows all environments to be described simply with points, lines, and polygons. However, good design systems extend this limited set so that more useful layout can be done. Given a good set of primitives for a particular style of design, any circuit can be built, usually by aggregating the components hierarchically. With a weak set of primitives, the resulting circuits are limited in complexity and clarity. Therefore the proper selection of design primitives is a significant aspect of the programming of a design system.

One effect that the selection of primitive components has on the design style is that it establishes the granularity of the lowest levels of the circuit. Given that objects can be infinitely subdivided, it is not interesting to select very tiny objects as primitives. A design made of components that are too elemental will have great complexity, possibly characterized by a very deep hierarchical description. For example, it is nonsensical to design an IC chip by using protons, neutrons, and electrons as the primitive components. Instead, areas of semiconducting material make more sense. Even better than that would be to use higher-level combinations of the semiconducting material that perform specific circuit operations. However, a selection of components at too high a level will restrict design flexibility. If the primitives are too complex, the designer will be forced to tailor special ones whenever the existing components are inappropriate for a given task. An optimal set of primitives allows easy and unrestricted design without introducing distracting detail.

In addition to providing sensible building blocks for design, the selection of primitive components determines the nature of the interaction between the designer and the design. When a circuit is specified, it is the primitive components that provide the fundamental interface. If the components do not match the designer's notions of the circuit being built, the interaction will become strained.

To provide properly for the designer's expectations, it should be understood that there are several different kinds of designers. Each designer requires a somewhat different set of primitives in order to match his or her mental model. For example, a **circuit designer** is usually concerned with the logical interactions of a circuit and is less concerned with its physical layout. A **mask designer**, on the other hand, is responsible for the physical layout and may be unaware of the overall nature of the circuit. A **machine architect** is a designer who is not concerned with the details of logical or physical layout. This person is interested in only the abstract floor-plan of large systems. Modern VLSI tools and techniques have merged all these design functions so that one person can specify a machine, its structure, and its layout. This person is the **VLSI designer**. It is important that each type of designer feel comfortable with the primitives of the design environment.

As an example, assume that the primitive components of MOS design are rectangular areas of metal, polysilicon, and diffusion. Then, a MOS transistor can be viewed as the overlap of a diffusion primitive and a polysilicon primitive (see Fig. 2.1a). Given such an environment, a VLSI or circuit designer may not notice that a change to one primitive affects the whole device (see Fig. 2.1b). These designers need a single, indivisible component that implements a transistor, because they think of the circuit in terms of such transistors. Mask designers can use the

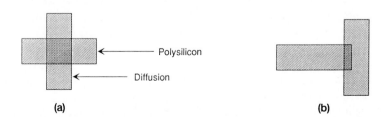

(a)   (b)

FIGURE 2.1 Incorrect primitives make design difficult: (a) MOS transistor composed of two primitive components (b) Motion of one primitive destroys the implicit transistor.

polysilicon and diffusion primitives because their view of a circuit is one of polygons on different material layers. However, even mask designers can benefit from a single transistor primitive, provided that it is flexible enough to describe all the necessary geometries.

Another example of primitive components that fail to match a VLSI or circuit designer's mental model are ones that make no use of connectivity. It has already been established that circuits are always interconnected, so the primitive components for VLSI design should reflect this fact. A system that provides unconnected regions of layout material is not as intuitive as one that distinguishes between components and connecting wires. The former system will not be able to capture networks as well as the latter system and will be less easy to use for many designers. Thus the set of primitives must match the designer's own ideas of the proper way to build a circuit.

As a final consideration, the number of primitive components should be kept small. This allows the entire set to be learned and used effectively by the designer. Any collection of primitives that is too large will not be used uniformly, which can degrade the quality of designs. Also, the pieces of multidesigner circuits will not match well if each designer has become familiar with, and used, a different subset of the primitive components.

In summary, a good collection of primitives must satisfy a number of conditions. For circuit design, there must be primitive components that can connect. The set of components must be flexible enough to generate any circuit without being so general that the layout has distracting detail. The components must also express the nature of the circuit in terms understandable to the designer. Finally, the collection must be small enough to be learned completely and used properly. This chapter will illustrate different sets of primitive components that were developed to meet these criteria in various areas of VLSI design.

2.1.2   Space of Environments

There are many design environments, each having different purposes in the realm of circuit production, and each making different demands on the necessary set of primitives. Some environments are considered to be very high level because their primitives are further defined by lower-level environments. Sometimes there are families of similar but not identical environments. The proliferation of environments within a family can be caused by new design methodologies, commonalities of some components, or variations in manufacturing processes.

All the different circuit design environments form a space that can be organized in many ways (one way is shown in Fig. 2.2). There is no correct ordering of design environments, but there is a general

Level	Environment	Parts
System	PMS	Processors, memories, links
Algorithm	Schematics	And, or, negation
	Temporal logic	Henceforth, eventually
	Flowchart	Test, compute, I/O
	Dataflow	Select, merge, function
Component	Register Transfer	Control, arithmetic, memory
	ISP	Registers, ALUs
Layout	MOS	Metal, polysilicon, diffusion, transistor
	Bipolar	Base, emitter, collector
	Packages	SSI, MSI, LSI
	Artwork	Rectangle, line, spline

FIGURE 2.2 Electrical design environments.

idea that the higher levels are more abstract and the lowest levels are more manufacturing-specific. Algorithmic environments are arbitrarily located between system and component environments, but they represent a different class that really spans all levels. Environment families also appear in this chart. For example, the **MOS** (metal oxide semiconductor) family includes **nMOS** (*n*-channel MOS) and **CMOS** (complementary MOS), which function differently but share many design attributes.

The rest of this chapter will outline some typical VLSI design environments and will describe a set of primitive components for each. This will illustrate the nature of the environments and also will indicate how a CAD system can be programmed to deal with them.

2.2   System Level

The environments with the greatest abstraction are at the system level. These environments view designs as collections of large functional units. The units can be arbitrarily complex, and they typically form the major components of a computer system (see Fig. 2.3). Connections in these designs also can be arbitrarily complex, carrying unspecified numbers of signals that use unspecified protocols. The reason for this is that, at a system level, the design is done by machine architects, so the precise method of communication is not important. All that matters is whether or not a particular function block connects to another. Sometimes the abstract nature of a connection is specified, such as whether it carries data or control. Often, however, a system-

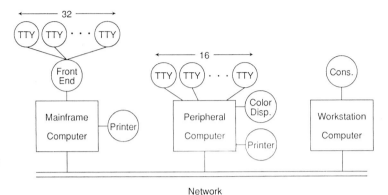

FIGURE 2.3 System design. Major components are informally drawn and connected.

level environment exists only to show design topology without providing excessive detail.

2.2.1 PMS Environment

The best example of a system-level environment is PMS [Bell and Newell]. PMS stands for "processor-memory-switch," three of the components in this environment (see Fig. 2.4). In addition to processors, memories, and switches, this environment has components for links, data operations, transducers, and control units. Solid lines are used to connect two components tightly, whereas dashed lines indicate secondary, control connections. Hierarchy can be used by defining new components that are collections of existing ones. The following paragraphs describe the seven primitives of the PMS environment.

The processor component (P) is used to represent the central processing element of a computer. This includes the instruction fetch, decode, and execution but does not include the instruction or data memory, nor any of the input/output (I/O) devices. This distinction can be difficult because some processors have built-in memory for stacks, caches, or microinstruction sets and, in these cases, the memory is considered to be part of the processor component. Also, all internal state such as machine registers and status bits are part of the processor.

P_t	Processor, name is "t"
M_t	Memory, nature is "t"
S	Switch
L	Link
D	Data operation
T_t	Transducer, nature is "t"
K	Control

FIGURE 2.4 PMS components.

The intention is to encapsulate a component that is typically indivisible and the same in any less abstract environment.

Memory components (M) are generally connected to processors. Although the precise amount of memory need not be specified, the nature of the memory is given as a subscript on the component letter. For example, M_p indicates primary memory, which might distinguish it from secondary memory, fast memory, double-ported memory, or any other scheme in a given PMS diagram.

A switch (S) is a general component for joining multiple objects in a controllable fashion. Switches can be used to select among multiple I/O devices on a processor, multiple memories, or even multiple processor components. Their presence in a PMS diagram indicates that there is a choice in the use of the connecting components.

The link (L) is used to indicate an unusual connection configuration. For example, the connection of two separate processors is frequently done with some sort of link that moves information across physical distances. Buses, networks, and telephone lines are all links. Some processors do not need links to communicate if, for example, they share memory or are directly connected with common internal state.

A data operation (D) is any functional block that generates data. The category includes processors but is really meant to cover those devices that perform input, such as digitizers and keypads. The opposite of a data operation is a transducer (T), which takes data and transforms them for some other use. All output devices are transducers, including printers, displays, and robots.

The final primitive component of the PMS environment is the control unit (K). A control unit is one that causes another component to function. Thus, to make a transducer or data operation work, a control unit is needed. Processors are the only other components that have control and so they can replace these control units in situations that demand more intelligence.

The PMS notation has been used to describe the configuration of computer systems [Bell and Newell; Siewiorek, Bell and Newell]. It is particularly useful in describing computer networks because it shows the components concisely. Figure 2.5 shows a PMS diagram for the computer network of Fig. 2.3. Notice the informal use of subscripts to qualify components and superscripts to indicate repetition.

PMS is a formal example of a systems-level environment. Many less formal environments are used for system description, such as the example in Fig. 2.3. There are also textual hardware-description languages that are able to express abstract structural descriptions [VanCleemput]. The important aspect of systems level environments is that design is independent of physical layout, describes gross functionality, and can be used to capture easily entire VLSI systems.

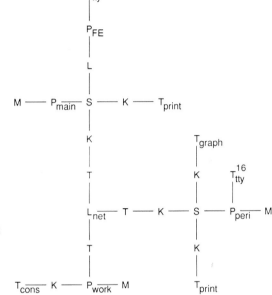

FIGURE 2.5 PMS description of the computer network in Fig. 2.3. Subscripts indicate component attributes and superscripts indicate repetition.

2.3 Algorithm Level

Somewhere in the continuum of environments that ranges from layout-specific to system-general are a set of environments that are more detailed than are the previously described machine architectures, but that have no circuit layout associated with them. These environments specify the precise function or algorithm of a design without providing an indication of how the algorithm is realized.

The most popular algorithm-level environment is the schematic, which uses logic components to implement Boolean functions. Temporal logic is an extension of standard logic that is particularly well suited to electronic circuit specification. Another way to specify an algorithm is with a flowchart that passes either flow of control or flow of data. Of course, the most common way of specifying algorithms is with programming languages. However, textual languages can always be implemented graphically by flowcharting them. Therefore, for the purposes of this discussion, programming languages will not be considered separately from the control-flow and dataflow environments.

2.3.1 Schematic Environment

A **schematic** is a stick-figure drawing of an electronic circuit. The process of drawing a schematic is sometimes called **schematic capture**

A	B	And $A \wedge B$	Or $A \vee B$	Implication $A \rightarrow B$	Negation $\neg A$	Exclusive Or $A \oplus B$
F	F	F	F	T	T	F
F	T	F	T	T	T	T
T	F	F	T	F	F	T
T	T	T	T	T	F	F

FIGURE 2.6   Truth tables.

because it graphically represents the circuit. It is so flexible and abstract that many circuit designs start this way and are subsequently translated into some other layout-specific environment for manufacturing. Synthesis tools (described in Chapter 4) can translate a schematic into layout, producing printed circuits, integrated circuits, and more. Eventually, these tools will be able to translate reliably from higher-level input (for example system-level environments) into lower-level layout (optimized circuitry as opposed to highly regular layout). Until such time, the schematic environment will be very popular.

In digital electronic design, the schematic components are strictly Boolean and the wires that connect these logical elements carry truth values. For every CAD system that can do schematic design, there is a different set of logic gates available to the designer. In pure logic there are four operations: NOT ($\neg$), AND ($\wedge$), OR ($\vee$), and IMPLICATION ($\rightarrow$), but common practice also includes EXCLUSIVE OR ($\oplus$). These operations are presented in a truth table in Fig. 2.6. Designers often combine these functions into more complex symbols that are treated like primitive components. For example, NAND ($\overline{\wedge}$) is simply the NOT of AND (the result is negated, not the input terms). Figure 2.7 shows some typical logic symbols and their variations. Notice that there is only one connection in the schematic environment: the Boolean that carries either *true* or *false*.

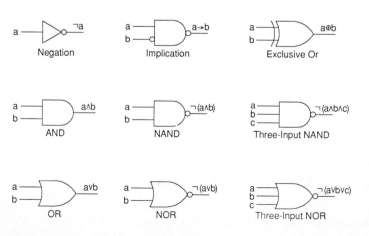

FIGURE 2.7   Logic gates.

In some situations, the number of different logic gates is too great and it is desirable to reduce the number of components by expressing all of them in terms of a few. For example with just one component, the NAND gate, all the previous logical operations can be derived:

$$\neg\, a \equiv a \barwedge a$$
$$a \wedge b \equiv (a \barwedge b) \barwedge (a \barwedge b)$$
$$a \vee b \equiv (a \barwedge a) \barwedge (b \barwedge b)$$
$$a \rightarrow b \equiv a \barwedge (b \barwedge b)$$
$$a \oplus b \equiv (a \barwedge (a \barwedge b)) \barwedge (b \barwedge (a \barwedge b))$$

These manipulations are aided by sets of equivalence rules that explain how one logic symbol can be expressed in terms of another. To prove these logical equivalences, mathematicians have developed sets of **axioms**. Given these axioms as truths, any logical manipulations can be done. For example, the following nine axioms are always true, regardless of the values of a, b, or c [Bell and Slomson]:

$$a \rightarrow (b \rightarrow a)$$
$$(a \rightarrow (b \rightarrow c)) \rightarrow ((a \rightarrow b) \rightarrow (a \rightarrow c))$$
$$(\neg a \rightarrow \neg b) \rightarrow (b \rightarrow a)$$
$$(a \wedge b) \rightarrow a$$
$$(a \wedge b) \rightarrow b$$
$$(a \rightarrow b) \rightarrow ((a \rightarrow c) \rightarrow (a \rightarrow (b \rightarrow c)))$$
$$a \rightarrow (a \vee b)$$
$$b \rightarrow (a \vee b)$$
$$(a \rightarrow b) \rightarrow ((c \rightarrow b) \rightarrow ((a \vee c) \rightarrow b))$$

These expressions are sufficient to define all logic because any true statement can be manipulated into one of them.

In addition to logic gates, there must be some accommodation for storage of data. Figure 1.11(b) shows the design of a flip-flop using AND, OR, NOT, and NAND gates. This circuit can be built electronically so that it correctly stores a single bit of data. With the addition of such a memory facility, any functional design can be done.

For the builder of CAD tools, it is useful to know how to draw all these schematic elements. The U. S. government has standardized the drawing style in a document called Military Standard 806B [Department of Defense]. The standard provides proper ratios for graphics and explains how to extend symbols when there are many input lines (see Fig. 2.8). In addition, the standard provides the meaning of abbreviations (C is ''clear,'' FF is ''flip flop,'' RG is ''register,'' and so on) and the proper location of annotation on bodies (top line is the

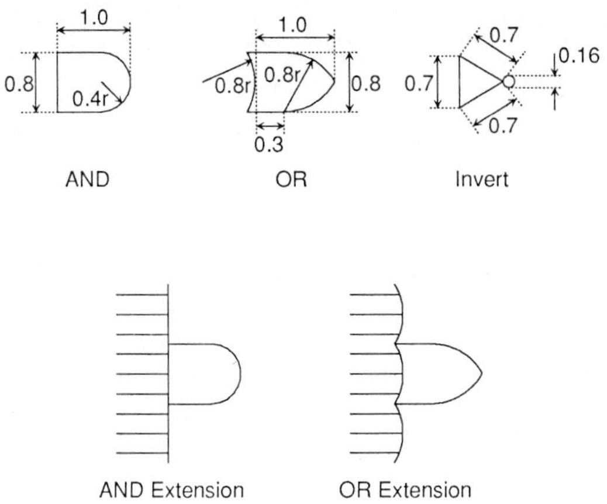

FIGURE 2.8 Military standard graphics for schematics.

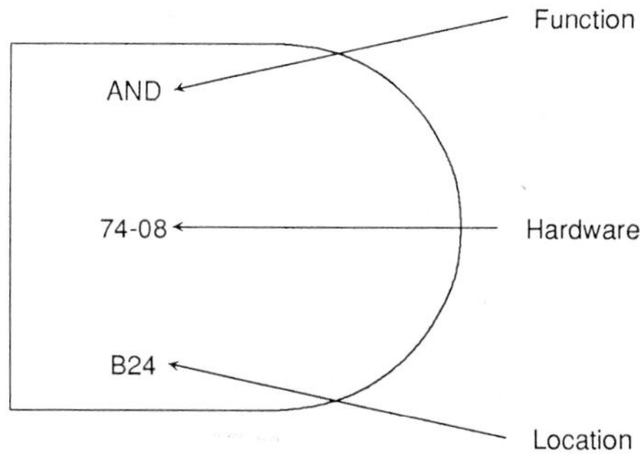

FIGURE 2.9 Military standard annotation locations on schematics bodies.

function, middle line is the type of hardware, lower line is the physical location; see Fig. 2.9). Although this standard is not followed rigorously and is incomplete in some areas, it does provide a starting point for the programmer of schematics systems.

2.3.2   Pseudolayout Environments

Although schematics provide a good way of abstractly specifying a circuit, layout must be done at some time. To help with the conversion, there are a number of design environments that have the feel of schematics yet are much closer to layout. These environments are typically specific about some aspect of layout while abstract about the rest.

The **sticks** environment is specific about the relative placement of components but abstract about their size and exact dimensions [Williams]. Transistors are drawn as stick figures with connecting wires that have no dimension (see Fig. 2.10). This design is much more

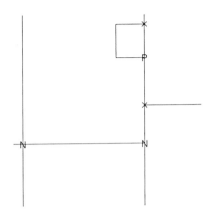

FIGURE 2.10   Sticks nMOS inverter layout. The "N" is a pulldown transistor, the "P" is a depletion transistor; and the "X" is a contact.

specific than a schematic because abstract logic gates are not allowed. Rather, only those components that can be implemented directly in silicon are present. To convert this environment into actual layout, the components are expanded to their proper size, wires are given actual silicon layers, and the distances are adjusted to be design-rule correct. Sticks environments must be tailored to a particular type of layout. Although most are for MOS layout, there are some for bipolar design [Szabo, Leask and Elmasry].

Another abstract environment used to approach layout is **virtual grid** [Weste]. As with sticks, only those components that will appear in the final layout may be used. Virtual grid, however, draws all components with a nominal size that, although not true to the final dimensions, do impart a feeling for the geometry (see Fig. 2.11). Another important feature of virtual grid design is that the relative spacings are preserved when true layout is produced. Thus two transistors that line up horizontally or vertically will stay in line, even if they are not connected. This is because the conversion to layout is

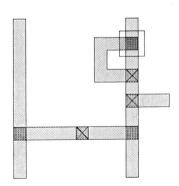

FIGURE 2.11   Virtual grid nMOS inverter layout.

simply a process that assigns actual coordinates to the virtual grid coordinates. This makes conversion quite simple yet retains the abstract nature of the design environment.

Yet another pseudolayout environment is SLIC [Gibson and Nance], which describes layout component placement using text letters. An array of these letters, displayed on a video terminal, shows component placement and connecting wires. Special letters distinguish horizontal from vertical wires, and allow correct connectivity to be specified.

2.3.3   Temporal Logic Environment

One problem with traditional logic is that it does not inherently deal with time in any way. One solution is to parameterize logic values with the current time, resulting in expressions like:

$$W(t) \wedge I(t) \rightarrow O(t + 1)$$

which means "if working at time t, and inputs are available at time t, then there will be outputs at time $t + 1$." However, because this structure can be awkward to use, recent work has turned to using **temporal logic**, a modal logic that has operators applying to sequences of states over time. Although many formulations of temporal logic have been used, this section will discuss one method that has been applied to electronic circuit descriptions [Malachi and Owicki].

Four basic operators can be used to describe state properties over time. The symbol $\diamond$ means "eventually" and its use in the expression "$\diamond p$" means that p will be true at some future time (or is true in the present). The symbol $\square$ means "henceforth" and indicates that its operand is true now and at all future times. The symbol $\bigcirc$ means "next" and indicates that its operand will be true in the very next moment of time. Finally, the symbol **U** means "until" and is used in a two-operand sense "p **U** q" to mean that p is true at all times from now until the first instance that q is true.

As with pure logic, a set of axioms relate these operators and show how they are used to define each other. For example,

$$\square p \rightarrow p \wedge \bigcirc p \wedge \bigcirc \square p$$

states that "henceforth" p implies that p must be true now, at the next interval, and henceforth beyond that interval. A simpler tautology of temporal logic is:

$$\square p \rightarrow p \text{ } \textbf{U} \text{ } q$$

In addition to basic operators, it is common to define new operators that are derived from the original set. These new operators are better suited to circuit design because they behave in ways that are more intuitive to designers. Thus the "latched while" operator is defined as:

$$pLWq \equiv p \rightarrow (p \ U \ \neg q)$$

meaning that, when p is true, it will remain that way as long as q remains true. Another derived operator that is important in defining the liveness of a system is "entails" ($\rightsquigarrow$):

$$p \rightsquigarrow q \equiv p \rightarrow \Diamond q$$

meaning that the truth of p demands the eventual truth of q.

Temporal logic is well suited to the description of digital circuits over time. This is because it can describe the behavior of a single component or a collection of parallel activities. In fact, some form of temporal logic can be found wherever the behavior of circuits is specified. Variations on basic temporal logic exist to improve its expressiveness. For example **interval temporal logic** generalizes the basic scheme so that subintervals of time can be conveniently described [Moszkowski].

2.3.4   Flowcharting Environment

Flowcharting is a general-purpose algorithm environment that is often used to represent textual languages graphically. Typically, the components of a flowchart operate on data and the connections in a flowchart indicate flow of control. Each component in a flowchart can be arbitrarily complex because it contains an English-language description of its operation. Typically, however, there are a limited set of components that perform distinct classes of operations.

Figure 2.12 shows the four basic primitives that are most often used to do control-based flowcharting. The computation component is used for all data manipulations, with the actual function written in the box. The decision component is used to make arbitrary tests and modify the flow of control. The I/O component indicates control of devices. The connector component is used for special control-flow cases such as are necessary in multipage designs or for program termination. The layout in Fig. 2.17 (shown later in this chapter) is an example of a control-based flowchart specification, but it does not differentiate the component types because it is a different environment (register transfer).

FIGURE 2.12   Control-flow components.

A somewhat different style of control-flow specification is the use of **state diagrams**. The components of a state diagram perform actions that may alter a global state. The connections between components are each labeled with a state, such that only one connection will match and direct the control to another component. State diagrams are a very formal way of expressing an algorithm, as opposed to flowcharting, which has imprecise rules for actions and tests.

2.3.5   Dataflow Environment

Dataflow programming is a twist on the commonly used flowcharting technique because the connections between components carry data values rather than control invocation [Rodriguez]. This means that parallel activity is easier to specify and that the operation of an algorithm is driven by the presence of data rather than a "start" signal.

The components of a dataflow program are called **actors** [Dennis] and there are many possible sets that can be used. Figure 2.13 shows one typical set of dataflow actors [Davis and Drongowski]. Dataflow actors function only when a predetermined set of input lines have data values waiting on them. This set of inputs is called the **firing set**. When the firing set becomes available, the actor executes and produces output values for other actors. A dataflow network that guarantees no more than one data value will be pending on any input line is **safe** [Patil].

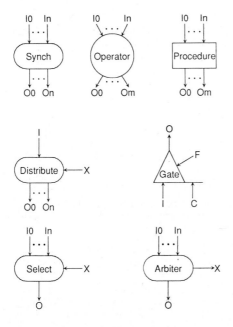

FIGURE 2.13   Dataflow primitives. Activity occurs when data are ready on the proper input lines.

The seven actors outlined in the following paragraphs are sufficient to describe any computation. In fact, some of them are redundant and exist purely for convenience.

The initiation of parallel computation is done simply by having one actor's output go to the input of multiple other actors. Then all the actors that receive input data begin operation, creating parallel execution sequences. Synchronization of these parallel paths is done with the "synch" actor that waits for all of its inputs to have data and then passes each input value to an output value at the same time. Therefore this actor synchronizes parallel data paths that complete at different times.

The "operator" actor is a general-purpose function primitive that can perform any arithmetic or logical operation. When the operation name is placed inside the actor symbol and the correct number of inputs and outputs are used, the operator functions in the proper manner. For example, the " + " actor with three inputs and one output will wait for all three data values to be present and then produce their sum on the output.

The "procedure" actor is similar to the "operator" actor except that it is used for programmer-defined functions. The name of a dataflow subroutine is placed inside the symbol as an indication that this code is to be run when the appropriate inputs appear. Both this actor and the "operator" actor are subroutine calls, but the former is for higher-level procedures whereas the latter is for low-level library functions.

The "distribute" actor is used to select among multiple output paths according to an index value. When the input and the index lines have data, this actor sends the input value to the output line specified by the index value. This actor is somewhat like a "case" statement in traditional programming languages and is also used for simpler "if"-type branching.

The "select" and "arbiter" actors are conceptually the opposite of "distribute" because they reduce multiple input values to a single output value in a controlled manner. In the case of the "select" actor, an index is used to determine which input will be used. When the index and the indexed input are present, that input is passed to the output. The "arbiter" does not need any particular input; rather, it executes as soon as any input appears, generating both the input and its index as outputs. Both "select" and "arbiter" reduce many inputs to a single output: One requires a predetermined choice of inputs and the other reports its own choice.

The final actor of this particular dataflow environment is the "gate," which is used for loop control. This actor has three inputs and an output. The condition input acts as a switch to select between the initial input and the feedback input. When the loop is not active, the condition is false, which causes the initial value to be passed to the output. Once this initial value is passed to the output, the loop is activated. Active "gate" actors ignore the initial value and pass the feedback value to the output. This continues while the condition input is true. When this condition becomes false, the loop terminates, the feedback values are ignored, and a new initial value is awaited to restart the loop.

The dataflow actors described here can be used to write arbitrarily complex algorithms. The design in Fig. 2.14 shows the computation of Fibonacci numbers in a recursive and parallel style. Notice that this subroutine starts at the bottom with its input parameters and filters up to the top where the function value is returned. This layout uses only the "operator," "procedure," and "distribute" actors and thus does not make use of the powerful concurrency control available in the environment. Nevertheless, graphically specified dataflow is a useful environment for the design of algorithms, both serial and parallel.

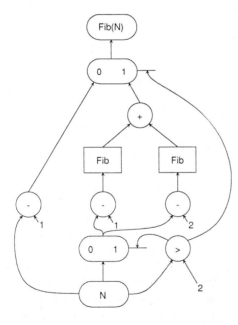

FIGURE 2.14 Dataflow layout for Fibonacci numbers. Flow begins at the bottom and filters upward.

2.4 Component Level

The next level of abstraction that defines a class of design environments is the component level. Environments of this type are usually composed of primitives that perform a small number of arithmetic or logical operations. For example, an adder or a small block of memory are acceptable component-level primitives, but an entire processor or a fully interfaced bank of memory are not.

A number of different environments exist at the component level. **Register transfer** is an environment that is used to describe the design of processors. **ISP** is a language that is used to describe the instruction set of a computer and how those instructions are implemented. Even **integrated-circuit packages**, although quite variable in their functionality, form a component-level environment in common use. However, this is also a layout environment and so it will be discussed later.

2.4.1 Register-Transfer Environment

The first component-level environment of interest is **register transfer** [Bell, Grason and Newell]. In this environment, the components are specific hardware manipulations of control and data with connections showing the flow for the desired algorithm. This environment can be used to describe the instruction set of computers because it shows the flow of control in a processor that effects the machine's operation. It is also useful in the design of algorithms. Actual implementations of these components were used to teach hardware design to university students.

The components of the register-transfer environment are named after the components of the PMS environment. Their level of detail is much greater and fewer assumptions are made about the contents of a component. For example, register transfer has no processor but instead constructs one from control, memory, and data operations.

The set of register transfer components is shown in Fig. 2.15. There are four classes of components: control (K), memory (M), data/memory (DM), and transducers (T). The control and memory components are similar to their counterparts in the PMS environment. The data/memory component is similar to the PMS processor, although much less powerful. The transducer component combines the functions of the PMS transducer and the PMS data operation.

Two PMS components are missing from this list: the switch and the link. The switch is a type of control component in register transfer, as will be discussed. The link does not appear in the register-transfer environment because of the connection distinctions. Rather than describe a connection in terms of physical distance, register transfer distinguishes connections by their contents.

Register transfer has four types of connections between components: bus, control, Boolean, and miscellaneous. The bus is a general-purpose data and control connection that passes synchronized information between components. Control connections are used to link the control

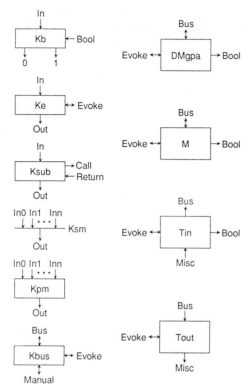

FIGURE 2.15 The components of the register-transfer environment. "K" is control; "D" is data; "M" is memory, and "T" is input/output.

components and also to invoke activity in other components. Boolean connections are special-purpose signals that aid the flow of execution by linking conditions with decision-making control components. Finally, miscellaneous connections exist to link these register-transfer components with external devices.

Every register-transfer design has exactly one K_{bus} component. This controls the bus connections for all other components and is the source of initial and terminal control that links the design with the outside world. In student hardware laboratories, where register-transfer modules were used most, this component provides the manual interface to the running circuit.

Flow of control is done with four components: K_b, K_{sub}, K_{sm}, and K_{pm}. The K_b component does a branch, conditional on a Boolean input. Only one of the two output connections will be invoked, causing that control module to execute next. The K_{sub} component executes a subroutine of control modules by invoking the first module, awaiting a completion signal on the return line, and then invoking the next sequential control component. The K_{sm} does a serial merge, invoking the next control component when any one input line is raised. This is useful in loops where multiple execution paths enter a single control stream. It is also useful for parallel execution control. Such a situation can exist if a single control-invocation line connects to multiple control components. Then the synchronization of these control paths can be accomplished with the K_{sm} or the K_{pm} components. The K_{pm} does a parallel merge that waits for all input lines to be invoked before it generates its own output.

All of these control components are used to construct a flow of execution in a register-transfer design. However, none of them cause any data to be manipulated. All data operations are controlled with the evoke component, K_e. The K_e acts like a subroutine call to a data component: It initiates the data operation, awaits completion, and then invokes the next sequential control component. All data components communicate between each other on the bus and communicate back to the control components via Boolean connections.

There are two internal data components: the DM_{gpa} and the M. The DM_{gpa}, or "general-purpose-arithmetic" component, is able to perform the simplest of arithmetic and logical operations. It holds two registers that it can move, add, subtract, complement, logically AND, logically OR, and shift. It generates Boolean signals that indicate numeric test results. This component is the workhorse of any algorithm done in the register-transfer environment. The M is the other internal component and it stores memory. To use it, both a memory address and a data value must pass across the bus on separate control cycles. The Boolean output signal is used in slower memories to indicate the completion of the operation. This allows other work to proceed in parallel with the memory access. The size and speed of the memory varies, yielding a subclass of these components.

Transducer components are used to convert between bus data and analog signals for external use. Input transducers can handle analog input, switch-bank sensing, and keyboards. Output transducers can print, plot, and produce general analog signals.

Figure 2.16 shows an example of register-transfer layout that normalizes a floating point number. The two DM_{gpa} components store and manipulate three values: a sign flag called Flag, the mantissa (Man) and the exponent (Exp). The top six components set the sign flag and ensure a positive mantissa. Following that, the mantissa is shifted until there is a "1" in its high bit (a normalized fraction is between one-half and one). Each shift of the mantissa requires an increment of the exponent to keep the overall value correct. When the number has been normalized, the sign flag is used to restore negative values.

The bus-control component links the data buses and controls the execution of the algorithm. Many register-transfer designs are done

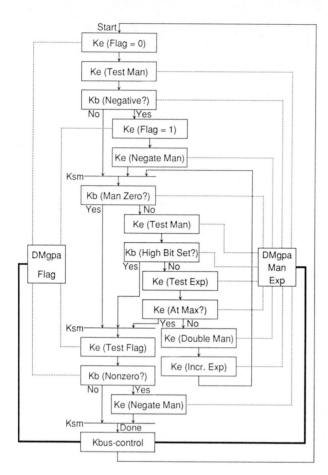

Start
Ke (Flag = 0)
Ke (Test Man)
Kb (Negative?)
No   Yes
Ke (Flag = 1)
Ke (Negate Man)
Ksm
Kb (Man Zero?)
Yes   No
Ke (Test Man)
Kb (High Bit Set?)
Yes   No
Ke (Test Exp)
Ke (At Max?)
Yes   No
Ke (Test Flag)   Ke (Double Man)
Ke (Incr. Exp)
Kb (Nonzero?)
No   Yes
Ke (Negate Man)
Ksm   Done
Kbus-control
DMgpa Flag
DMgpa Man Exp
Ksm

FIGURE 2.16 Register-transfer layout to normalize a floating-point number. Shown complete with bus connections (thick) and data-evocation lines (dotted).

without explicitly showing the data and K_{bus} components or any bus connections. Figure 2.17 shows the same algorithm without these datapath complexities. This makes the design simpler without sacrificing the detail of this environment, because the data components and their connectivity can be inferred from the control components.

In addition to being a useful environment for algorithm specification, register-transfer designs are able to be built in hardware. Every component in the environment has associated electronic modules that can be wired together. Thus a working register-transfer machine can be built directly from these designs. Although these register-transfer modules are no longer used, the concepts in this environment are still valid and can be realized with other implementation methods.

Many hardware-description languages textually capture the same information as the register-transfer environment does. One language

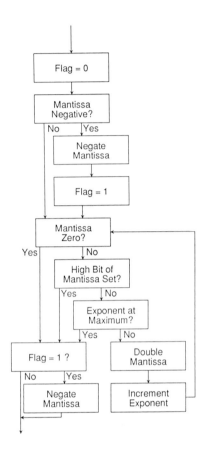

FIGURE 2.17 Register-transfer layout to normalize a floating-point number. Shown without bus or data-evocation connections.

separates data structure from control [Baray and Su], similar to the distinction made in register transfer.

2.4.2 ISP Environment

Bell and Newell also defined a component-level environment for describing computers [Bell and Newell]. **ISP**, or **instruction-set processor**, is a language for describing the instructions in a processor and how they are implemented. Notation is provided to specify the machine organization, to manipulate machine registers, and to specify parallel or pipelined execution. For example, the statement:

$$(OP = 1) \rightarrow (M[OPERAND] \leftarrow AC; AC \leftarrow 0);$$

states that if OP is equal to 1 then two actions will occur in parallel: The value of the accumulator register AC will be deposited at memory location OPERAND and the accumulator will be cleared. This statement handles the decoding and execution of a "deposit and clear accumulator" instruction. Prior to such a statement, the meanings of every term must be defined. For example, the OP field may be defined as:

$$OP := M[PC]<0:2>$$

which means that the three high bits of memory-location PC define this field. Similarly, the meaning of PC, AC, and even the memory M must be explicitly defined. Thus ISP resembles a programming language, complete with declarations and executable statements.

Nonparallel sequencing is specified with the "next" clause, as in the following example:

$$(AC \leftarrow AC + T; \text{ next } M[D] \leftarrow AC)$$

This ensures that the addition of T to AC is completed before the result is stored at memory location D. Determination of when an operation can be done in parallel is dependent on the architecture of the processor. This architecture is specified implicitly by the various ISP components and the operations that can be performed on them.

ISP is often used to describe computer instruction sets. It is a highly algorithmic environment, which leads one to believe that it should not be a component-level environment. However, there are direct associations between ISP statements and the machine registers that implement them because the text can be used to infer a diagram of the target machine. Thus ISP is another example of a component-level environment. These environments illustrate detailed electronic design without completely precise specification.

2.5 Layout Level

Layout-level environments exist primarily for the generation of final manufacturing specifications. Every component in these environments has precise dimensions, and its location in the design corresponds exactly to its final position in the manufactured circuit. Although some layout-level environments have information about higher levels of design, this information is not mandatory. For example, nMOS design can include connectivity information, behavioral descriptions, and links to equivalent circuits in other environments. Ultimately, however, all that matters is the geometry of the layout, so it is acceptable for the nMOS environment to contain only rectangles as primitives. The needs of the designer determine the complexity of these layout environments.

Three different types of layout environments are worth mentioning here. First, the integrated-circuit domain includes the MOS, bipolar, gallium arsenide, and other microsilicon environments. MOS devices have received much attention lately as a desirable medium in which to build large systems [Glasser and Dobberpuhl]. Second, the macro electronic environments include printed-circuit-board design, wire-wrap-board design, and others of that ilk. These environments have been used to build large systems for many years. Finally, the pure artwork environment exists for the generation of graphics, not electronics, and is useful in its own right as well as in combination with circuit layout.

Besides providing different layout environments, a good design system should allow these components to be intermixed. Mixed environment designs are useful for process experimentation, in which new structures are being combined. They are also valuable in design when partial layout is used with schematics or other nonspecific components. Finally, mixed environments are useful in large system designs that demand interconnection of many components that are not of uniform origin.

2.5.1 nMOS Environment

The *n*-channel metal oxide semiconductor environment is one of two MOS environments that will be discussed. The distinguishing characteristic of MOS is its use of a **field-effect transistor** to do switching. In this type of transistor, the switching element runs over the switched line, causing a field to be produced at the intersection (see Fig. 1.18b). This field conditionally inhibits the flow of current in the switched line. An algorithm that has been reduced to a collection of logic gates can be implemented in the MOS environments. This is because all logical operations can be implemented with combinations of these transistors.

There are different layers of conducting and semiconducting material that are available for the implementation of nMOS. The switching layer is **polysilicon** and is drawn with diagonal parallel lines in this book. The switched layer is **diffusion** and is drawn with a textured area of perpendicular lines. These two layers cannot cross each other without forming a component: either a transistor or a direct contact. To make general-purpose connections that do not interfere with other layers, there is a **metal** layer, drawn with a texture of dots, which can cross polysilicon or diffusion without connecting. Special contacts can be used, however, to connect metal to the other layers.

These three layers are the only ways to run wires in an nMOS circuit, although some fabrication processes have an extra layer of metal or polysilicon. There are, however, many other layers that are used to make a chip. An **implant** region, drawn with light texturing, is used over the switching area of a transistor to set it permanently "on." This alternate kind of transistor is useful for obtaining the appropriate voltage levels needed to implement logical functions. The direct connections between metal, polysilicon, and diffusion use intermediate layers such as the **contact-cut** and the **buried-contact** layers. The entire chip is typically covered with a layer of protective coating called **overglass**, which must be removed at the bonding pads to allow the chip to be externally wired. And to make things even more complicated, some nMOS fabrication processes have special layers, such as light implant, hard or light enhancement, and secondary contact, which are rarely, if ever, used during layout.

Given that there are many layers and that geometry is the only important result, one is tempted to provide a purely geometric set of design components for the nMOS environment. In fact, any implementation of the nMOS environment that does not have components for specifying arbitrary shapes on all the layers is likely to fall short of the designer's needs. However, it is possible to provide more complex components that will make much of the design phase easier. If these components form a consistent view of nMOS at some higher level of design, then the problems of conversion in and out of this environment will be lessened. Also, layout environments will be able to cater to circuit designers if these components include connectivity.

Figure 2.18 shows a set of high-level primitives that can be used for nMOS design. These components, used by the Electric system (see Chapter 11), have been adequate for the layout of medium- to large-scale ICs. There are, of course, the two types of transistors in this environment: enhancement (the normal polysilicon over diffusion)

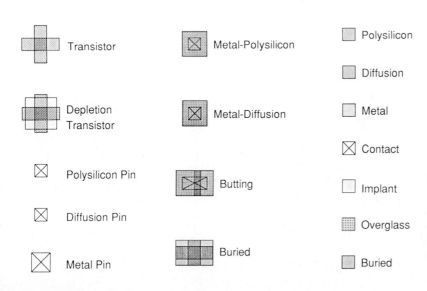

FIGURE 2.18 The primitive components of the nMOS environment. Transistors and intralayer pins are on the left; interlayer contacts are in the center; basic-layer components are on the right.

and depletion (with the implant layer). In addition, there are four basic layer-to-layer contacts: a metal-to-polysilicon, a metal-to-diffusion, a triple-layer butting contact, and a polysilicon-to-diffusion buried contact. In order to connect these components, there are wires that can run in all three layers. These wires can make junctions to themselves with the three-pin components. Finally, the right side contains basic-layer components that can be used to make any geometric structure. The design of an inverter with these primitives is illustrated in Fig. 2.19.

In order to encourage their use, these nMOS primitives must be more flexible than they appear in Fig. 2.18. For example, the transistor component should be able to scale properly in either dimension and even to bend (see Fig. 2.20). Also, the contacts should be available in other configurations, either by having extra primitive components or by parameterizing the components that are there.

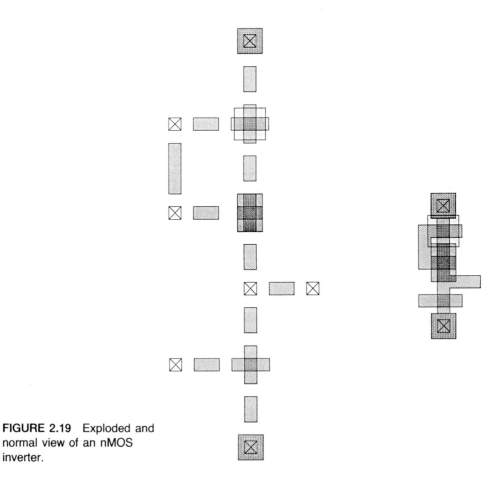

FIGURE 2.19 Exploded and normal view of an nMOS inverter.

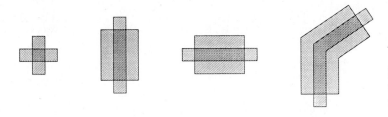

FIGURE 2.20 Varying configurations for an nMOS transistor.

The use of advanced design components in a layout-level environment such as nMOS does not hinder the flexibility of circuits that can be built. Circuit and VLSI designers prefer to deal with such a set of primitives because it agrees with their connectivity-oriented view of circuitry. For mask designers, the transistor and contact primitives can be ignored so that design can be done solely with geometric components. However, many mask designers understand enough about circuitry to appreciate these higher-level primitives. In fact, some designers prefer to use **cell libraries** with even-higher-level primitives that perform complex circuit functions [Newkirk and Mathews]. All that is needed for any designer is a small but powerful set of components that produces correct layout while providing information relevant to the task at hand.

2.5.2 CMOS Environment

The complementary metal oxide semiconductor (CMOS) environment is very similar to the *n*-channel MOS devices previously discussed. As in all MOS environments, a field-effect transistor is used that is simple enough to be modeled with a few components. Also, as in all MOS environments, there is a polysilicon layer that gates a diffusion layer, and one or more metal layers that can run anywhere. There are also contact structures, special implant layers, optional additional polysilicon layers, and an overglass layer for chip protection. In other respects, however, CMOS and nMOS are quite dissimilar.

CMOS design is a wholly different way of thinking about circuits [Weste and Eshraghian]. In nMOS, the enhancement transistor is a gate and the depletion transistor is used to obtain a "weak" logical high or becomes logically low by being gated to ground. Thus nMOS design centers around the operations that make a signal either weakly high or strongly low. This asymmetry means that certain transitions take longer to complete (setting a signal high takes longer than setting it low) and that certain states take much more power (holding a signal low consumes more power). The amount of space used in an nMOS design is small, but the power and speed characteristics are limiting.

In CMOS there are two transistors that either open the gate when on, or open the gate when off. There are no weak truth values and so the speed of any operation is as fast as the fastest nMOS action. Power requirements are also lower. The only problem is that every operation must be duplicated and this uses larger amounts of space. Fortunately, the space costs are not always of major concern and there are dynamic design styles that reduce the area [Weste and Eshraghian]. For all these reasons, CMOS is rapidly becoming the most popular integrated-circuit environment.

Because the two CMOS transistors perform logically opposite functions, they and all of their connecting diffusion wires must be enveloped in implant areas that dictate the sense of the function. The CMOS designer must therefore consider not only metal, polysilicon, and diffusion, but also implant regions: **well** and/or **native substrate**.

Designing an environment for CMOS can take many paths. In addition to having a set of geometric primitives for the mask designer, the environment can provide contacts and transistors for use by chip designers. Given that all diffusion paths are implant sensitive, however, these higher-level transistor and contact primitives will fall short of completing the layout, because the implant areas will have to be specified with lower-level primitives. The solution, shown in Fig. 2.21, is to define two sets of diffusion-related primitives for the two implant regions. This means that, in addition to having two transistors, there are two types of diffusion wire, two wire junction pins for diffusion, and two metal-to-diffusion contacts. This allows the design to be done

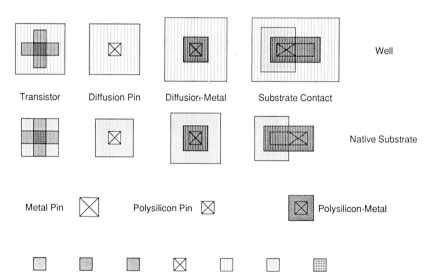

FIGURE 2.21 The components of the CMOS environment. Top two rows are implant-sensitive components. Bottom row contains basic-layer components.

without undue considerations for the implant regions, because most of these layers will appear as a byproduct of using the proper components and wires. Figure 2.22 shows a CMOS inverter made from these components.

CMOS illustrates the problem of having too many layers to manipulate conveniently. In typical environments, the design is so confusing that a set of **standard cells** is made and used for all layout, thus avoiding

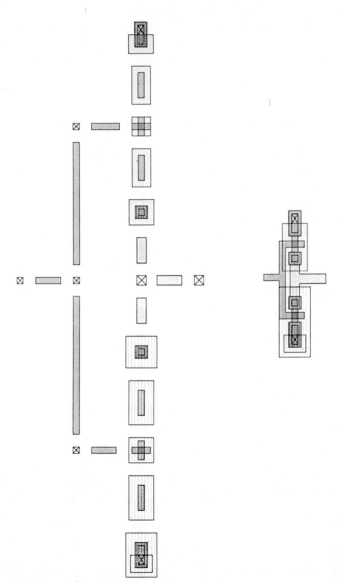

FIGURE 2.22 Exploded and normal view of a CMOS inverter.

the need to deal with the complexities of this environment [Feller]. Standard-cell packages change the nature of the design environment by providing high-level primitives. In addition, they can impose their own design methodologies that totally redefine the nature of the layout task [Schediwy]. However, direct CMOS layout can still be accomplished by simplifying the design components to a manageable and consistent set, so that confusion is reduced and design is more easily done.

2.5.3   Bipolar Environments

The bipolar environments form a family, much like the MOS family. Many notions can be shared among the members of this family, and many aspects are different. One commonality is that the basic switch for the bipolar environments is a **junction transistor**, which has a base, an emitter, and a collector. This transistor is much different in structure from, and is more complex than, the field-effect transistor used in MOS (see Fig. 2.23). Also, bipolar environments rarely make component connections with semiconducting material: Usually metal wires are the only ones used.

A sample bipolar environment is shown in Fig. 2.24. The transistor

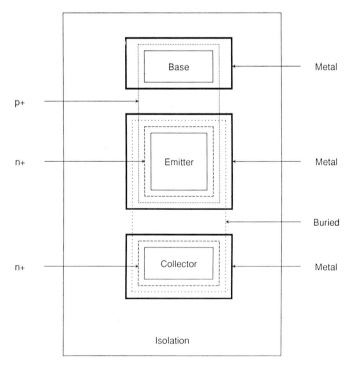

FIGURE 2.23 A typical *npn* bipolar transistor.

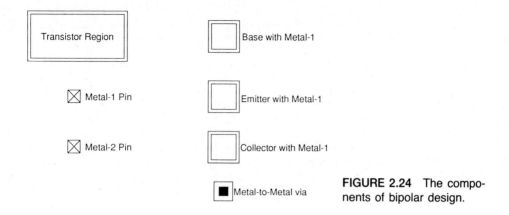

FIGURE 2.24   The components of bipolar design.

is an empty implant region into which base, emitter, and collector components can be placed. These components have metal surrounding them because metal is the only way to make a connection. However, the many other layers needed to specify a bipolar circuit precisely are not shown, because they can be automatically derived from these simpler components [Szabo, Leask and Elmasry]. The bipolar environment shown here has two layers of metal, so it needs two metal pins and an intermetal contact (called a **via**). Some bipolar environments also have polysilicon or a third metal layer.

The flexibility of designing custom transistors, resistors, and diodes is exactly what is needed in a bipolar environment. The limited set of transistors available in MOS is not sufficient. Instead, the transistor must be broken down into its basic components in order to satisfy the designer. Note, however, that some sensible combinations of layers can be made as in the case of the bases, emitters, collectors, and the via. By insisting that the components of a transistor all connect to the transistor implant area, the design system can still treat transistors as single objects and capture more information for the circuit designer. Thus even in a very low-level environment such as this, it is possible to provide more than purely geometric design primitives.

2.5.4   Integrated-Circuit–Packages Environment

Integrated-circuit (**IC**) packages are the component-level environment of choice today. Most algorithm design is converted to this for implementation, because this is the medium that is commonly available. Standardized voltage levels allow these packages to be directly interconnected and standard pin spacings allow them to be physically connected.

FIGURE 2.25 74-00 quadruple two-input NAND gate package.

A typical IC package has two rows of wires running on either side of a plastic or ceramic body (see Fig. 2.25). These chips, called **dual inline packages** (DIPs) always have the wire leads spaced 0.1 inch apart or multiples thereof. DIPs can have as few as two pins but typically have 14 or 16 pins and occasionally have over 100. Some packages have a single row of pins and are **single inline packages**. Recently, packages have appeared with pins on all four sides, called **quad packages**. Finally, there are packages with pins spaced uniformly across the underside, and they are called **pin grid arrays**.

Package connections depend on the surface to which the packages are attached. A wire-wrap board is one that has pin sockets on one side attached to long terminal posts on the other. The ICs plug into the side of the board with sockets and the wires run between the terminals on the other side.

Printed-circuit boards are a less flexible method of connecting IC packages because they are more permanent. The packages are soldered to the board and the wiring is etched along the board surface, making all changes difficult. Because they are more permanent, printed-circuit boards are used in production designs that need no debugging, but do need to be produced inexpensively. Many printed-circuit boards contain multiple layers of wire paths. This allows complex connectivity without the problems of path crossing. Simple multilayer boards have wiring on two sides but some printed-circuit boards have six or more layers.

The integrated-circuit packages have a wide range of complexity. Early packages were called **SSI**, or **small-scale integration**. This meant that a handful of gate-level components could be built into a single package. For example, the contents of the package in Fig. 2.25 is of SSI complexity. The power and ground pins must be connected to standard voltages (TTL designs use +5v and 0v respectively) and the levels on the logic gate pins also use standard voltages (higher than a threshold for true, lower for false). This particular SSI package has four NAND gates, each with two inputs. Typically, anything with up to 40 logic gates is within the realm of SSI.

More advanced packages are in the **MSI** family (**medium-scale integration**). The logic diagram in Fig. 1.11 is typical of the complexity of MSI packages. In this family, the complexity limit is about 500 gates, although this number and all other family distinctions are somewhat arbitrary.

The most densely packed integrated circuits today are **LSI**, or **large-scale integration**. The capabilities of a single LSI chip include large memories (one million bits in a package), multipliers (16-bit parallel multiplication), and processors (Z80 or MC68000 single-chip processors). LSI chips can be as complex as one million gates, which covers all packaged chips that have been manufactured to date.

Beyond LSI is the promise of **VLSI** (**very-large-scale integration**) and even **ULSI** (**ultra-large-scale integration**). Such chips will have so much information that the management of their design presents difficulty for present-day design systems. Currently, complex systems are being built in **WSI** (**wafer-scale integration**), in which hundreds of integrated circuits interconnect on a single wafer. Entire architecture-level descriptions can be embodied in a single VLSI system, causing a paradox in the notion that integrated-circuit packages are at the component level of design. This only confirms the futility of trying to categorize environments: What is an entire architecture to one designer is merely a component in a layout to someone else.

2.5.5 Artwork Environment

Artwork is not electrical. An artwork environment is simply a set of components for the design of arbitrary graphics. These graphics can be used to embellish circuit layouts (which is common in PC-board and chip layout) or they can be used independently of electronics; for example, to make figures for papers and slides. In any case, an artwork environment must be able to do any kind of layout, so a first reaction is to make this environment have a few basic geometric components such as a line, a polygon, a circle, and some text.

A look at any sketchpad editor, however, will reveal that complex primitives are often provided to the artist. Arrowheads, triangles, splines, and many more shapes appear in design menus. Even circuit and other scientific symbols can be found. Thus the existence of this environment in conjunction with the other electronic environments described here can be beneficial, given that components from all environments can be mixed in a single design. Since a CAD system must provide the graphical human interface, I/O, database facilities, and design primitives, it may as well throw in a few purely graphical components and be truly useful.

2.6   Summary

This chapter has illustrated a wide range of environments that can be provided for digital electronic design. Each environment consists of primitive components and their connections. It is important that an environment model the designer's notions of circuitry and provide a comfortable means of communication with the design system. Environments exist at many levels of specification to help designers from the machine architect to the IC mask designer. There are even some environments that have nothing to do with electronics. All together, it should be seen that a well-planned set of components can be a great aid to design, and a good collection of environments can work together to provide a powerful design facility.

Questions

1 How would you modify PMS for modern computer architectures?

2 What is the most difficult aspect of translating from dataflow to control flow?

3 Propose an alternative organization of design environments to the one in Fig. 2.2.

4 How would you change the schematics environment to include wires with multiple signals (such as buses or transmission lines)?

5 How does the sticks environment differ from the schematics environment?

6 How would you graphically describe the ISP environment?

7 Why is CMOS design more difficult than nMOS?

8 Convert the AND, OR, NOT, and IMPLICATION to equivalent logic using only NOR.

References

Baray, Mehmet B. and Su, Stephen Y. H., "A Digital System Modeling Philosophy and Design Language," Proceedings 8th Design Automation Workshop, 1-22, June 1971.

Bell, C. Gordon; Grason, John; and Newell, Allen, *Designing Computers and Digital Systems Using PDP 16 Register Transfer Modules*, Digital Press, Maynard, Massachusetts, 1972.

Bell, C. Gordon and Newell, Allen, *Computer Structures: Readings and Examples*, McGraw-Hill, New York, 1971.

Bell, J. L. and Slomson, A. B., *Models and Ultraproducts: An Introduction*, North-Holland and American Elsevier, New York, 1971.

Davis, A. L. and Drongowski, P. J., "Dataflow Computers: A Tutorial and Survey," University of Utah UUCS-80-109, July 1980.

Dennis, J. B., Fosseen, J. B., and Linderman, J. P., "Data Flow Schemas," Proceedings International Symposium on Theoretical Programming, 187-216, 1972.

Department of Defense, "Graphic Symbols for Logic Diagrams," MIL-STD-806B, Washington, D.C., February 1962.

Feller, A., "Automatic Layout of Low-Cost Quick-Turnaround Random-Logic Custom LSI Devices," Proceedings 13th Design Automation Conference, 79-85, June 1976.

Gibson, Dave and Nance, Scott, "SLIC—Symbolic Layout of Integrated Circuits," Proceedings 13th Design Automation Conference, 434-440, June 1976.

Glasser, Lance A. and Dobberpuhl, Daniel W., *The Design and Analysis of VLSI Circuits*, Addison-Wesley, Reading, Massachusetts, 1985.

Malachi, Yonatan and Owicki, Susan S., "Temporal Specifications of Self-Timed Systems," Proceedings C-MU Conference on VLSI Systems and Computations (Kung, Sproull, and Steele, eds), Computer Science Press, 203-212, 1981.

Moszkowski, Ben, "A Temporal Logic for Multilevel Reasoning about Hardware," *IEEE Computer*, 10-19, February 1985.

Newkirk, John and Mathews, Robert, *The VLSI Designer's Library*, Addison-Wesley, Reading, Massachusetts, 1983.

Patil, Suhas S., "An Asynchronous Logic Array," Project MAC tech memo TM-62, Massachusetts Institute of Technology, May 1975.

Rodriguez, Jorge E., *A Graph Model for Parallel Computations*, PhD dissertation, Massachusetts Institute of Technology, Report MAC-TR-64, September 1969.

Schediwy, Richard R., *A CMOS Cell Architecture and Library*, Masters thesis, University of Calgary Department of Computer Science, 1987.

Siewiorek, Daniel P.; Bell, C. Gordon; and Newell, Allen, *Computer Structures: Principles and Examples*, McGraw-Hill, New York, 1982.

Szabo, Kevin S. B.; Leask, James M.; and Elmasry, Mohamed I., "Symbolic Layout for Bipolar and MOS VLSI," *IEEE Transactions on CAD*, to appear, 1987.

VanCleemput, W. M., "An Hierarchical Language for the Structural Description of Digital Systems," Proceedings 14th Design Automation Conference, 377-385, June 1977.

Weste, Neil, "Virtual Grid Symbolic Layout," Proceedings 18th Design Automation Conference, 225-233, June 1981.

Weste, Neil and Eshraghian, Kamran, *Principles of CMOS VLSI Design*, Addison-Wesley, Reading, Massachusetts, 1985.

Williams, John D., "STICKS—A graphical compiler for high level LSI design," Proceedings AFIPS Conference 47, 289-295, June 1978.

REPRESENTATION

3.1   Introduction

The most fundamental part of a VLSI design system is the representation that it uses to describe circuits. Proper representation is crucial; it can make or break the entire program. A good representation allows the circuit to be conveniently stored, accessed, and manipulated, but a bad representation limits the system by making it wasteful, hard to debug, and slow. This chapter is about representations for VLSI design databases.

The most important concept is to make the representation similar to the mental model in the user's head. This ensures that simple changes to the system requirements can be implemented with simple changes to the representation. It also results in more intuitive interaction when dealing with lower levels of the system. For example, internal error messages will make more sense if the representational context of the error is familiar to the user.

Another point about representation is that it must be tailored to produce the best response to the user. This means that the classic tradeoff of space and time must be evaluated and used to plan the representation. If the design system is destined to run on a very fast computer with limited memory, then the representation must be compact, storing less and letting the system compute more. If the code has to run on a slower machine with large physical and virtual memory, then the representation should keep more information to speed the immediate data-access needs. This large memory configuration is more common today and will be presumed throughout the chapter.

The first part of this chapter contains a general discussion of representation properties that should be found in any system. Such properties include object orientation for maximum extendibility, and techniques for improved response. Also covered are the basic functions of a database that make demands on the representation. These functions include memory allocation, disk formats, change, and undo control.

The second part of this chapter addresses itself specifically to the four aspects of VLSI design discussed in Chapter 1: hierarchy, views, connectivity, and geometry. Hierarchy is represented by having complex components that are actually instances of other cells lower in the hierarchy. Views are represented with highly flexible data structures that can hold descriptions of any kind and can link to other views. Connectivity is represented with networks of components and their connections. Geometry is represented with shapes, graphic attributes, and transformations. As this chapter will show, it is not a simple task to include all these representations in a design system.

3.2   General Issues of Representation

3.2.1   Objects and Links

A very important feature of database representation is flexibility. In VLSI design, there seems to be no limit to the analysis and synthesis tools that can be used. All these tools manipulate different information, which must be represented. Therefore the representation must be flexible enough to contain new information wherever and whenever needed. To achieve this flexibility, the representation must not have any inherent limits. For example, fixed-sized arrays are generally a bad idea because they cannot adapt to unusual design needs. Experience has shown that no preconceived limits to the design activity are going to be acceptable to every designer. Just when it appears that one million gates is enough for a layout, there will be a demand for ten million. Just when it seems that only 50 layers are needed for integrated circuit fabrication, some process engineer will propose 60. Even simple user-interface attributes, such as the number of characters in a cell name, cannot be determined in advance.

The best way to achieve flexibility is to throw out the fixed-array construct and use **linked lists** [Knuth]. Linked lists are a better way to describe varying-length information because they consume less space for small tasks and extend cleanly for the big jobs. The list contains **objects**, which are records of data that are scattered randomly in memory. Ordering of the objects is done by having a list head pointing to the first object and a link in each object that points to the next object. A special "null" pointer in the last object identifies the end of the list.

Figure 3.1 shows the difference between linked lists and arrays. In this figure, four records of data are represented with three data items per record. The array representation in Fig. 3.1(a) stores the records in sequential array order. Objects are referenced by index so the addressing of the Data B value of record N requires access to location Base $+ N \times 3 - 2$ of memory. The linked-list technique shown in Fig. 3.1(b) references objects by direct memory pointers so addressing of the Data B value of the record at address N only requires access to location $N + 1$. To allow for an unlimited number of data, objects can be allocated from the freespace outside the extent of the design program rather than taken from preallocated memory inside the system.

The advantages of linked lists are their shorter computation when addressing, their unlimited expandability, and the ability to insert or

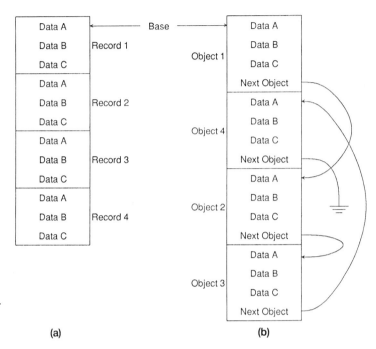

FIGURE 3.1 Arrays of records versus linked lists: (a) Arrays are difficult to reorganize (b) Linked lists are better.

(a)

(b)

delete records in the list without having to move memory. The disadvantages of linked lists are the extra storage space that the links consume and the fact that finding an entry by its position in the list requires search. Extra storage should not be a concern, since time is usually more important than space. The problem of searching for an entry by its index in a list is rarely encountered in a pointer-based system because there are few if any index values.

One decision that must be made when using linked lists of objects is whether to make the lists doubly or singly linked. A singly linked list is one that has pointers in only one direction (to the "next" object). A doubly linked list points in both directions (to the "next" and the "previous" objects). Doubly linked lists are faster when insertions or deletions are being made, because more information is available for pointer manipulation. However, they take up more space and do not always produce a noticeable speed improvement. The design system programmer must determine the needs of each situation and decide whether or not to use doubly linked lists.

3.2.2   Attributes

Every object in the database is a record that contains data items. These items are **attributes** that have names, values, and often specific types associated with them. Figure 3.2 shows typical attributes for a

Attribute	Value
Center	(55,102)
Size	2×6
Orientation	$0°$
Name	Transistor
Cell	Inverter
Environment	nMOS
Connections	Polysilicon, Diff In, Diff Out

FIGURE 3.2 Attribute/value pairs for a transistor.

transistor object. In practice, there is a never-ending set of attributes that must be represented, caused by the never-ending set of synthesis and analysis tools. Each tool is used to manipulate a different aspect of the object, so each introduces new attributes. Given that a representation should place no limits on the design activity, it is necessary that the creation of object attributes be similarly flexible. Since it is impossible to determine in advance exactly which data entries are going to be needed when planning a record organization, the program must be able to create new attribute entries during design without any prearrangements with the database.

The most general scheme for expandable storage of object attributes is to use a linked list of attributes and their values. In such a scheme, an object consists of nothing but a list head that links its attribute/value records (see Fig. 3.3). This scheme is used in LISP-based design systems [Batali and Hartheimer; Brown, Tong, and Foyster]. It has the disadvantage of making the system very slow because search is required to find an object's attribute values. To speed up attribute

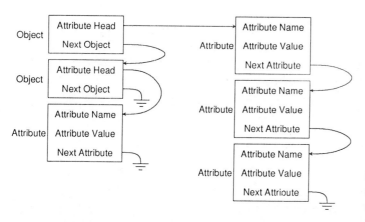

FIGURE 3.3 Object consists solely of linked lists of attribute/value pairs.

access, a hashing table can be stored in the object. Such a table is an array that gives initial guesses about where to look in the attribute list. Although this technique speeds access, that speed will degrade when the number of attributes grows larger than the size of the hash table. Also, hashing requires some algorithm to produce the hash key and that, too, will add delay.

A good compromise between speed and flexibility is to use a combination of fixed record entries and additional hashed attributes (see Fig. 3.4). Records with preassigned attribute slots are the very fastest, so they should be used for the most frequently accessed entries. One of those preassigned slots can then be the hash table that references the less frequently accessed attributes. Inside the design system, the code to access attributes can usually be tailored to use the proper method. Those pieces of code that cannot know the storage method of attributes in advance will have to search the hash table first and then step through the names associated with the fixed attributes. To speed such programs, hashed entries can be created for fixed attributes so that the hash table is the only place that need be consulted. In order to save memory, these additional pointers need be added to the hashed table only after they are referenced for the first time. Many of these facilities are available in modern object-based programming environments [Weinreb and Moon].

The scheme described here works well for any analysis or synthesis tool the attributes of which are mostly in fixed positions. However, as the number of tools grows, efficiency may decline. To prevent new tools from operating very slowly because their attributes are all in hashed entries, the representation can provide a small amount of "private" memory for each analysis and synthesis tool. This memory is a fixed attribute that is dedicated to the particular tool. The analysis

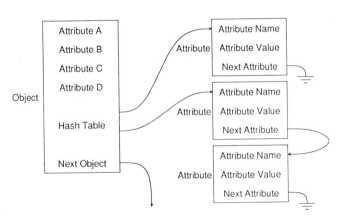

FIGURE 3.4 Fixed and hashed-extendible attributes.

and synthesis tools can use that entry for faster access, either by placing the most important attribute(s) there or by using it as a pointer to their own private attribute information. In general, the private attribute needs to be only one word in size to be useful.

Attribute flexibility is very important in a design representation. But so is efficiency. Therefore it is worthwhile to incorporate a collection of storage techniques to achieve the best possible results. This permits the design system to be powerful and fast, even at the expense of code complexity.

3.2.3 Prototypes

In any useful design environment, much information will be stored on an object. Some of this information is specific to each object, but much of it is general to all objects of a given class. In such circumstances it is advantageous to place the common information in a single common location. This location is called a **prototype** (see Fig. 3.5).

Prototypes should exist for every object in a design. A single prototype defines a class of objects and describes every actual object

Prototype for NAND Gate

Name = "NAND"
Default Size = 10 x 6
Environment = Schematics
Connections = In1, In2, Out
Behavior = "Out = In1 NAND In2"

Prototype for In2 Connection

Name = "In2"
Location = Lower-Left
Type = Input

Prototype for In1 Connection

Name = "In1"
Location = Upper-Left
Type = Input

Prototype for Out Connection

Name = "Out"
Location = Center-Right
Type = Output

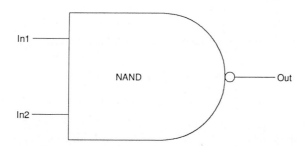

FIGURE 3.5 Prototype information.

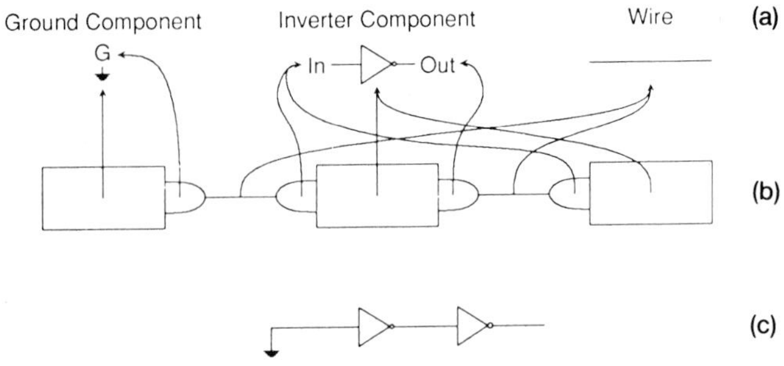

Ground Component   Inverter Component   Wire   (a)

G   In ──▷── Out

(b)

FIGURE 3.6 Instances and
prototypes: (a) Prototypes (b)
Instance objects (c) Represented design.

(c)

in that class. Attributes on the actual objects point to their prototypes (see Fig. 3.6) and, if it is desired, attributes on the prototypes can head linked lists of the actual objects. New attributes to be added to the database can be placed in the prototype object if they are common to everything in the class, or they can be placed in the individual objects if they are specific to them.

3.2.4   Object Marking

Many applications work with a subset of the objects in a design. Examples include selection of objects in a particular area, on a given net, or on a certain layer. To make this selection, some criterion is applied to the database and those objects in the subset are marked for future analysis. This intermediate marking step is necessary because the selection algorithm may be very time consuming and the analysis code repeated often. For this reason it is important to be able to mark efficiently a selected subset of objects for rapid subsequent identification.

The obvious marking technique is to set a **flag** on an object when it has been selected. This flag is a Boolean attribute on the object and so it consumes little space. The analysis phase needs only to examine the flags to determine the state of the object. The problem with marker flags is that their use requires multiple passes of every object in the design: one pass to clear all the flags, one pass to mark the desired objects, and one pass for every list of selected objects that is requested. The flag-clearing pass is the only one that will necessarily require a complete scan of the design, because many selection and examination algorithms work locally. However, they all require that the flags on every object be properly set. If the flag clearing is done after the analysis rather than before the marking, then the final analysis pass can be used to clean up the flags. This might save time but it is not a safe technique and results in a less robust program. Thus the use of flags as object markers is acceptable but potentially slow in large design databases.

A better marking technique is the use of **time stamps**. A time stamp is an integer attribute that is stored in an object when that object is selected. The value of the stamp is changed before every marking pass so that old stamp values become automatically invalid. The analysis pass then looks for time-stamp values equal to the current one that was used in marking. In addition, time-stamp values can be manipulated in such a way that extra information is encoded in the stamp value. The advantage of time stamps is that they do not have to be cleared and so they do not require that extra complete database scan. The disadvantage is that they consume an entire word of 32 or more bits to represent (be warned: small integer time stamps such as 16-bit values can overflow), whereas flags require only one bit.

When the analysis phase finds itself making complete passes of the design to find a small set of marked objects, then it is time to start accumulating the marking information outside of the database. Rather than storing marking attributes on the objects, pointers to the marked objects should be stored in external lists. These lists are as long or short as they need to be and are not dependent on the size of the design. Analysis does not have to search the entire circuit: It has its selected set in a concise list. Although this method consumes extra memory for the marking list, it actually may save memory because it does not require marker attributes in the database. The technique will take more time because the allocation, construction, and freeing of these lists is much more time consuming than is the simple setting of a flag or a time stamp, but for many applications this is worthwhile. The final decision about which marking method to use depends heavily on the nature of the selection and analysis algorithms.

3.2.5   Memory Allocation

Memory allocation is required in all phases of design. It typically consists of only two operations: the allocation of a specified amount of unused memory and the freeing of an allocated block. Some systems reduce this to a single operation that allocates on demand and frees automatically when the object is no longer used. Special **garbage-collection** operations find the unused memory by counting the number of references to each block [Deutsch and Bobrow].

Not only are objects and their attributes allocated, but temporary search lists and many other variable-length fields are carved out of free memory. It is important to deal with the memory properly to make the system run efficiently. Nevertheless, many design systems foolishly ignore the problem and trust the operating system or language environment to provide this service correctly.

One basic fact about memory allocation is that it takes time to do. Most programming environments perform memory allocation by invoking the operating system and some even end up doing disk I/O to complete the request. This means that a continuous stream of allocations and frees will slow the design system considerably. If, for example, memory is allocated to store a pointer to a marked object and then freed when the marker list is no longer needed, then every selection step will make heavy use of a general-purpose memory-allocation system. This approach is not very efficient. A better technique is to retain these allocated blocks in an internal list rather than freeing them when the analysis is done. Requests for new objects will consult this internal free list before resorting to actual memory allocation. Initially, the internal free list is empty, but over time the size of the list will grow to be as large as the largest number of these objects that is ever used.

The reason for not freeing objects is that they can be reallocated much more quickly from an internal free list. The memory allocator in the operating system cannot organize freed blocks of memory by their object nature because it knows nothing about the design program. Therefore it will mix all freed blocks, search all of them when reallocation is done, and coalesce adjacent ones into larger blocks whether or not that is appropriate. All this wastes time. The design system's own free list is much more intelligent about keeping objects of a given class together and can respond appropriately. Thus the speed of the program is improved.

Another time-saving technique is to allocate multiple objects at a time, in larger blocks of memory. For example, when the internal list of free marker objects is empty, the next request for a new object should grab a block of memory that can hold 50 such objects. This memory is then broken down into 49 free objects on the internal list and one that is to be used for the current request. This results in fewer calls to the operating system's memory allocator, especially for objects that are used heavily. In addition, some space will be saved since the operating system typically uses a few words per allocated block of memory as bookkeeping. The drawback of this block-allocation technique is that the program will typically have allocated memory that is unused, and this can be wasteful.

The biggest problem with memory allocation is that the program will run out of memory. When it does on a virtual-memory computer, the operating system will begin to page the design data onto disk. If the use of memory is too haphazard, the contents of an object will be fragmented throughout the virtual address space. Then the operating system will thrash as it needlessly swaps large amounts of memory to get small numbers of data. To solve this, it is advisable for design systems that run on virtual machines to pay special attention to the paging scheme. For example, all objects related to a particular cell should be kept near each other in memory. One way to achieve this is to allocate the objects at the same time. Unfortunately, as changes are made the design will still fragment. A better solution is to tag pages of memory with the cells that they contain. Memory used for a different cell will be allocated from a different memory page and all cell contents will stay together. By implementing virtual memory as "clusters" of swappable pages, the design activity will remain within a small number of dedicated pages at all times [Stamos].

In addition to aggregating a design by cells, clusters of virtual memory can be used for other special-purpose information such as the attributes of a particular design environment, analysis, or synthesis tool. The resulting organization will provide small working sets of memory pages that swap in when needed, causing the design system to run faster. For example, if all timing-related attributes are placed together, then those memory pages will remain swapped out until needed and then will swap in cleanly without affecting other data.

When the computer does not support a virtual address space, the design system must do its own swapping. This internal paging is typically done by having only one cell in memory at a time. The code becomes more complex because all references to other cells must explicitly request a context change to ensure that the referenced data are in memory. In virtual-memory systems, all data are assumed to be in memory and are implicitly made so. In an internally paged system, the data must be loaded explicitly. Many design operations span an entire hierarchy at once and will run more slowly if paged one cell at a time. For example, it will be harder to support hierarchical display, network tracing, and multiple windows onto different cells. Although the design system could implement its own form of virtual memory, that would add extra overhead to every database reference. It is better to have a virtual computer that implements this with special-purpose hardware. Also, letting the operating system do the disk I/O is always faster than doing it in the application program, because with the former there is less system overhead. Thus the best way to represent a design is to allocate it intelligently from a large, uniform memory arena.

3.2.6 Disk Representation

If a design is going to be represented by placing it in virtual memory, then a very large design will consume much virtual memory. To modify such a design, the system will have to read the data from disk to memory, make changes to the memory, and then write the design back to disk. This is very time consuming when the design files are large and the virtual memory is actually on disk. Even nonvirtualmemory systems will waste much time manipulating large design files. What is needed for better speed is an intelligent format for disk representation.

By placing a "table of contents" at the beginning of the design file, the system can avoid the need to read the entire file. Individual parts can be directly accessed so the design can be read only as it is needed. This is fine for efficient perusal of a design but there will still have to be a total rewrite of the file when a change is made. If each cell is stored in a separate disk file, then the operating system's file manager will perform the table-of-contents function. This allows rapid change to single cells but opens up the problem of inconsistent disk files since they can be manipulated outside of the design system.

An ideal disk format would replicate exactly what is in memory. When this is done on a virtual-memory machine with flexible paging facilities, it is not necessary to read or write the entire design file at once. Instead, the file is "attached" to virtual memory by designating the disk blocks of the file to be the paging space for the design program. Initially these memory pages are all "swapped out" but, as design activity references new data, the appropriate disk blocks are paged in. Changes to the design manipulate the disk file in only those places corresponding to the modified memory. Allocation of more memory for a larger design automatically extends the size of the disk file. Of course, writing the design simply involves updating the changed pages. This scheme requires the smallest amount of disk I/O since it accesses only what it needs and does not require a separate paging area on disk.

The only problem with storing the precise contents of memory on disk is that object pointers do not adjust when written to disk and then read back into a different address in the computer. If relocation information is stored with the file, then each page must be scanned after reading and before writing to adjust the pointers. This requires extra time and space in addition to creating the problem of having to identify every pointer field in memory. Relocation could be done with special-purpose hardware that relocates and maintains pointer information in memory and on disk. Unfortunately, no such hardware exists. The only way to avoid these problems and still obtain a single representation for disk and memory is to stop using pointers and to use arrays and indices [Leinwand]. This makes I/O faster but slows down memory access during design.

3.2.7 Database Modification

The last general issue in representation to be discussed here is the control of changes. These changes come from the synthesis tools, from the analysis tools, and from that most advanced tool of all, which does both synthesis and analysis, the user interface. Once a change is requested, it might be further transformed by side effects within the database that produce more modification. These side effects are constraints that tie together different pieces of the database so that a change to one results in a change to the others. Such constraints belong in the database as an integral component of change so that the database can guarantee their uniform and consistent application. The original change, combined with the effects of constraints, forms a constrained set of changes that should be retained for possible reversal. Undo control also belongs in the database because there may be many sources of change in the design system, all of which wish to guarantee that undo will be possible. Thus database modification actually involves a change interface, constraint satisfaction, and undo control.

The interface to database modification should be a simple set of routines. When the representation consists of objects with attributes, it is sufficient to have routines to create and delete objects and to create, delete, and modify attributes. In practice, however, there are different types of objects with differing attributes, so special interfaces will exist to manipulate each object. Also, some attributes come in sets and should be modified together. For example, a single routine to change the size of an object will affect many coordinate attributes, such as the bounding area, the corner positions, and the aspect ratio. Although these side effects can be implemented with constraints, efficiency considerations dictate that separate routines exist to implement the frequently used functions.

In a VLSI design system, manipulation routines should be provided for the basic aspects of circuit elements. This means that there should be interfaces for manipulating hierarchy, views, connectivity, and geometry. Hierarchy manipulation requires routines to create and delete cell definitions and their instances. View manipulation requires routines to deal with correspondence pointers throughout the database. Connectivity manipulation means having routines to create and delete components and wires. Finally, geometry manipulation needs routines to alter the physical appearance of an object.

3.2.8   Constraint

Constraint can appear in a number of different forms. Typically, an object is constrained by placing a **daemon** on it, which detects changes to the object and generates changes elsewhere in the database. When two objects are constrained such that each one affects the other, then a single daemon can be used that is inverted to produce a constraint for both objects. For example, if two objects are constrained to be

Object	Change	Constraint
A:	ΔA	$C = C + \Delta A$
B:	ΔB	$C = C + \Delta B$
C:	ΔC	$A = A + \Delta C/2$
		$B = B + \Delta C/2$

FIGURE 3.7 Constraint rules for "$A + B = C$."

adjacent, then moving either one will invoke a constraint to move the other. A more complex constraint can involve three or more attributes. An example of this is simple addition, in which one attribute is constrained to be the sum of two others. If any attribute changes, one or both of the others must adjust.

Constraint specification can be made simple by demanding certain restrictions. One way to do this, when constraining more than one object, is to require a separate constraint rule on each object so that there is always a rule waiting for any change [Batali and Hartheimer; Steele]. For example, when three attributes are constrained, a rule must appear for each one specifying what will happen to the other two when a change occurs (see Fig. 3.7). If this is not done, the database will have to figure out how to satisfy the constraints by inverting one rule to get the other. Another way to simplify constraint specification is to restrict the allowable constraints to a simple and solvable set. The Juno design system allows only two constraints: parallelism and equidistance [Nelson]. The Electric design system has the limited constraints of axis orthogonality and rigidity (see Chapter 11). These constraints are simple enough that no complex rule inversion is needed because all rules and their inverse are known a priori.

More complex constraint systems have the advantage of allowing the user to build powerful design facilities. Total constraint flexibility can be found in Sketchpad [Sutherland] and ThingLab [Borning]. These constraints are generalized programming-language expressions that are solved by precompiling them for efficiency and by using error terms to indicate the quality of satisfaction. Solution of such constraints is not guaranteed and can be very time consuming (see Chapter 8, Programmability, for a discussion of constraint solving).

Representation of constraint can take many forms. In limited-constraint systems such as Juno and Electric, the constraints are simply sets of flags on the constrained objects. In more complex systems, the constraints are subroutines that are executed in order to yield a solution. These database daemons detect changes to constrained objects and pursue the constrained effects. In addition to methods for storing and detecting constraints, there must be special structures to aid in constraint application. For example, there should be attributes to prevent constraint loops from infinitely applying the same set of actions. A time stamp on an object can be used to detect quickly when a constraint has already executed and is being run for the second time. When such a loop is detected, it indicates that the database may be overconstrained and that special action is necessary.

3.2.9   Undo

When a requested change and all its resulting constraints have been determined, the system is left with a set of constrained changes to the database. Some of these were explicitly given and others were the implicit results of constraints. The implicit results should be returned to the subsystem that invoked the change. However, these results may be intermingled with the explicit changes and may make no sense when out of order. For this reason, changes must be collected in the proper sequence, regardless of whether the change came directly from the application program or from internal constraints. The complete set of changes should then be presented to the application subsystem, which can then see the total effect. A uniform subroutine interface can be established for broadcasting changes, and it should mimic the one that was used to request the changes.

The collected set of changes should also be retained by the database for possible undoing. In order to enable this, the complete nature of the change must be held so that the original can be restored. This is best accomplished by having a "change" object associated with every database modification, showing what was changed and how. The change object may contain old values from before the modification or it may describe the nature of the change in such a way that it can easily be reversed. One complication is that deletion changes will have to store the entire contents of the deleted object to allow for proper reconstruction when the deletion is undone. This is not very efficient, so deletions usually just describe the operation, point to the deleted object, and delay its actual destruction while there still is a chance that the deletion will be undone.

Every change requested by an application subsystem results in a batch of constrained change objects, organized in an ordered list. The database should be able to undo such a batch of changes by tracing it backwards. A truly powerful design system will allow undoing of more than the last change and will retain an ordered history of change batches so that it can undo arbitrarily far back. Implementation of undo can be done by removing the change objects from the history list and replacing them with the undone set. This undone set of changes may be more compact because it may combine multiple changes into a single one. In such a scheme, users must understand the extent of each undo change. Also, because this scheme allows redo by undoing an undo, care must be taken to prevent indiscriminate use of the facilities that may produce an inconsistent database.

A better implementation of undo control is to back up through the history list without creating new entries. Redo is then a simple matter of moving forward through the list, and further undo moves farther backward. When a new change is introduced, the history list is truncated and extended at the current point, thus losing the ability to redo. This scheme provides a cleaner change mechanism for both the user and the implementor.

Change control is time and space consuming. It can take arbitrary amounts of time to solve constraints and produce a consistent database. Also it can take arbitrary amounts of space to represent database changes, especially when these changes are multiplied by constrained effects and saved for possible undo. However, the resources consumed by such a change mechanism are well worth the effort because they make the design system responsive and forgiving.

3.2.10   Summary of Basics

This chapter has shown so far that a design consists of objects with attributes. For maximum flexibility and efficiency, these should be implemented with fixed record entries, linked lists, and attribute hash tables. The objects are allocated from unused virtual memory and are represented on disk in a variety of ways best to suit the nature of the computer system. Special techniques can be used to scan and mark the database. Also, the representation of change, constraint, and undo control is a fundamental aspect of a design database.

The rest of this chapter will discuss techniques for dealing with the four aspects of VLSI design: hierarchy, views, connectivity, and geometry.

3.3   Hierarchy Representation

The first aspect of VLSI design that must be represented is hierarchy. Hierarchical layouts have entire collections of circuit objects encapsulated in a cell definition. Instances of these cells then appear in other cells, which means that their entire contents exists at each appearance.

The representation of cell instances can be done with instance objects. These objects, which point to their cell definitions, are actually **complex** components, as opposed to **primitive** components such as the NAND gate. Complex components can use the same object structure as primitive components use, but their prototype objects have different attributes. For example, a primitive prototype may have attributes that describe it graphically, whereas a complex prototype will contain a list head that identifies the subobjects inside the cell. Although it is tempting to create a new object type so that design can be done with components and instances, the representation is much cleaner if only components are used because then there are fewer database objects and they can be treated uniformly.

Given this uniform representation of hierarchy, every cell is a component prototype. In Fig. 3.8, the design of Fig. 3.6 is shown in its proper perspective as a complex prototype called "Bignothing." Note that the "Out" connection on the rightmost inverter component in "Bignothing" is exported and called "Final." Other cells may contain instances of the "Bignothing" cell, thus including its contents. The "Something" cell in Fig. 3.8(c) has two components: one that is a primitive component and one that has a complex prototype. The complete layout is shown at the bottom of the figure.

Because complex component prototypes are objects, the question

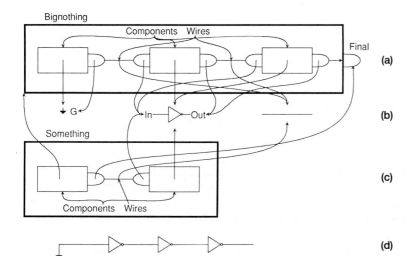

FIGURE 3.8 Hierarchy: (a) Complex prototype for "Bignothing" (b) Primitive prototypes (c) Complex prototype for "Something" (d) Represented layout.

of where to store their subobject list heads is resolved. These pointers are simply attributes in the complex-component prototype objects. However, a new issue is raised: how to represent the lists of component prototypes. To do this, two new object types must exist: the **environment** and the **library**. The environment is a collection of primitive-component prototypes, organized to form a design environment such as was discussed in Chapter 2. A library is a collection of complex component prototypes, or cells, presumably forming a consistent design. A good design system allows a number of different environments and permits multiple libraries to be manipulated. Figure 3.9 shows an overall view of the objects in such a design system. Environments provide the building blocks, which are composed into cells. Cells are then hierarchically placed in other cells, all of which are represented in libraries. A collection of library objects therefore contains everything of value in a particular design.

Although libraries provide convenient ways to aggregate collections of cells, a further level of abstraction may be used to aggregate libraries by designer. In multiperson designs, a **project** can be defined to be a collection of works from many people [Clark and Zippel]. Subprojects identify the work of individuals and eventually arrive at libraries and cells to describe the actual design. Thus hierarchy can be used to describe both circuit organizations and human organizations.

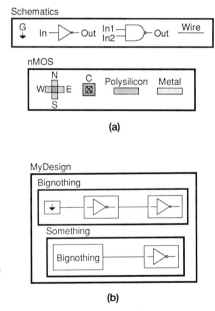

FIGURE 3.9 Environments and libraries: (a) Environments (b) Libraries.

3.4 View Representation

The next aspect of VLSI design that must be represented is views and their relationships. Representation of views makes use of different methods for each view. Layout views require geometric representations and schematic views require topological representations, both of which are discussed in the following sections. Temporal views make use of relative event times and signal values [Arnold; Borriello]. All of these views should be able to link together to create a single object [Zippel; Clark and Zippel]. A good view-correspondence system allows these view links to be established uniformly and thoroughly in the database.

There are no truly good view-correspondence systems today, so their representational needs are ill-defined. What is known is that arbitrary information must be available to do proper correspondence maintenance. Combined with the other representational facilities described in this chapter, a few additional attribute types should fill the need. In particular, the database must be able to link two designs done in different views and to describe their relationship fully.

Linking of cells is done with pointer attributes that can relate the various objects in the database. For example, two cells that represent the same circuit in different views will each have attributes that identify the cell in the other view. Components and wires inside the cells will have attributes that link them to their counterparts in the other view. This pointer information links the database across hierarchies, showing the same design in different layout styles. If the views are so different as to be unable to associate common cells, then further pointer information must be represented. For example, the link between an algorithm and a layout will have to identify textual equations and their equivalent components. Thus the capability demanded by views is the ability for attributes to contain pointers to any other object in the database.

Besides storing pointers, it is useful to have classes of pointers that carry meaning. One database system for CAD has three different pointer types that can describe a rich set of relationships [Atwood]. The **a-part-of** pointer is for composition ("contact 7 is a-part-of the Inverter cell"). The **an-instance-of** pointer is for type qualification ("component 12 is an-instance-of an Inverter cell"). Finally, the **a-kind-of** pointer describes relationships between categories ("Superbuffer is a-kind-of Inverter"). By allowing multiple a-kind-of links and selecting common attributes among the linked objects, the database can store a complex yet compact description. For example, Superbuffer is a-kind-of MOS cell, but it is also a-kind-of Inverter and a-kind-of Driver.

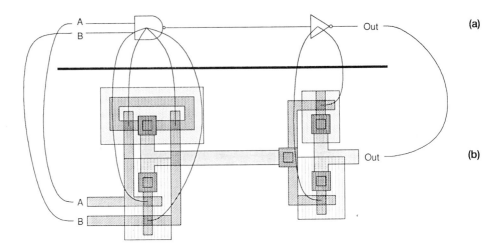

FIGURE 3.10   View-correspondence links: (a) Schematic (b) CMOS.

The primary a-kind-of link is to MOS cell, with weaker links to Inverter and Driver, selecting only those properties that are relevant.

The most important feature of view correspondence is its many-to-one mapping of components. A single object may have a pointer to multiple other objects (see Fig. 3.10). Simple pointer attributes will not be sufficient for this, so an array construct will be needed to provide one attribute with many pointers. Arrays should have the capability of being either fixed length, with the length stored in the attribute, or variable length, being delimited by a null termination pointer much like that in a linked list. One-dimensional attribute arrays are sufficient because more complex data structures can all be stored as vectors.

The aspect of design that makes view correspondence hard to maintain is the fact that designers are constantly rearranging the objects that are linked between views. When this is done, some of the links are temporarily broken. The attributes must be able to track this information so that the design can be checked and reconstructed. Therefore every view link needs additional information for state (the pointer is valid/invalid, the designer is aware/unaware of the change, the data of the change is———, and so on), and even a textual field in which notes and comments are made about the view. A proper software approach to this problem would use structured records as attributes so that the varying kinds of information could be kept in one place. Even without this structuring of attributes, it is necessary to have all these data types. A text-message attribute is useful for design documentation, and integer arrays can hold arbitrary information such as design parameter and status tables. All of this is useful in the representation of views.

3.5   Connectivity Representation

3.5.1   Nodes, Arcs, and Ports

Networks provide a suitable representation for connectivity. Most commonly, **nodes** of a network are electronic components and **arcs** are connecting wires. Some systems reverse this to represent wires as nodes and components as arcs [Karplus]. Without significant loss of generality, the discussion here will adopt the former assumption to show how networks can represent connectivity.

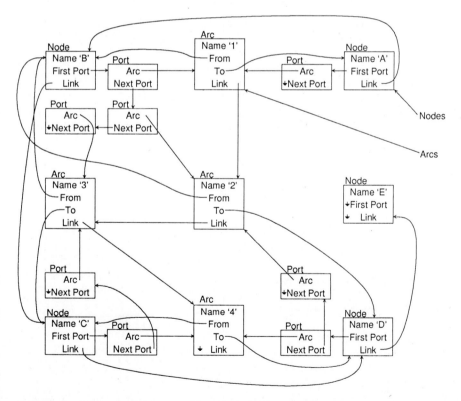

FIGURE 3.11   Node, arc, and port objects form a network.

In network representations, there are two types of objects: nodes and arcs. Each arc object connects exactly two nodes and so it has attributes that identify these two objects. However, it is not as easy to list every arc that connects to a node because there is no limit to the number of connections that a node can have. In addition to this need for unlimited arc pointers, there may be other attribute information associated with these node-arc connections. For this reason, a connection object is used to link nodes and arcs, called a **port**. A port is the site of an arc connection on a node. By having explicit port objects in a linked list, it is possible to represent an unlimited number of connecting arcs on a node and to store any information about the connection.

The most basic connectivity representation has node, arc, and port objects (see Fig. 3.11). The node and arc objects in this figure are named and so have ''name'' attributes, which identify them. These objects also have ''link'' attributes, which collect them into two lists that form the circuit. An external list head for these lists is all that is needed to identify the entire collection. To link arcs with nodes, each arc object has a ''from'' and ''to'' attribute that points to node objects. Also, each node object has a ''first port'' attribute that heads a list of port objects. If there are no arcs connected to a node (as in

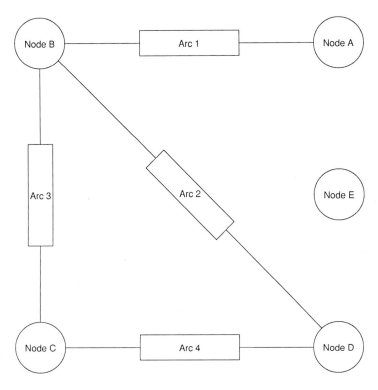

FIGURE 3.12 The network of Fig. 3.11.

node E of the figure), then this field is null because there are no ports. For every arc that does connect, there is a port object that contains a pointer to that arc. The port object also has a "next port" attribute that points to the next port object in the linked list on the node. When there are no more arc connections, and therefore no more ports, the "next port" attribute contains a null value to terminate the list. Figure 3.12 shows the circuit in a more natural way.

When representing a VLSI circuit both geometrically and as a network, there is a question about how the geometry of components and wires will relate. Since the wire attaches onto the component, its

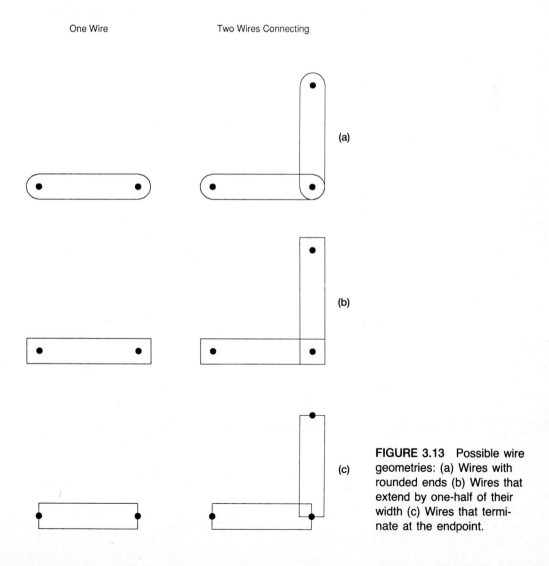

FIGURE 3.13   Possible wire geometries: (a) Wires with rounded ends (b) Wires that extend by one-half of their width (c) Wires that terminate at the endpoint.

geometry will overlap that of the component. Also, when two wires meet, their geometry must be correct at the intersection point. Three popular approaches are for wires to be rounded, for them to extend beyond the endpoint by one-half of their width, and for them to terminate cleanly at their endpoint (see Fig. 3.13). With no extension, some additional geometry must appear at wire intersections to make a correct connection. If this additional material is round, the wires can meet at all angles. With an extension of one-half of the width, wires will meet correctly when at orthogonal orientations. However, extended wires that meet at nonorthogonal angles will have to extend somewhere between zero and one-half of the wire width [Mead and Conway]. Yet another problem in the one-half-width extension scheme is that the two wires may have differing widths. In this case, wire extension should be by one-half of the width of the narrowest wire.

The representation of connected networks is somewhat complex. Part of this complexity is a response to the need for arbitrary circuit-size capacity and part is a result of cooperation with other representational needs.

3.6   Geometry Representation

The last aspect of VLSI design to be represented is its spatial dimensionality. There must be convenient ways to describe flat polygonal objects that are aggregated in parallel planes. In addition, the objects must be organized for efficient manipulation and search.

3.6.1   Shape, Transformation, and Graphics

Actual representation of geometry consists of shape, transformation, and graphical attributes. The **shape** attribute contains information about the actual drawing of the object and can be described in many ways (see Fig. 3.14). Given a shape-type attribute and an array of shape

Shape Type	Shape Parameters
Rectangle	Center, size
Polygon	Point count, point coordinates
Line	Begin, end, thickness
Circle	Center, radius
Arc	Center, start, end
Text	Location, message, typeface

FIGURE 3.14   Shape description.

data parameters, any figure can be described. These shape parameters should be represented with integers because integer arithmetic is faster on many computers and does not suffer from uncertain precision. However, the programmer must beware of integer overflow, which goes unnoticed on most computers. Also, when using integers, the unit of spacing should be chosen to allow maximum accuracy, both now and in the future. For example, if a 1×1 square in the database represents 1 square micron, then submicron IC layout cannot be done with integers. The smallest representable unit should match the smallest unit size of the manufacturing device. Most IC mask-making equipment works at centimicron detail, so one unit should represent one-hundredth of a micron. In PC design, one unit should represent one micron. Be warned that very small units of representation will result in very large values, which can overflow even 32-bit integers, common on today's computers.

A common abstraction in design systems is to use a scalable layout unit that is independent of actual spacing. Mead and Conway coined the term **lambda** (λ) to describe a single unit of distance in the design. As the value of λ varies, the entire circuit scales. Spacing is usually expressed in integer multiples of λ, although half and quarter λ units may be used. Besides a λ grid, it is also convenient to have coarser spacings for certain tasks. Several researchers have independently found that an 8λ grid is appropriate for the placement of cells [Schediwy; Sproull and Sutherland].

One issue that arises in the description of polygon shape is how to handle complex polygons such as those that intersect themselves. Although there are many solutions to this problem, one stands out as the most reasonable, especially for VLSI design [Newell and Sequin]. A point is defined to be inside a polygon if it has a positive (nonzero) **winding number**. The winding number is the number of times that another point, traveling around the perimeter of the polygon, wraps around the point in question. Besides this formal definition, Newell and Sequin show how complex polygons can be drawn and decomposed into simpler polygons.

Transformation attributes indicate modifications to the shape data. Transformations are typically linear, which means that the shape can be rotated, translated, or scaled. In a two-dimensional domain, this information can be represented with a 3×3 matrix of floating-point values. Since homogeneous coordinate transformations are not used in circuit design, only the left six elements of this matrix are valid [Newman and Sproull]. However, square matrices are easier to manipulate, so they continue to be used.

Another transformation that is rarely used in VLSI design is scaling. This is because object scaling is mostly done by modifying the shape

Attribute	Parameters
Color	Red, green, blue
Texture	Pattern
Filled-in	Boolean
Mask layer	Layer number

FIGURE 3.15   Graphical attributes.

coordinates rather than the transformations, and screen scaling is separate from the object representation. If scaling is eliminated, then the array elements will all fall into the range of zero to one and can be stored as fixed-point integers. This makes transformation operations faster on most computers. Thus a transformation matrix containing translation and rotation can be represented with six integers. Chapter 9, Graphics, discusses these transformations more fully.

The final aspect of geometric representation is graphical attributes that describe the visual qualities of an object. Examples are shown in Fig. 3.15. Note that this is different from shape because these graphical attributes are independent of coordinate values and can apply to any geometric object. Graphics attributes describe an object so that it can be distinguished from others that it overlaps. With the three geometric attributes of shape, transformation, and graphics any geometric object can be described.

3.6.2   Orientation Restriction

One limitation that is often found in VLSI design is a restriction on the possible orientations that an object may have. Since arbitrary rotation requires transcendental mathematics, systems that allow them are usually more complex than those that allow only a fixed set of orientations. A limit to the possible orientations will make the representation more compact and faster. However a limit should be used only if the design environment lends itself to such restrictions.

In some VLSI layout, all geometry must be **Manhattan**, which means that the edges are parallel to the x and y axes. Every polygon is a rectangle and can be oriented in only one of eight ways (see Fig. 3.16). If the rectangle is uniform in its contents, no transformation is necessary since all orientations are merely variations on the shape. It is only when transforming a detailed component or a cell instance that these eight orientations are used.

In other VLSI design environments it is necessary to provide 45-degree angles. This is double the number of orientations in Manhattan geometry and isb essentially the limit at which such layout is done. Although some synthesis tools generate geometry at arbitrary angles,

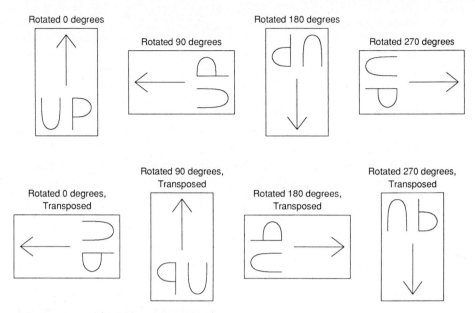

FIGURE 3.16 Manhattan orientations.

sometimes called **Boston** geometry [Bryant], designers rarely use such a facility. The reason is that arbitrary-angle design rules do not exist for many IC processes and, if they did, would be so difficult to check that only a computer could produce error-free layout.

In non-IC environments such as schematic design, the geometry can appear at arbitrary angles. However, even Manhattan rules can be used and are often preferred because they make the layout look cleaner (see Fig. 3.17). If total flexibility of orientation is desired, the

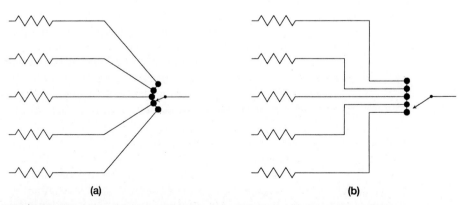

FIGURE 3.17 Geometry restrictions dictate the appearance of the layout: (a) Arbitrary angles (b) Manhattan.

Manhattan rotations can still be coded as special cases. Full transformations can then be represented with 3×3 matrices, but one of the unused entries in the matrix indicates the special cases of Manhattan orientation.

3.6.3 Geometry Algebra

To go beyond the basic attributes of geometric representation, it is necessary to consider the kinds of operations that will be performed on these objects. One such operation is geometry algebra. This set of operation performs union, intersection, expansion, and contraction on object shapes. Union operations are typically used to merge two adjoining shapes into one. Intersection is used to find the common area of two shapes, often for the purpose of eliminating redundancy. Expansion and contraction operations push the edges of a shape away from or toward the center.

Combinations of these operations can be very powerful. For example, design-rule checking is often done by expanding an object by its minimum spacing distance and then looking for areas of intersection with other objects. These areas are the design-rule violations. Expansion and contraction are used in compensation: a preprocessing step to fabrication that accounts for inaccuracies in manufacturing. There are even VLSI databases that require these algebra functions. For example, Caesar [Ousterhout 81] represents rectangles in its database and uses intersection tests to keep the rectangles nonredundant whenever changes are made. Nonredundant databases are occasionally required and often preferred. This is because they store less information and their analysis takes less time.

There are many forms of shape representation and each has its own methods of being algebraically manipulated. Manhattan rectangles are simple to modify. Polygons with 45-degree sides are more complex [Lanfri]. Arbitrary-angle polygons are the most complex but are often approximated with stair-stepped Manhattan rectangles [Batali and Hartheimer]. Circles, arcs, and conics demand a separate set of geometric algorithms, and become even more complex when mixed with polygonal representations. Here again, many curved shapes are approximated polygonally [Whitney]. The user should be allowed to have control over the grain of curve approximation so that the necessary smoothness can be produced.

The Polygon Package [Barton and Buchanan] performs all the geometric algebra functions and, in addition, allows both straight and curved lines to be used in the description of polygon boundaries. **Sheets**, which are multiple collections of these lines, can be aggregated to define polygons with holes. The most interesting representational aspect of this system is the organization of pairs of adjacent lines, sets of two pairs, sets of four pairs, and so on into a binary tree. The top of each tree encloses an entire polygon and each lower level of the tree has a bounding box that encloses its lines. This allows a search to rapidly find the specific lines involved in polygon intersections.

Another useful technique for storing polygonal information is **winged-edge** representation [Baumgart]. This representation captures the interrelation of polygons by sharing common edges and vertices. Thus a line on the border of two polygons will point to each one and will also be linked to the previous and subsequent edges on both polygons. This abundance of information is particularly useful in three-dimensional polygonal representations and can be simplified somewhat in the two-dimensional CAD world.

3.6.4   Search

Search is the operation that places the greatest demands on the geometric representation. Two operations that are important are the search for the neighbors of an object and the search for all objects at a given point or area. The way to make this efficient is to store sorted adjacency in the representation so that this information does not have to be determined during the search. Such adjacency will take time to compute when changes are made but will save much time later.

The problem is to determine what form of adjacency to represent. If the geometric objects are sorted by their center, then there is no information about their extent, and the relational information will not properly convey object location. For example, Fig. 3.18 shows two polygons the center coordinates of which do not indicate a spatial ordering in the x axis. The correct information to represent is the boundaries. By storing sorted edges, a walk through the edge list will precisely identify the objects in a given neighborhood.

The simplest representation of sorted edges is a linked list of objects according to edge order. One list can order the edges along only one axis. Therefore two lists are needed: one to cover horizontal edges and a second list to cover vertical edges. In some circumstances, the lists may need to distinguish between the two sides of an object. This means that the horizontal-edge list is actually two lists: one with

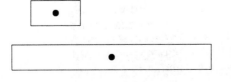

FIGURE 3.18 Center-based object ordering is useless for a search.

the top edges and one with the bottom edges. For nonrectangular objects, a minimum bounding rectangle is used. These lists can be used to delimit a neighborhood because, when the near edge of an object extends beyond the bounds of the search, then the remaining objects on that direction's list are also out of range. For example, when searching for all objects in the range of $5 \le x \le 10$, the program searches the left list until all objects bounds are less than 5 and then searches the right list until all objects bounds are greater than 10. The problem with this technique is that every object is represented multiple times and so it will be examined repeatedly during the search. Thus when a search for all objects in an area is conducted, the left and right lists will give no information about the vertical position of an object, so a scan of these lists will not limit itself in the up and down directions. Only by clever use of these lists can waste be prevented. For example, if the area of search is taller than it is wide, then the left- and right-edge lists should be used since they will do a better job of selection.

The tree is an obvious and well-used representation. Binary trees divide the represented objects in half and then divide each half into subhalves until each individual object is at a leaf of the tree. For two-dimensional representation, the **quad-tree** divides space into four parts (in half along each axis) and then each quadrant is further divided into subquadrants until each object is placed properly in the tree. In general, a rectangle to be stored in a quad-tree will be placed at the lowest level that completely contains its area. This means that a rectangle that is centered along either axis will not be wholly in any quadrant and will be stored at the top of the tree. This loss of efficiency occurs throughout the quad-tree because of its inability to adapt to the needs of the data. One quad-tree implementation augments the spatial organization with x and y binary trees inside each quadrant to speed the search of objects once the quadrant is selected [Kedem]. The only significant advantage of quad-trees is their ability to be addressed efficiently using the binary properties of coordinate values [Reddy and Rubin; McCreight]. The high bit of the x and y coordinate are concatenated into a two-bit number that indexes one of the four top-level quadrants; the next bit down is used to select the next subquadrant. Given proper quad-tree organization, addressing can be done exclusively with bit manipulation.

Another representation for spatial organization is called **corner stitching** [Ousterhout 84]. In it, all space in a layout is stored as a perfectly tiled area of nonoverlapping rectangles. Even the empty spaces have rectangles delimiting them. These area rectangles then have pointers that connect their corners to the adjacent area rectangles: The lower-left corner points to the rectangles below and to the left;

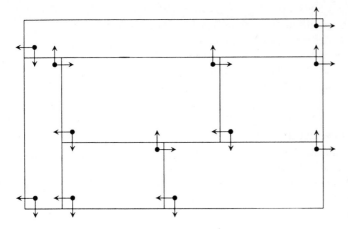

FIGURE 3.19   Corner stitching.

the upper-right corner points to the rectangles above and to the right (see Fig. 3.19). With this representation, it is fairly simple to find the neighborhood of an object and to walk through the pointers in search of an object at a selected coordinate. However, it is necessary to use multiple "planes" of these rectangles in order to represent layout that overlaps on different layers. Also, the representation is mostly intended for Manhattan geometry and becomes complex when used otherwise.

Perhaps the best representation for the storage and search of spatial objects is the use of **R-trees** [Guttman]. This structure is a height-balanced tree in which all leaf nodes are at the same depth and contain spatial objects. Nodes higher in the tree also store boundary information that tightly encloses the leaf objects below. All nodes hold from M to $2M$ entries, where M is typically 25. The bounding boxes of two nodes may overlap, which allows arbitrary structures to be stored (see Fig. 3.20). A search for a point or an area is a simple recursive walk through the tree to collect appropriate leaf nodes.

Insertion and deletion can be complex operations for R-trees. To insert a new object, the tree is searched from the top; at each level, the node that would expand least by the addition is chosen. When the bottom is reached, the object is added to the list if it fits. When there is no more room, the bottom node must be split in two, which adds a new element to the next higher node. This may recurse to the top, in which case the tree becomes deeper. Splitting a node is done by finding the two objects farthest apart along any dimension, and then clustering the remaining objects according to the rule of minimal boundary expansion.

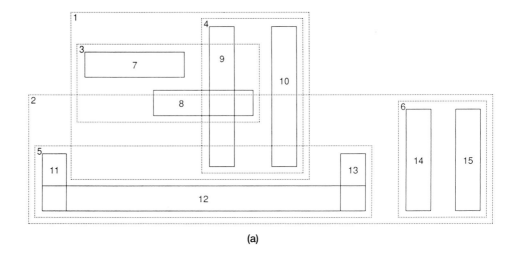

(a)

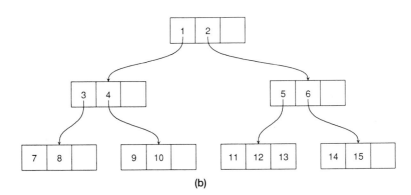

(b)

FIGURE 3.20 R-tree organization: (a) Spatial ordering (b) Tree representation.

R-tree deletion is done simply by removing the entry. If, however, the removal causes the node to have too few objects (less than M), then the entire node is deleted and the remaining valid objects are reinserted with the algorithm described in the previous paragraph. Deletion of a node may cause the next higher node also to contain too few objects, in which case the deletion recurses and may result in a shortened tree. Object-size updates require deletion and reinsertion.

Regardless of the spatial representation, search can be aided by providing an initial guess as to which objects are in the search area. When the search involves finding the neighbors of a particular object, that object is useful as the initial guess. When the object at a specified coordinate is sought, the most recently selected object can be used as the initial guess for the search; designers often work within a small area, so it is safe to assume that the next object selected will be close to the last object selected. This is the principle of **locality** and it applies to many tasks.

Speedy search requires that large amounts of information be available and rapidly accessible. This information must be organized well and the initial conditions for search must be favorable in order to achieve that speed. Efficient search is needed by many aspects of design and is crucial to a good design system.

3.7   Summary

This chapter has discussed the representation of VLSI design. The first half of the chapter covered basic representational needs for any design task. The remainder of the chapter concentrated on the specific representations for circuitry: hierarchy, views, connectivity, and geometry. Implementation of a complete VLSI design database is complex but provides the necessary basis for the algorithms described in the rest of this book.

Questions

1 When is the array a useful structure?

2 Besides relieving the problem of fragmented memory, what is another advantage of clustering memory allocation into separate arenas?

3 How could a pointer-based representation remain identical on disk and in memory?

4 What is the disadvantage of having the database manage change control?

5 To implement wires the ends of which do not extend beyond their connection point, some systems pull the connection points toward the center and then compute a normal wire that extends its ends by one-half of its width. What problems might this approach cause?

6 Would sparse matrix representations work for VLSI geometry? Why or why not?

7 How would you extend corner stitching to handle arbitrary angles?

References

Arnold, John E., "The Knowledge-Based Test Assistant's Wave/Signal Editor: An Interface for the Management of Timing Constraints," Proceedings 2nd Conference on Artificial Intelligence Applications, 130-136, December 1985.

Atwood, Thomas M., "An Object-Oriented DBMS for Engineering Design Support Applications," Proceedings Compint Conference, Montreal, 299-307, September 1985.

Barton, E. E. and Buchanan, I., "The Polygon Package," *Computer Aided Design*, 12:1, 3-11, January 1980.

Batali, J. and Hartheimer, A., "The Design Procedure Language Manual," AI Memo 598, Massachusetts Institute of Technology, 1980.

Baumgart, Bruce Guenther, *Geometric Modeling for Computer Vision*, PhD dissertation, Stanford University, August 1974.

Borning, Alan, "ThingLab—A Constraint-Oriented Simulation Laboratory," PhD dissertation, Stanford University, July 1979.

Borriello, Gaetano, "WAVES: A Digital Waveform Editor for the Design, Documentation, and Specification of Interfaces," unpublished document.

Brown, Harold; Tong, Christofer; and Foyster, Gordon, "Palladio: An Exploratory Environment for Circuit Design," *IEEE Computer*, 16:12, 41-56, December 1983.

Bryant, Randal, "Preface," Proceedings 3rd Caltech Conference on VLSI (Bryant ed), Computer Science Press, v-viii, March 1983.

Clark, G. C. and Zippel, R. E., "Schema: An Architecture for Knowledge Based CAD," *ICCAD '85*, 50-52, November 1985.

Deutsch, L. P. and Bobrow, D. G., "An Efficient Incremental Automatic Garbage Collector," *CACM*, 19:9, 522-526, September 1976.

Gosling, James, *Algebraic Constraints*, PhD dissertation, Carnegie-Mellon University, CMU-CS-83-132, May 1983.

Guttman, Antonin, "R-Trees: A Dynamic Index Structure for Spatial Searching," *ACM SIGMOD*, 14:2, 47-57, June 1984.

Karplus, Kevin, "Exclusion Constraints, a new application of Graph Algorithms to VLSI Design," Proceedings 4th MIT Conference on Advanced Research in VLSI (Leiserson, ed), 123-139, April 1986.

Kedem, Gershon, "The Quad-CIF Tree: A Data Structure for Hierarchical On-Line Algorithms," Proceedings 19th Design Automation Conference, 352-357, June 1982.

Knuth, Donald E., *The Art of Computer Programming, Volume 1/Fundamental Algorithms*, Addison-Wesley, Reading, Massachusetts, 1969.

Lanfri, Ann R., "PHLED45: An Enhanced Version of Caesar Supporting 45 degree Geometries," Proceedings 21st Design Automation Conference, 558-564, June 1984.

Leinwand, Sany M., "Integrated Design Environment," unpublished manuscript, April 1984.

McCreight, E.M., "Efficient Algorithms for Enumerating Intersecting Intervals and Rectangles," Xerox Palo Alto Research Center, CSL-80-9, 1980.

Mead, C. and Conway, L., *Introduction to VLSI Systems*, Addison-Wesley, Reading, Massachusetts, 1980.

Nelson, Greg, "Juno, a constraint-based graphics system," *Computer Graphics*, 19:3, 235-243, July 1985.

Newell, Martin E. and Sequin, Carlo H., "The Inside Story on Self-Intersecting Polygons," *Lambda*, 1:2, 20-24, 2nd Quarter 1980.

Newman, William M. and Sproull, Robert F., *Principles of Interactive Computer Graphics*, 2nd Edition, McGraw Hill, New York, 1979.

Ousterhout, J. K., "Caesar: An Interactive Editor for VLSI Layouts," *VLSI Design*, II:4, 34-38, 1981.

Ousterhout, John K., "Corner Stitching: A Data-Structuring Technique for VLSI Layout Tools," *IEEE Transactions on CAD*, 3:1, 87-100, January 1984.

Reddy, D. R. and Rubin, Steven M., "Representation of Three-Dimensional Objects," Carnegie-Mellon University Department of Computer Science, Report CMU-CS-78-113, April 1978.

Schediwy, Richard R., *A CMOS Cell Architecture and Library*, Masters thesis, University of Calgary Department of Computer Science, 1987.

Sproull, Robert F. and Sutherland, Ivan E., *Asynchronous Systems II: Logical Effort and Asynchronous Modules*, to be published.

Stamos, James W., "A Large Object-Oriented Virtual Memory: Grouping Strategies, Measurements, and Performance," Xerox PARC SCG-82-2, May 1982.

Steele, G. L. Jr., *The Definition and Implementation of a Computer Programming Language Based on Constraints*, PhD dissertation, Massachusetts Institute of Technology, August 1980.

Sussman, Gerald Jay and Steele, Guy Lewis, "CONSTRAINTS—A Language for Expressing Almost-Hierarchical Descriptions," *Artificial Intelligence*, 14:1, 1-39, August 1980.

Sutherland, Ivan E., *Sketchpad: A Man-Machine Graphical Communication System*, PhD dissertation, Massachusetts Institute of Technology, January 1963.

Weinreb, Daniel and Moon, David, "Flavors: Message Passing in the Lisp Machine," MIT Artificial Intelligence Lab Memo #602, November 1980.

Whitney, Telle, *Hierarchical Composition of VLSI Circuits*, PhD dissertation, California Institute of Technology Computer Science, report 5189:TR:85, 1985.

Zippel, Richard, "An Expert System for VLSI Design," Proceedings IEEE International Symposium on Circuits and Systems, 191-193, May 1983. tation, Massachusetts Institute of Technology, January 1963.

Weinreb, Daniel and Moon, David, "Flavors: Message Passing in the Lisp Machine," MIT Artificial Intelligence Lab Memo #602, November 1980.

Whitney, Telle, *Hierarchical Composition of VLSI Circuits*, PhD dissertation, California Institute of Technology Computer Science, report 5189:TR:85, 1985.

Zippel, Richard, "An Expert System for VLSI Design," Proceedings IEEE International Symposium on Circuits and Systems, 191—193, May 1983.

SYNTHESIS TOOLS

4.1 Introduction

VLSI design has reached the point at which most circuits are too complex to be specified completely by hand. Although computers can be used to record layout, thus making design easier, they can also be programmed to take an active role by doing synthesis or analysis. **Synthesis tools** generate circuitry automatically and **analysis tools** attempt to verify existing circuitry that has been designed by hand. This chapter covers synthesis tools and the next two chapters are devoted to analysis tools.

There is nothing magical about synthesis tools; in fact many of them are simple applications of machine iteration. When design is done manually, it is easy to see areas of high repetition and regularity that lend themselves to automation. It is the combination of many of these techniques that begins to make computers appear intelligent in circuit design. Ultimately, it would be valuable to have a single synthesis tool that produces a complete layout from a high-level description of a circuit. This is the goal of today's **silicon compilers**, which are collections of simpler tools. However, since silicon compilers have not yet advanced to the stage at which they can replace all other tools, it is important for a design system to provide the individual synthesis tools so that they can be used selectively.

This chapter will illustrate a number of synthesis tools for VLSI design. These techniques are all based on the premise that a circuit consists of cells and their interconnection. Some tools generate and manipulate the contents of cells (PLA generators, gate-array generators, gate-matrix generators, compacters); some tools work with the layout external to cells (routers, placement systems, pad-frame generators); and some tools work with the interrelation of the cells and their environment (aligners, pitch matchers). Of course, there are silicon compilers that combine these techniques to produce cells and complete systems, ready for fabrication. Finally, there are synthesis tools that help to convert completed layout into suitable form for manufacturing (compensation, mask-graphics generators). These are just some of the tools that should be provided by a good design system.

4.2 Cell Contents Generation and Manipulation

The first type of synthesis tool to be discussed are those that work at a local level, within the cells of a chip. These tools generate the cell layout from a more abstract specification of the cell's functionality.

To accomplish this task, the layout must typically be restricted to some regular pattern into which the original specification can be mapped. Thus PLA designs have rigid AND and OR planes of transistors, gate-arrays have fixed transistor pairs, and gate matrices have columns of gating material crossing rows of switchable interconnect.

All cell design has regularity; even layout produced by humans makes use of complex but regular patterns in the designer's mind that translate function to layout. These patterns guide the designer during layout and help to determine when the design is complete. Some artificial intelligence systems have attempted to codify these rules [Kim, McDermott and Siewiorek; Kowalski and Thomas; Kollaritsch and Weste; Mitchell, Steinberg and Schulman; Simoudis and Fickas], but the results often have noticeable patterns that are no better than the regular layout methods used by traditional generators. This simply shows that manual design is at the far end of a continuum of layout methods, used to build circuits from high-level functional specifications. The remainder of this section discusses a few of the regular styles of layout that are available today for cell synthesis.

4.2.1   PLAs*

Combinational circuit elements are an important part of any digital design. Three common methods of implementing a combinational block are **random logic**, **read-only memory (ROM)**, and **programmable logic array (PLA)**. In random-logic designs, the logic description of the circuit is directly translated into hardware structures such as AND and OR gates. The difficulty in this method is that the placement and interconnection cost is high. In a large system, this cost could be prohibitive. The ROM is useful for tabular data that has little regularity, but it is very wasteful of space for data that could be algorithmically derived. The PLA combines features of both other methods by allowing the designer to realize combinational design with programming taps on a logic array.

The PLA is made up of an AND plane and an OR plane. The AND plane produces product terms of input variables selected by the programming taps and the OR plane produces the sums of these product terms selected by a different set of programming taps. The symbolic representation of the places where the programming taps are placed is known as the **personality matrix** of the PLA. Figure 4.1 shows the generic structure of a PLA that programs these logic functions:

$$f = (a \wedge b \wedge (\neg\ c)) \vee ((\neg\ b) \wedge (\neg\ a))$$
$$g = (a \wedge c) \vee ((\neg\ a) \wedge (\neg\ c))$$

---

* This section was contributed by Sundaravarathan R. Iyengar, Case Western Reserve University.

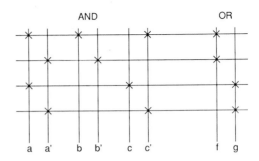

FIGURE 4.1 PLA for the example equations: X marks the programming taps.

The PLA occupies less area on the silicon due to reduced interconnection wire space; however, it may be slower than purely random logic. A PLA can also be used as a compact finite state machine by feeding back part of its outputs to the inputs and clocking both sides. Normally, for high speed applications, the PLA is implemented as two NOR arrays. The inputs and outputs are inverted to preserve the AND-OR structure.

PLAs are popular because their generation can be automated, which frees the designer from spending valuable time creating random-logic gates. Since the PLA generator fixes the physical structure of the PLA, there arises the problem of accommodating the designer's special requirements, if any. The first requirement would be to reduce the area occupied by the PLA. Usually the personality matrix is so sparse that a straightforward mapping of the matrix to the silicon will result in wasted silicon area. Instead, the PLA is **split** and **folded** to reduce the area required for the physical implementation. Instead of having one AND plane and one OR plane, the PLA can be split into many more AND and OR planes. Also, the input and output columns can be moved and folded such that there are two to a column instead of the usual one. The rows can be similarly moved and folded. There have been many folding algorithms published [Hachtel, Newton, and Sangiovanni-Vincentelli].

The second important requirement is speed of operation. Switching speed is affected by parasitic capacitance and the driving capacities of input and pullup stages. Options to change the driving capacities of these stages would be a convenient addition to the PLA generator. This would involve estimating the parasitic capacitance and automatically sizing the driver transistors.

There are several more options that a programmer can provide for the user. For instance, there could be an option to correct electrical problems such as voltage buildups on the power and ground lines of the PLA. Sometimes, due to the interconnection requirements of the surrounding circuitry, it might be helpful to have the input and output entry points for the PLA appear on different sides of the layout. In a MOS PLA, for example, it is very simple to provide the outputs at the top of the PLA instead of at the bottom.

When a PLA is used as a finite-state machine, the input and output stages need to be clocked using two different phases or two different edges of the clock. Thus it is convenient to provide an option that generates clocked input and output stages when required. The clock, when off, would simply prevent the corresponding stages from loading. Regardless of clock usage, it is good to estimate the worst-case delay through the PLA (see Timing Verification, in Chapter 5). Since the timing of the clock phases is critical for proper operation of the PLA, this feature is important.

In conclusion, a PLA generator is a useful tool that forms an important part of many silicon compilers. Given a high-level algebraic or a truth-table description of the PLA, it generates the mask-level description, which can be fabricated directly. Since the generator uses a predetermined structure for the layout, it is important to be able modify it through high-level parameters. A parameterized generator is more useful than one that supports only a rigid structure. A simple way to implement the PLA generator with parameters is to first draw a sample PLA on a piece of paper (or on the screen) and to identify major building blocks. Typically there are input and output stages, pullups, programming taps, and the connection primitives that connect the planes together. Parameterizing each of these blocks allows more high-level options to be provided.

4.2.2   Gate-Arrays

The **gate-array** is a popular technique used to design IC chips. Like the PLA, it contains a fixed mesh of unfinished layout that must be customized to yield the final circuit. Gate-arrays are more powerful, however, because the contents of the mesh is less structured so the interconnection options are more flexible. Gate-arrays exist in many forms with many names; for example, **uncommitted logic arrays** [Ramsay] and **master-slice** [Freeman and Freund].

A typical gate-array is built from blocks that contain unconnected transistor pairs, although any simple component will do. An array of these blocks combined with I/O pads forms a complete integrated circuit and offers a wide range of digital electronic options (see Fig. 4.2). These blocks are internally customized by connecting the components to form various logical operators such as AND, OR, NOT, and so on. The blocks are also externally connected to produce the overall chip.

Designing a circuit using gate-arrays consists of specifying the

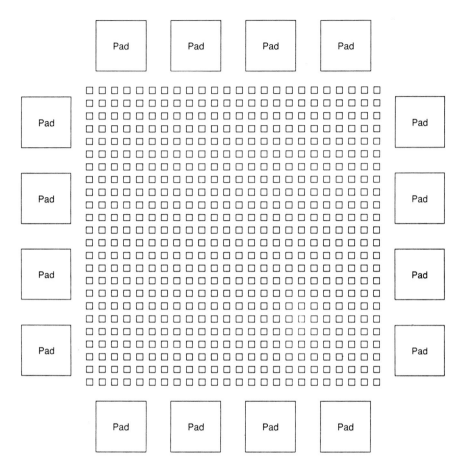

FIGURE 4.2   A typical gate-array. Blocks are in the center, pads are at the periphery.

internal and external connection requirements and then routing wires to produce the layout. There are high-level languages that help with the production of these connections. In these languages, designs can be easily specified by aggregating multiple blocks into larger functional units called **macros**. This is similar to macros in programming languages and it gives a semblance of hierarchy to the gate-array design. In addition to these macros, some gate-arrays have special-purpose circuitry, such as memory cells, so that these popular functions can be implemented more efficiently.

To aid in making connections, wire routing space is typically left between the gate-array blocks. Greater density is achieved in **channelless arrays**, which leave no wiring space and then using an ''etching'' layer to eliminate circuitry and create space where it is needed. Another way to save the space between blocks is to have an extra layer of metal that is reserved exclusively for interconnect. This lets routing run across blocks rather than having to work around them. However, these connections can cross each other only in the area between blocks where it is possible to change layers.

Gate-array generation is another form of silicon compilation because a completed chip can be produced without any concern for the layout environment. Higher-level languages specify a circuit in terms of logic functions and the computer translates it into gate-array connections. Also, gate-arrays can be fabricated quickly because the blocks and pads are on a different layer than the interconnect. This results in many fewer fabrication masks to generate: Only one or two interconnect masks are needed for each new gate-array. These interconnect layers are also usually the last layers processed during fabrication, so they can be placed on top of nearly completed wafers that have been prepared in advance. Another advantage of gate-arrays is that they can take advantage of special manufacturing processes that produce high-speed parts.

The disadvantage of gate-arrays is that they are not optimal for any task. There are blocks that never get used and other blocks that are in short supply. Since block placement is done in advance, interconnect routing can become complex and the resulting long wires can slow down the circuit. Also, the design will not be compact since interblock spacing is fixed to allow worst-case routing needs. One estimate is that gate-arrays are three to six times less efficient than a custom layout [Heller].

4.2.3   Gate Matrices

The **gate matrix** (not to be confused with the gate-array) is the next step in the evolution of automatically generated layout from high-level specification [Lopez and Law]. Like the PLA, this layout has no fixed size; a gate matrix grows according to its complexity. Like all regular forms of layout, this one has its fixed aspects and its customizable aspects. In gate matrix layout the fixed design consists of vertical columns of polysilicon gating material. The customizable part is the metal and diffusion wires that run horizontally to interconnect and form gates with the columns.

It is a simple fact that layout is composed of transistors and the wires that connect these transistors. Weinberger observed that for MOS circuitry it is possible to transpose these transistors so that they reside on top of the connecting wires [Weinberger]. This is because the polysilicon wire in MOS is also the gate of a transistor. Converting a circuit to this form saves space and achieves regularity. The **Weinberger**

array treats each column as a single NOR gate and requires all logic to be reduced to combinations of this element.

A gate matrix also makes the assumption that many of the vertical polysilicon lines are forming common circuit elements but the type of logic gate is unspecified (see Fig. 4.3). For example, each vertical line can be a NAND gate in nMOS by placing a load at the top of the line and ground on the bottom. Each gate on the column then acts as a pulldown that combines serially to form a multi-input NAND function. CMOS can also implement NAND gate matrices fairly easily [Weste and Eshraghian]. There are automatic systems to produce gate matrices from high-level circuit descriptions [Liu; Wing, Huang, and Wang]. Liu's system begins with logic gates, converts them to NANDs, reduces some of the redundant logic, and produces CMOS layout (see Fig. 4.4). The final step of assigning transistors to columns is prone to inefficiency since incorrect placement can result in long horizontal interconnect wires, and too many of these wires will cause the gate matrix to be slow and large.

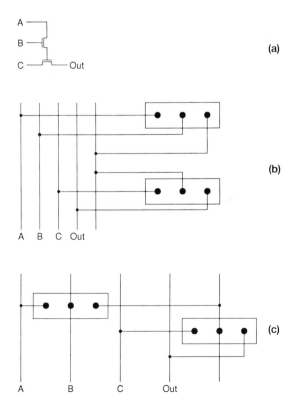

FIGURE 4.3 Gate matrices: (a) Sample circuit (b) View as gates and interconnect (c) Gates placed on interconnect.

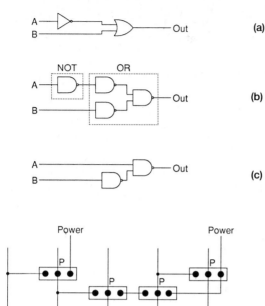

FIGURE 4.4 Gate-matrix generation: (a) Original logic (b) Translation to NAND gates (c) Optimization of NAND gates (d) Gate-matrix layout.

Wing, Huang, and Wang have solved the transistor-placement problem in two steps. Each column is first ordered such that its transistors connect directly to both adjoining columns. This is essentially the **greedy algorithm** [Balraj and Foster]. A simple connectivity graph can be built and traversed for this task. There will certainly be gaps in the adjacency, but this initial placement is refined in the second step, which examines each transistor and computes the range of columns to which it is connected. This range is used to reorder the columns so that each transistor is closer to the center of its activity. The actual column and row spacing is then determined according to the interconnect needs. Rows are rearranged to obtain complete wiring and columns are widened if more space is needed.

Another automatic system for gate matrix optimization uses high-level rules to guide the process [Kollaritsch and Weste]. These rules examine the circuit, rearrange components, and compact. As with all rule-based systems, this one can be easily adjusted to provide better results in unusual cases.

4.2.4   Other Regular Forms

In addition to gate matrices, there are many other ways to generate layout automatically from higher-level specifications. **Diffusion line tracing** is a method that strings together the transistors of a MOS design, connecting sources to drains as far as possible, and then routing the remaining polysilicon and metal connections [Nogatch and Hodges]. The **storage/logic array** [Goates *et al.*] is an enhanced PLA that contains memory elements so that state can be captured internally rather than having to loop around the outside of the array. The Tpack system allows designers to identify areas of a small layout that will be repeated according to a personality matrix [Mayo]. This allows large but regular circuits to be easily designed. Other methods are constantly being invented for the construction of layout. As these methods improve, they will approach the quality of hand layout and earn their place as the key component of silicon compilation.

4.2.5   Compaction

The last synthesis tool to be discussed under the category of cell contents manipulation is **compaction**, which removes unnecessary space from a design by moving everything to its closest permissible distance. This step can be done on a completed chip but is usually done on a cell-by-cell basis to recover unused space as soon as possible.

The simplest compaction method, called **one-dimensional compaction**, works by squeezing space along the x axis and then along the y axis. This is repeated until no more space can be recovered [Williams; Hsueh and Pederson; Weste; Entenman and Daniel]. One-dimensional compaction is simple to do, runs fairly quickly, and will recover most of the savable space. More advanced techniques attempt to save additional space by jogging wires or moving components; thus they are performing placement, routing, and compaction all at once. These methods have combinatorial work to do and produce only slightly better results than simple compaction methods.

One-dimensional compaction uses information about components and their connecting wires. The wires that run perpendicular to the direction of compaction link a set of components into a single **track** that will adjust together. Components and tracks are typically represented with a graph that shows the movable objects and their movement limitations. For example, in Fig. 4.5 the track with components A, B, and C will move together when x-axis compaction is done, as will the track with components E, F, and G. Although a track may run for the entire length or width of the cell, it may also be treated as a

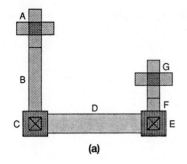

(a)

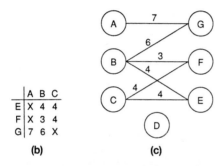

	A	B	C
E	X	4	4
F	X	3	4
G	7	6	X

(b)   (c)

FIGURE 4.5 One-dimensional compaction (along *x* axis): (a) Layout (b) Minimum-spacing matrix (c) Minimum-spacing graph.

locally connected piece of layout. This allows all unconnected parts of the design to compact properly, independent of other layout that may originate in the same row or column.

Compaction is done by adjusting the coordinates for each track. However, when virtual grid design styles are used, the initial spacings are all on a dimensionless grid, so the compaction process becomes one of assigning real coordinates to the tracks [Weste; Entenman and Daniel]. Regardless of the initial coordinate system, one-dimensional compaction must determine the minimum distance between two tracks of layout. This can be done by scanning all combinations of components in both tracks and computing the greatest design-rule spacing that is necessary. In Fig. 4.5, for example, transistors A and G must be 7 units apart from centerline to centerline, based on a 3-unit minimum diffusion spacing; 3 units between the centerline of A and its diffusion edge; and 1 unit between the centerline of G and its diffusion edge.

Each component must remain at a particular distance from its neighboring components, and this collection of distance constraints is found in the graph. Construction of the graph can be augmented by connectivity information so that electrically equivalent components are allowed to touch or overlap [Do and Dawson]. There can also be weighting factors in the graph to address nongeometric considerations such as the desire to keep diffusion lines short [Mosteller]. The graph, which has nodes for each component and arcs for each spacing rule, can be used to determine quickly the allowable separation of two tracks. Transistor A does not interact with objects E and F because they are not on the same level. Nevertheless, it is necessary to check for objects that are diagonally spaced because they may interact. Scanning the rest of these two columns shows that the 7-unit spacing is the worst case and thus defines the limit of compaction.

The previous example can also be used to illustrate the need for repeated application of one-dimensional compaction. The first x compaction places the two columns 7 units apart which is limited by the location of transistors A and G. It is possible, however, that during y compaction transistor G will move down but transistor A will not, due to other constraints. Then, the second pass of x compaction will bring the two columns to 6 units apart, defined by the minimum distance between wire B and transistor G. One-dimensional compaction is usually fast enough to be done continuously until there is no improvement made in either direction.

Although this method seems simple, there are pitfalls that must be observed. Weste observed that it is not sufficient to check only the next adjacent track when compacting. Pathological cases found in some environments cause layout two tracks away to present more severe constraints than does the layout in the immediately adjacent track [Weste]. Therefore the construction of the constraint graph must extend beyond the boundary formed by tracks of components, and must evaluate the worst-case spacing among all combinations of objects.

Another problem that can arise in alternating axes during one-dimensional compaction is that motion along one axis can block more significant motion along the other. With an interactive system, this can be controlled by having the designer specify areas and directions of compaction [Williams; Scott and Ousterhout; Mori]. Some systems allow designers to augment the spacing graph with their own constraints [Do and Dawson; Mosteller]. When these additional user constraints create an overconstrained situation, Do and Dawson leave the layout uncompacted with the hope that the orthogonal compaction phase will remove the blockage. Another feature that their interactive system provides is the notion of a **fence** that designers place to prevent unwanted mingling of components. Regardless of interaction flexibility, the problem remains that one-dimensional compaction has no global information to help it achieve the best results. What is needed is a way to relate the x and y constraints so that the sequencing of the process uses all of the available information.

Both x- and y-axis constraints can be incorporated into a single network to enable more global compaction [Watanabe]. This network has arcs between all components that may interact during compaction. These arcs are tagged to indicate whether they can act alone or can work in conjunction with a dual constraint that runs in a perpendicular direction. This notion of a dual constraint enables diagonally spaced components to compact in either direction. For example, two contacts that must remain 3 units apart can stay 3 units away in x, or 3 units away in y, but when they are placed on a diagonal it is not necessary to enforce both constraints. This is because diagonal constraints are spaced farther apart than is actually necessary.

Optimizing a two-axis constraint network involves a **branch-and-bound** technique that iteratively finds the longest path in either direction and removes one-half of a dual constraint to shorten that path. This can take much time on a large network, so it is useful to reduce the initial graph size as much as possible. Watanabe found a property of uncompacted layout that helps to reduce this network: If every component is viewed as a vertex and every wire as an edge, the initial design forms an uninterrupted area of polygons. A constraint between two components is not necessary if those components do not border on the same polygon.

Humans often compact by jogging wires so that components can tuck into the tightest space. This can be done automatically by splitting wires at potential jog points so that the two halves can move independently. One observation is that good jog points occur where two parallel wires are connected by an orthogonal wire [Dunlop]. The orthogonal wire can be jogged to save space (see Fig. 4.6). Of course, jogs can be placed anywhere and the most general jog-insertion technique is to remove all wires and reroute them [Maley]. Routing methods will be discussed later in this chapter.

As a final consideration, it is possible to use **simulated annealing** as a model of compaction [Sechen and Sangiovanni-Vincentelli]. This technique, which is useful for many synthesis tasks, makes severe adjustments to the layout at the beginning and subsequently damps its action to make smaller changes, thus mimicking the annealing process of cooling molecules [Metropolis *et al.*]. By making component placement and orientation changes between compaction passes, each change can be seen either to help or to hurt the compaction. Major

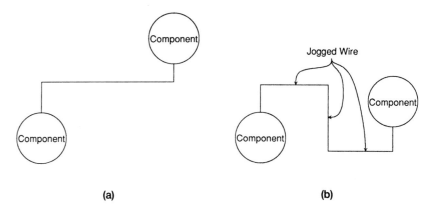

(a)   **(b)**

FIGURE 4.6   Jog insertion during compaction: (a) Before (b) After.

changes are tried first so that the overall layout configuration can be established. The minor changes are made last, in an attempt to save the small bits of space. This method consumes vast amounts of time but achieves very good results.

A more global issue in compaction is how to do it in a hierarchical environment. Typically, leaf cells are compacted first and then composition cells are adjusted to squeeze space around the leaf instances [Ackland and Weste]. This can cause problems when leaf-cell compaction realigns connection points, causing intercell connections to be broken. Although a hierarchical-layout constraint system, such as that in Electric (see Chapter 11), will keep these connections safe, most systems must do pitch matching (discussed later in the chapter) to ensure that hierarchical compaction will produce valid results [Weste; Entenman and Daniel].

Compaction is useful in improving productivity because it frees the designer from worries about proper spacing. With any compacter— even a simple one-dimensional compacter—layout can initially be done with coarse spacing so that the circuit can be better visualized. The designer is free to concentrate on the circuit and need not waste time scrutinizing the layout for unused space. Also, the existence of a compacter makes it easier to produce cell-contents generators because precise layout issues can be deferred. Finally, a compacter can be used to convert layout from one set of geometric design rules to another [Schiele].

4.3   Generators of Layout Outside the Cells

The next class of synthesis tools deals with layout external to the cell. Since most active circuitry is inside the cells, these tools handle the tasks of intercell wiring (routing), cell placement, and pad layout. Since routing can be very complex and can consume more area than the actual cells do, it is important to plan the connections in advance. Although proper cell planning and placement can make the routing simpler, tools that do this automatically are rare.

4.3.1   Routers

A router is given sets of points on unconnected cells that must be connected. The space between these cells can be as simple as a rectangle or as complex as a maze (see Fig. 4.7). When the routing space is rectangular and contains connection points on two facing sides, it is called a **channel**. When routing is done on four sides of a rectangular area, it is called a **switchbox**. More complex routing areas usually must be decomposed into channels and switchboxes. Many different tech-

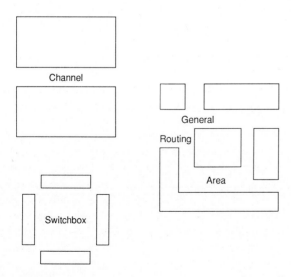

FIGURE 4.7   Routing areas.

niques exist for routing, depending on the number of wires to run, the complexity of the routing space, and the number of layers available for crossing.

Maze Routing   Maze routing is the simplest way to run wires because it considers only one wire at a time. Given the connection points and a set of obstacles, the maze router finds a path. The obstacles may be cells, contacts that cannot be crossed, or wires that have already been run.

The first maze-routing technique was based on a grid of points to be traversed by the wire [Moore; Lee]. Now known as the **Lee-Moore algorithm**, it works by growing concentric rings of monotonically increasing values around the starting point. When a ring hits the endpoint, a path is defined by counting downward through the rings (see Fig. 4.8). This method is so simple that hardware implementations have been built [Blank]. It can also employ weighting factors that increment by more than one and use the lowest value when two paths merge. This can prevent wires from running outside of an area when they could run inside.

One problem with Lee-Moore routing is that it demands a work array that is very large. Moore observed that the integers 0, 1, and 2 can be used cyclically since the adjacent values are always plus or minus one. However, even a two-bit work array can be very large. To avoid this problem, maze routing can be done with traditional

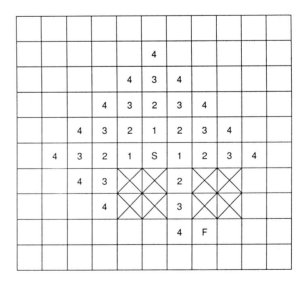

FIGURE 4.8 The Lee-Moore routing algorithm: Beginning at "S," integers radiate toward "F"; the shortest path counts backward from the end.

maze-solving techniques that walk along the obstacle walls. Figure
4.9 illustrates this technique: The router moves from the starting point
toward the finish until it hits an obstacle. It then chooses to walk to
the left (although it could choose to walk to the right, so long as it
is consistent). When the obstacle is cleared, it heads for the endpoint—
but once again hits an obstacle, this time a wire. Crawling along these
obstacles to the left makes the path go all the way around the cell to
get to the endpoint. Note that the extra loop made near the wire
obstacle can be optimized away so that there are only seven turns
rather than nine. The **Hightower** router is a variation of the maze
technique that constructs paths from both ends until they intersect
and complete a connection [Hightower].

The basic maze router is inadequate for many tasks, as the figure
illustrates. For example, if the router chooses to walk to the right
instead of to the left when it hits an obstacle, then it will quickly loop
back to its starting point. This is an indication that it should change
directions and start again. If the router then returns for the second
time after walking both to the left and to the right, the wire cannot
be routed. This is the case in Fig. 4.9 when an attempt is made to
connect "start" to "aux." The only solution in such a situation is to
change layers so that the wire can cross other wire obstacles.

Multilayer routing is an obvious extension to this basic technique.
Typically the wire changes layers only long enough to bridge other
wires that form an obstacle. Some schemes have one layer that runs
horizontally and another layer that runs vertically [Doreau and Koziol;
Kernighan, Schweikert, and Persky]. Routing on each layer is not

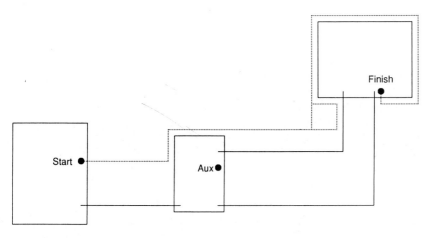

FIGURE 4.9 Maze routing.

difficult because all wires are parallel, so the only question is how to order sets of parallel paths. Since horizontal and vertical paths are already known, the location of layer contacts can be fixed so that the remaining problem is how to wire between them. This is done by swapping wires until they all fit through the available spaces and around the layer contacts. This scheme employs a constraint network that indicates how many wires can pass between any two obstacles. The network simplifies the computation of wire width and design-rule spacing. It ensures that wires will never form obstacles, but it creates a large number of layer contacts.

Other considerations must be addressed when doing multilayer routing. The most important is to minimize the number of layer changes. Contacts to other layers take up space on both layers and form obstacles for other wires. Also, excessive layer switching can delay a signal in the final circuit. Therefore the router should optimize the layer switches by refusing to make a change if it is not necessary for obstacle avoidance. A useful option to consider in multilayer routing is that there may be different starting and ending layers desired for a wire so that, once a layer change is made, the wire can remain in the layer to which it was switched. Of course, an alternative to switching layers is to route around the outside of an obstacle, which can consume much space. To allow for this without making absurd detours, some routers have parameters to choose between extra run length and extra layer changes for obstacle avoidance [Doreau and Koziol].

Another multilayer-routing consideration is the priority of layers. Since not all layers are the same, the speed of signals on a given layer should bias that layer's use. Also, some signals are restricted to certain layers. In IC layout, the power and ground wires must always run on metal layers [Keller]. When these wires cross other signals, it is the other signal that must change layers. One way to ensure this is to run the power and ground wires first; however, there must still be special code to ensure the correct placement of these wires. The PI system uses a **traveling-salesman** technique to find a path that runs through each cell only once in an optimal distance [Rivest]. Since this path never crosses itself, it divides the layout into two sides on which power and ground can be run without intersecting.

In printed-circuit routing, there are some signals that must run on specific layers. For example, those wires that are important in board testing must run on outside layers so that they can be probed easily. Therefore the multilayer router must allow special requests for certain wires' paths.

Multiple-Wire Routing Routing an entire circuit one wire at a time can be difficult, because each wire is run without consideration for others that may subsequently be routed. Without this global consideration, total consistency is hard to achieve. Preordering of the wires to be routed can help in some circumstances. For example, if the wires are sorted by length and are placed starting with shortest, then the simple and direct wiring will get done without trouble and the longer wires will work around them. Alternatively, if the longest wires are run first, then they will be guaranteed to run and the short connections will have to struggle to complete the circuit. Another solution is to run wires in the most congested areas first [Doreau and Koziol]. This is a good choice because it most effectively avoids the problem of being unable to complete all the wiring.

Another option when routing many wires is the use of alternative routes for a wire. If each wire that is routed is kept on a stack, then this information can be used to back up when a new wire cannot be run. By backing up through the stack and undoing wires up to the offending wire that causes the blockage, an alternate path can be made so that subsequent wires can have a different environment that may be better (see Fig. 4.10). The nature of this routing change must be saved on the stack so that the same choice will not be made again if the router backtracks to that point a second time. If repeated backtracking to redo a wire exhausts the possibilities for that wire, backtracking must then go beyond the offending wire to an earlier one that allows even more possibilities. This process can be extremely time consuming as it repeatedly routes wires one at a time. With a little bookkeeping, much of the computing can be saved so that wires that reroute after a backtrack are routed faster. Nevertheless, there are

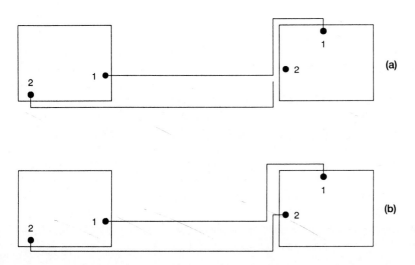

(a)

(b)

FIGURE 4.10   Backtracking in routing: (a) First attempt to route net 2 (b) Backtracking reroutes net 1.

no guarantees that recursive juggling of these wires will result in a good solution. One way to improve the chances for routing is to search the space of possible wire paths so that only those wires that promise no space increase are placed [Kernighan, Schweikert, and Persky].

Another way to manipulate single paths to obtain global consistency is with the simulated-annealing technique used in compaction [Kirkpatrick, Gelatt, and Vecchi; Sechen and Sangiovanni-Vincentelli]. Having a global measure of routing quality and making random changes allows each change to be seen either to help or to hinder the routing. Major changes, which are tried first, establish the overall routing configuration; minor changes are done last to get the details of the wiring correct. This method also consumes vast amounts of time, but achieves very good results.

Channel and Switchbox Routing   An alternative to single-wire routing is **channel routing**, in which all the wires in a channel are placed at once. This technique essentially sweeps through the routing area, extending all wires toward their final destination. Although this routing method can get stuck at the end of the channel and fail to complete many of the wires, it does make global considerations during the sweep so the results are often better than they are with maze routing. Also, it is faster because it considers all wires in parallel, examines the routing area only once, and avoids backtracking. Those routers that can handle connections on all sides of a routing area are called **switchbox routers** [Soukup] and they generally work the same way as channel routers. If the routing area is not rectangular, it must be broken into multiple rectangles and connection points assigned along the dividing lines (see Fig. 4.11). This step is called **global routing** or **loose routing** [Lauther].

Channel and switchbox routing typically make the assumption that wires run on a fixed-pitch grid, regardless of wire width or contact size. All contact points are also on these grid locations. Although this may consume extra space because of the need to use worst-case design rules, the area can always be compacted in a postprocessing step. Some routers are able to create new grid lines between the initial rows and columns if the new lines are necessary to complete the wiring successfully [Luk; Lauther].

Although there are many channel and switchbox routers [Deutsch; Rivest and Fiduccia; Doreau and Koziol], only one will be discussed; it combines good features from many others [Hamachi and Ousterhout]. This router works with a switchbox that contains connection points, obstacles, and obstructions and makes connections using two layers of wire (see Fig. 4.12). **Obstructions** are those areas that can be crossed

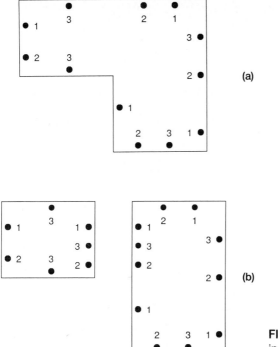

(a)

(b)

FIGURE 4.11   Global routing: (a) Before (b) After.

only on certain layers. **Obstacles** block all layers, forcing the router to work around them.

The routing task is viewed as follows: Each net is in a particular horizontal track and must eventually get to a different horizontal track to complete its wiring. The wiring is run one column at a time, starting from the left and sweeping across the area. At each column, there are rules to determine which nets will switch tracks. When the sweep reaches the right side of the area, the layout is complete and all nets have been routed.

In order to implement this model for switchboxes, the nets connected at the top and bottom must be brought into the first horizontal track, and any connections to nets that are not on track centers must jog to the correct place. Once the nets are all on horizontal tracks, routing can begin.

As each column is processed, priority is given to those vertical wirings that can collapse a net. This is the case with the first column on the left in Fig. 4.12 that collapses net 2. Notice that the layer changes when running vertical wires. This allows vertical and horizontal tracks to cross without fear of connection. To prevent excessive layer

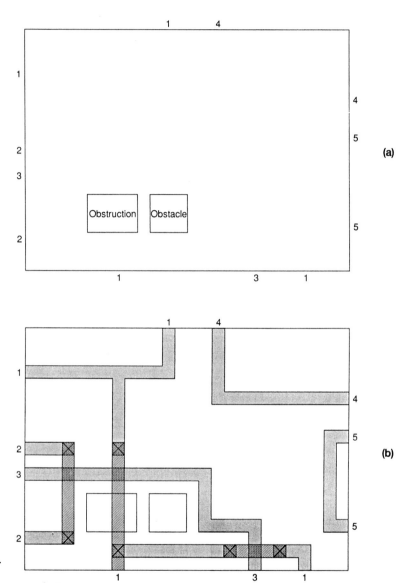

FIGURE 4.12 Switchbox routing: (a) Initial configuration (b) After routing.

switching, a postprocessing step eliminates unnecessary switches and moves contacts to maximize use of the preferred layer (if any).

The next priority in processing a column is to shift tracks toward a destination. Net 3 in the figure makes a downward jog as soon as it has cleared the obstacle so that it can approach its destination. Similarly, net 4 jogs down to its correct track immediately. Special rules must be used to handle the right side of the channel; otherwise

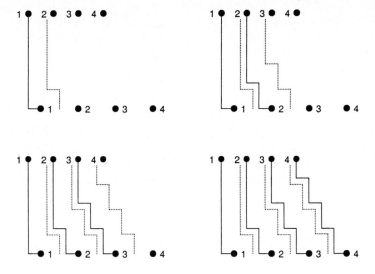

FIGURE 4.13   River routing. Dotted lines show areas of avoidance that guide subsequent wires.

wires like net 5 will not know their destination until it is too late. These rules reserve an appropriate number of columns to route the right edge.

In addition to these basic rules, there are much more complex ones that dictate when to create a contact, when to vacate a track in anticipation of an obstruction, and when to split a net in preparation for multiple end connections. When the rules are followed in the proper order, channel routing is achieved.

River Routing   In some situations the routing needs are specialized and can be solved more easily due to the problem's constraints. One such simplified technique is **river routing**, which connects many wires that are guaranteed not to cross, but instead run parallel to each other. River routing can be used to connect two cells that have inexact pitch and alignment [Wardle *et al.*]. The issue in river routing is how to adjust the spacing properly if one end is wider than the other. One method runs the wires of the bus serially and constructs areas of avoidance after each wire [Tompa]. Figure 4.13 shows this process: The dotted lines are the areas of avoidance that are created after the wires are placed. Special methods such as this allow unusual routing needs all to be met in the wiring of a circuit.

4.3.2   Placement

Placement, or **floor-planning** as it is sometimes called, is the process of positioning cells such that they connect well and do not waste space. Automatic placement tools therefore need to know the size of each cell and the complete set of routing requirements. Since this information is the same as what is required of routers, the two are often thought of as two parts of the same process. In fact, **placement and routing** comprise a single step in automatic printed-circuit-board layout. However, the two are often independent and have different algorithms driving them. Also, they must function separately as each produces results needed by the other: First, placement establishes an initial location of cells; next, routing connects the cells and indicates, by its success, how changes to the placement can help. The process is repeated until suitable layout is produced. Although an ideal placement and routing system would do these tasks uniformly, no such systems exist today. What is needed is to be able to share the placement and the routing goals in a common data structure so that the system can alternate between the two as the priorities change.

One system does approach this unification by dividing the overall placement and routing task into smaller pieces that are invoked alternately [Supowitz and Slutz]. This system observes that the vertical, lower channel of a T-shaped routing area needs to be routed before the horizontal, upper area. This T-shaped routing area is created by three cells on the left, right, and top. Routing the vertical, lower channel allows the cells on the left and right to be placed correctly. Then the upper area can be routed and the top cell can be placed correctly. The assumption is that a crude placement has already been done so that the placement and routing step can simply connect while fine-tuning the cell locations. Another assumption is that there is no cycle of T-shaped routing areas, which would prevent a good ordering of the placement and routing steps.

Automatic placement is difficult because of the extremely complex problems that it must satisfy. For example, if the wiring is ignored and only placement is considered, the task is simply one of tiling an area optimally, which is NP-complete [Garey and Johnson]. This means that the program for packing cells tightly can take an unreasonable amount of time to complete. If wiring considerations are added to this task, then every combination of cell placements that is proposed will also require seriously large amounts of time in routing. Because of these problems, placement cannot be done optimally; rather, it is done with heuristics that produce tolerably good results in small amounts of time.

One heuristic for automatic placement is **min-cut** [Breuer]. This technique divides the unplaced cells into two groups that belong on opposite sides of the chip. Each side is further subdivided until a tree-structured graph is formed that properly organizes all cells. Determination of the min-cut division is based on the wiring between the cells. The goal of min-cut is to make a division of cells that cuts the fewest wires (see Fig. 4.14). The number of cells should be divided approximately in two, such that each half has a high amount of internal connectivity and the two halves are minimally interconnected.

The min-cut technique cannot do the whole job of placement. It considers neither cell sizes nor the relative orientation of cells within a group. However, it is one method of reducing the size of the problem by breaking it up into smaller subproblems. Another way of doing the same thing, called **bottom-up** [Rivest; Heller, Sorkin, and Maling], is conceptually the opposite of the min-cut method because it grows clusters rather than cutting them down.

The bottom-up method of automatic placement presumes that all cells are distinct and must be grouped according to their needs. A weighting function is used to indicate how closely two cells match. This function, like the one for min-cut, counts the number of wires that can be eliminated by clustering two cells. It can also take cell

Groupings	Cuts
AB CD	10
AC BD	6 ***
AD BC	10

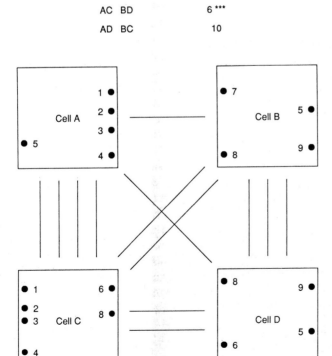

FIGURE 4.14   Min-cut placement method. The goal is to find a partitioning of the cells that results in the fewest wire cuts.

sizes and other weights into account. Clustering then proceeds to group more and more cells until there is no savings to be had by further work. Figure 4.15 illustrates this technique. Initially, there are four unrelated cells. Bottom-up planning finds the two that fit together best by eliminating the most number of wires. In addition to considering

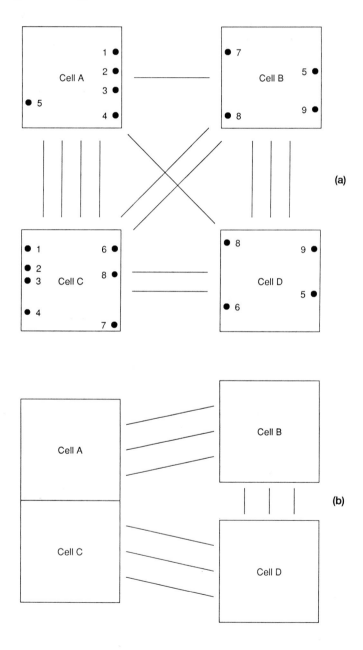

FIGURE 4.15 Bottom-up placement method: (a) Before (b) After.

wiring outside of cells, it is necessary to examine the complexity of wiring inside a cluster. If this were not done, then the optimal bottom-up clustering would be a single cluster with all cells and no external wires.

Once min-cut or bottom-up methods have organized the cells into their relative orientations, a more precise placement is needed. This can be done by converting the placement graph into a dual graph that includes position [Kozminski and Kinnen]. The method recursively picks off edge members of the original graph and creates a new graph with position and size information. By proceeding in adjacency order, the new graph is built in a "brickwork" style.

An alternative to converting the placement graph into a more precise placement is to compute this information as the initial placement graph is built. In one system, the initial layout is seen as a square, and each min-cut division divides the square proportionally to the area of cells on each side of the cut [Lauther]. When min-cut is done, the square indicates proper area and position, but incorrect aspect ratio. The problem then is to fit the actual cells into the idealized areas of the square. A number of approximate methods are used, including rotating cells and spreading open the area. Since routing requirements may spread the layout further, this approximation is not damaging.

Besides min-cut and bottom-up methods, those that find analogies to physical systems that behave in similar ways also can be used to do placement. For example, if a circuit is viewed as a resistive network, in which each cell is a constant voltage node and each collection of intercell wires is a resistor, then the problem becomes one of finding the correct voltage. The voltage values that are close indicate cells that should be near. To implement this technique, the square of the wire lengths becomes the objective function that is solved with matrix techniques [Cheng and Kuh]. The resulting graph can then be adjusted into actual placement using the techniques already described.

Another physical emulation that is used for placement is simulated annealing [Kirkpatrick, Gelatt, and Vecchi; Sechen and Sangiovanni-Vincentelli]. This method was shown to be useful in compaction and routing, so it should be no surprise that it also works for placement. Given an initial floor-plan, the annealer makes severe changes and evaluates the results. The severity of the changes decreases until there is nothing but detail to adjust. Like the other annealing tasks, this method consumes large amounts of time but produces good results.

Other information can be used to produce more intelligent cell placement. For example, there can be special code to detect cells that align well and should be placed together. This can be done before the general-purpose placement begins. An **idiomatic** floor-planner is one that recognizes these special situations and uses appropriate algorithms in each case [Deas and Nixon]. Determining which idioms exist in the layout is done by examining the routing needs and applying specialized rules that recognize arrays, trees, complex switchboxes, and so on.

Once a basic placement has been found, it is possible to use other methods to fine-tune the layout. The results of routing will indicate the amount of channel utilization so that the cell spacing can be adjusted. This information can be used to control another iteration of placement or it can simply indicate the need for compaction of the final layout. It has also been shown that the placement and routing cycle can be shortened by having routing take place after every step of min-cut division [Burstein, Hong and Pelavin]. Once again, the results of routing help with subsequent placement.

Besides routing, other tools can be used to help improve the results of placement. One system even integrates timing information to help placement and routing iterations [Teig, Smith, and Seaton]. This timing analysis, described more fully in Chapter 5, finds the worst-case circuit delay so that the placed modules and routed wires can optimize that path.

Automatic placement is a difficult problem that has not yet been solved well. However, it is one of the few tasks in which hand design excels over machine design, so it is worth trying to solve. With good placement, automatic layout can become a useful reality. Until such time, these systems will provide alternative layouts from which a designer can choose.

4.3.3   Pad Layout

The pads of an integrated-circuit chip are large areas of metal that are left unprotected by the overglass layer so they can be connected to the leads of the IC package. Although large compared to other circuit elements, the pads are very small and make a difficult surface on which to bond connecting wires. Automatic wire-bonding machines can make accurate connections, but they must be programmed for each pad configuration. The chip designer needs a tool that will help with all aspects of bonding pads.

The immediate objection of many designers to the offer of automatic pad layout is that they want to place the pads manually. Given that pads consume large areas, the designer often wants total placement control to prevent wasted space. Chips have not typically had many pads anyway, so manual techniques do not consume much time.

The answer to such objections is that any task, no matter how simple, must be automated in a complete design system. As the number of special-purpose chips increases and their production volume decreases, it will be more important to design automatically than to design with optimal space. Also, the programming of the bonding machine can become bottlenecked if it is not directly linked to the CAD system. In modern chips there can be hundreds of pads; placing them manually will lead to tedium and confusion. Finally, there are only a few considerations that must be kept in mind when doing pad layout and these can be easily automated.

One consideration in pad placement is that the pads should be near the perimeter of the chip. The bonding machine may spatter metal and do damage to the circuitry if it has to reach into the center. Some fabrication processes even require that no active circuitry be placed outside of the pads. Others, however, allow pads anywhere on the chip. Another reason to place pads on the edge is to keep the bonding wires from crossing. The pads must present an uncluttered view from the outside of the chip.

In addition to being located on the edge of the chip, pads should also be placed uniformly. This means that there must be approximately the same number of pads on all four edges and that they must be spaced evenly along each edge. On special chips that have rectangular layout, the pads may be evenly spaced along only two edges. Equal spacing makes automatic bonding easier, and uniform pad density keeps the bonding wires from cluttering and possibly shorting. The proper limitations of pad spacing must also be taken into consideration (currently no less than 200 microns between centers [Keller]).

One algorithm for pad placement is called **roto-routing** [Johannsen]. This technique starts by sorting the pads into the same circular sequence as are their connection points in the chip. The pads are then placed into a uniform spacing around the chip and rotated through all positions. The one position that minimizes overall wire length to the pads is used.

Another concern in proper pad layout is the location of power and ground pads. Although some applications demand multiple power and ground pads for high power consumption, there are typically only one of each, and they should generally be on opposite sides of the chip, because they are usually bonded to leads that are on opposite sides of the package. In addition, they must be correctly connected to the other pads in order to get proper distribution of power and ground. Figure 4.16 shows one configuration of power, ground, and pads. The ground runs outside of the pads because it is less critical than the power rail, which runs just inside. This arrangement allows each pad to have easy access to the supply voltages it needs to function properly. In order to get these voltages to the chip, however, there must be a gap in the inner power rail.

Routing of pad signals to the main chip circuitry can be done

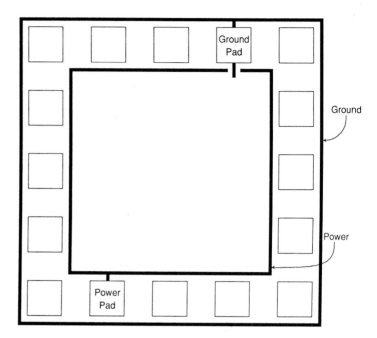

FIGURE 4.16 Pad frame showing power and ground rails.

automatically with standard switchbox techniques. It is important to try to place the pads close to their destination so that signal wires do not run the length and width of the chip. Techniques for automatically assigning pads to signals can take into account both wire length and the importance of that signal's timing. For example, a two-phase clock brought in from outside should be placed on two pads that have similar distances to the active circuit.

Although bonding-pad placement is a minor aspect of integrated-circuit design, it is a necessary part of any good CAD system. Changing design methodologies call for tools that can do all the work and prevent minor oversights from turning into major failures. There are detailed constraints in pad layout, and failure to observe any one of them can cause the chip to fail totally.

4.4 Cells and Their Environment

The next class of synthesis tools deals with the general environment of cells. These tools do not concern themselves with explicit layout, but rather ensure that the layout is properly used. Included in this category are alignment and pitch matching tools.

4.4.1   Alignment and Pitch Matching

When completed cells are placed together to form a circuit, there are two ways that they can connect: implicitly by abutting with composition rules or explicitly by connection rules. Abutment presupposes that the contents of the cells are precisely designed so that their mere presence near each other forms connecting wires. This has the advantage that less space is wasted between the cells because no complex routing of wires is needed. The disadvantage is that some space may be lost inside the cells in order to get the wires to the right place. This process of properly **aligning** wires for external connection is part of **pitch matching**.

Pitch matching is actually the stretching and shrinking of cells so that their bounding boxes line up. For example, if a multibit parallel subtraction unit is to be built from negation and addition cells, the aspect ratio of the two cells should be pitch matched so that both have the same height (or width, depending on how they are stacked). Since addition is more complex than negation, the negation unit will probably have to be stretched in order to pitch match. It is also necessary to know all of the connections into and out of each cell in order to do this properly.

Pitch matching can be done in a number of ways. The manual method requires that the most complex cell be designed first so that the largest dimensions are established. Subsequent cell layout will start with a fixed width or height according to the way in which the cells adjoin. The order of cell design will depend heavily on the floor-plan of the chip, with each newly designed cell obeying the dimension rules along one axis and forming new dimensions along the other axis for subsequent cells. The pitfall is obvious: Any single change can destroy all this careful planning and require redesign of major portions of the chip. The advantage of all this manual effort is that each newly designed cell that must meet one dimension can be optimized in the other.

Automatic pitch matching requires not only that the design system know the location of all connection points, but also that it understand how to stretch the layout between these points. One way this can be done is by designing on a stretchable or virtual grid [Weste]. Layout done in this style must acknowledge wires as special stretchable entities that can be adjusted within their particular design rules. Typically, only Manhattan wires can be used, since cells have rectangular shape and must stretch along these axes.

When stretchable wires are explicitly known to the design system, pitch matching of two cells is simple, since the connection points and cell edges can be adjusted to the greatest common spacings. In order to achieve space savings, the layout that results from these two matched cells can be compacted together (see Fig. 4.17). This will reduce each individual cell to its smallest layout within the limitations of its connection constraints. If the compaction is good, there will not be much wasted space in either cell.

The advantage of automatic alignment and pitch matching is that changes to the design can be handled easily. The disadvantage is that fine-tuning of area consumption cannot be done. However, given better tools for compaction and floor-planning, more efficient fabrication processes, and ever more complex designs, automatic techniques become very attractive.

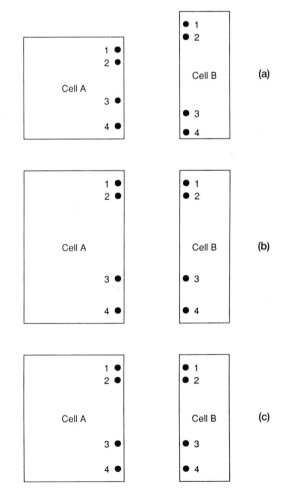

FIGURE 4.17   Pitch matching and alignment: (a) Initial layout (b) Pitch matched and aligned (2-to-3 on A stretched; 3-to-4 on B stretched) (c) Overall layout compacted.

4.5   Silicon Compilers

The meaning of the term "silicon compiler" has changed over the years as advances in CAD tools have been made. Initially it implied a glorious dream of specifying some extra switch to a normal programming language compiler so, for example, it would translate FORTRAN into layout, rather than into object code. Soon, however, this dream was reduced to the use of regular methods for converting logic into layout, such as those discussed at the beginning of this chapter. Although these methods do not use conventional programming languages as input, they do translate from a layout-independent structural or behavioral specification into complete circuitry and therefore are called "silicon compilers".

The current excitement about silicon compilation began with the work on a system called Bristle Blocks [Johannsen]. Here was a system that could take a high-level description of a processor (an ISP description, see Chapter 2) and produce complex MOS layout of the type often done by hand. Although the system never produced completely working chips, those that followed did [Southard], and they sparked a new meaning for silicon compilers: systems that construct custom layout from high-level languages. No longer can a PLA generator be seriously called a silicon compiler.

The secret to the success of Bristle Blocks is the fact that its regularity is in the floor-plan. Rather than have a small regular structure that is customized, this system has a processor floor-plan that is customized (see Fig. 4.18). All aspects of the ISP language can be mapped into parts of this processor: The data requirements specify the hardware datapaths and the control requirements specify the structure of the PLA microcode decoder. The datapath consists of columns of constant-width cells that transfer data between the buses. Each column is custom assembled to perform one of the specified processor functions. Special parameterized cells are textually described so they can adjust their width and connections to stack correctly. These cells are all designed to function with a two-phase clock for synchronizing data transfer with processing.

Modern silicon compilers translate structural descriptions into IC layout using many of the same steps as do traditional language compilers. A front-end parser reads the input description and converts it to a structural representation such as the processor floor-plan used by Bristle Blocks. Some more advanced systems can handle higher-level behavioral descriptions and reduce them to the structural level before producing layout [Mitchell, Steinberg, and Schulman]. The back end

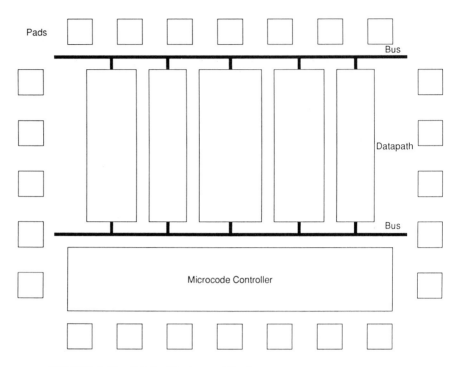

FIGURE 4.18   Bristle Blocks architecture.

then produces layout in two phases. First, a set of procedurally defined cells are invoked to produce the bulk of the circuitry for data operations. Control sequences are often built with traditional PLA techniques. Second, routers make the necessary connections between the major structural components. There are even optimization passes that can be invoked to compact the layout.

Most silicon compilers are specialized to produce one type of design. For example, FIRST always builds bit-serial chips for signal-processing applications [Denyer, Murray and Renshaw]. In specialized silicon compilers, the input language can be very high level, allowing nonprogrammers to specify chips for their own particular needs [Kahrs]. Another specialized silicon compiler, Miss Manners, compiles path expressions into the necessary layout for synchronizing circuit activity [Balraj and Foster].

Some silicon compilers have multiple architectures available to them [Pope, Rabaey and Brodersen], but they all are restricted to mapping a specific input language to a particular style of layout. The most generalized commercial compilers simply have a larger set of architectures available to them [SCI; Seattle Silicon; Buric and Matheson; VTI]. Ayres describes a general-purpose language that can be used to construct architectures, cell generators, and even analysis schemes for silicon compilers [Ayres].

One way to improve the flexibility of silicon compilers is to have them automatically select the correct architecture from the input specification. Even harder would be to generate the architecture automatically according to the needs of the specification. If a silicon compiler could accept a general-purpose hardware-description language and always produce a reasonable layout, then it would have achieved its glorious initial goal.

4.5.1 Expert Systems

Another approach to flexibility in silicon compilation is the use of **expert systems**, which are collections of high-level rules that interact to produce intelligent results. Each rule is expressed as an if-then pair that shows a precondition and an action to take. The expert system continuously scans the list of rules to find and execute those that are applicable. Typically, there are two kinds of rules: **control** and **design**. Control rules direct the overall task, for example:

> **if** need-an-inverter **then**
>> create-pullup,
>> create-pulldown,
>> connect-pullup-to-pulldown.

Design rules are more specific and generate the final results:

> **if** creating-pullup **then**
>> find-closest-power,
>> place-depletion-transistor,
>> connect-gate-to-drain,
>> connect-source-to-power.

Rules can be arbitrarily complex and may interact with each other in nonobvious ways. However, given enough rules with the right preconditions, fairly intelligent activity can be described.

Because the process of converting behavior to layout is so vast, most expert systems tackle only one piece. The Vexed system translates behavior to logic gates, seeking a structural organization without attempting complete layout [Mitchell, Steinberg, and Schulman]. By storing both actual behavior and required specifications, the system knows when an adequate design has been produced. Talib is an expert system that finishes the design task by translating logic gates to nMOS layout [Kim, McDermott, and Siewiorek].

Some expert systems are specialized for particular tasks. The Hephaestus system has rules for producing PLA and ring-oscillator layouts [Simoudis and Fickas]. What makes this system interesting is its ability to track each design decision and later to explain how a solution was achieved.

Expert systems may also be used exclusively for their control functions, directing other expert or nonexpert system tools. Such systems attempt to mimic the overall steps taken by human designers who hierarchically organize their circuit. In the Cadre system, multiple experts cooperate in the layout task [Ackland *et al.*]. Starting with a structural specification of relative cell locations, the system moves up and down the hierarchy to obtain a good layout. A cell-generation expert is invoked first [Kollaritsch and Weste]. Next, a floor-plan is made and an evaluation expert determines whether the cells or the floor-plan should be adjusted. The manager expert controls this iteration until a satisfactory layout is produced. The notion of moving up and down the hierarchy is also found in other expert systems [Gajski].

Silicon compilers are exciting because they raise the level at which design is done. It is no longer necessary to manipulate transistors in a layout. The specification is at a much higher level, generally structural. Silicon compilers must advance to the level at which they can handle any abstract description, convert it to layout, and be efficient enough to keep the designer from wanting to modify that layout. Tools will then develop for simulating and analyzing the high-level descriptions so that all design activity can take place at a comfortable distance from IC layout.

4.6   Postlayout Generators

Once a design is complete it must be manufactured or else it will remain just so much wallpaper. This process involves removing hierarchy by flattening the circuit, modifying the layout graphics for the manufacturing needs, making a template or **mask** for each layer, fabricating the actual circuit, and then packaging it for use. The set of synthesis tools to be discussed here is concerned with the layout modifications that must be done for this process.

4.6.1   Compensation

When a circuit is fabricated, differences are introduced between the mask data and the true circuit. For example, contact cuts involve the removal of material from previous layers, and because of the nature of the chemicals used, more area gets removed than is indicated on the mask. In order to get the correct dimensions on the final artifact, the mask dimensions must be **compensated**, in this case by shrinking.

Compensation is highly process-dependent—much more so than design rules. For example, most MOS processes have design rules that prevent polysilicon lines from being closer together than they are wide, but one process will compensate polysilicon by shrinking the geometry, whereas another process will bloat it. One reason that the direction of compensation is so variable is that some masks are negative and some positive, so it is never clear whether a layer's creation is the result of etching away unwanted material or protecting what is wanted.

Shrinking on the final circuit indicates the need to bloat on the mask, which is called **positive** compensation. When the mask is to be shrunk, that is **negative** compensation. This compensation is typically performed immediately prior to mask creation and is therefore never displayed to the designer.

Positive compensation is quite simple: Every polygon is made larger by the specified amount. This is done by moving all coordinates away from the center of the polygon. When two polygons overlap because they have each grown, there may be no need to remove the redundancy. This is because the mask-generation process sorts all the polygons prior to writing them and can merge overlaps easily.

Negative compensation is more difficult because two adjacent polygons that shrink will no longer adjoin. Polygon edges that border other polygons on the same layer therefore must not be shrunk. This requires the negative compensation process to spend much time exploring the neighborhood of each polygon. Once a common border is found, it may be necessary to break an edge into multiple edges in order to maintain connectivity while still shrinking (see Fig. 4.19).

The only ameliorating aspect of negative compensation is that it may be possible, in special circumstances, to do it as easily as positive compensation. If, for example, the contact-cut layer is to be shrunk, then there is no fear about loss of adjacency, because no two contact-cut polygons ever connect. This special case is not uncommon and should be available in the compensation system to save time. It presupposes, however, that no other part of the design system has divided a single contact cut into multiple polygons. If this assumption cannot be made, then more rigorous compensation must be done.

Compensation is one of the "hidden" operations in IC layout because the designer does not think about it at all. Nevertheless, it is a crucial step that enables the finished chip to resemble the specified design.

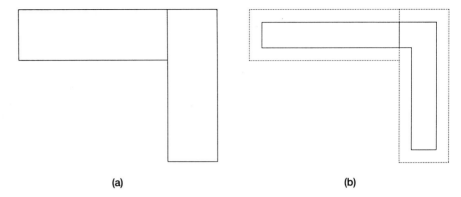

(a)   (b)

FIGURE 4.19 Negative compensation: (a) Before (b) After (dotted lines show before).

4.6.2 Mask Graphics

There are special markings that must be placed on the masks of an integrated circuit. These markings are used for mask alignment, process checking, and wafer cutting. In general, they are placed on a mask after compensation because their specified dimensions are indicative of desired mask markings and not wafer markings.

Alignment marks are needed on every layer so that the masks can be properly registered with each other. Typically the first mask to be used has a series of empty **target** boxes on it. Each subsequent layer to be aligned will then have a **pattern** that fits into one of these boxes. Almost any pattern and target will do, but they should be designed so that they can be easily spotted on the masks and do not obscure each other during alignment. If there are uniform gaps in the pattern-target overlap, it will be easier to check visually for alignment errors (see Fig. 4.20). Also, these gaps will help if one or both of the masks is produced as a negative image because of process requirements.

Maximum alignment accuracy is achieved by aligning all masks against the first one. This ensures that no masks are more than two alignment errors off from each other. In some cases, however, it is more important that a particular pair of masks align with each other than that they align with the overall wafer. In such cases, the set of masks will contain their own targets rather than use targets from the first layer. Decisions about which layers should be targeted against which other layers are generally made by process engineers, who best understand the important dependencies.

Another mark that is placed on a mask is a **fiducial** for gross

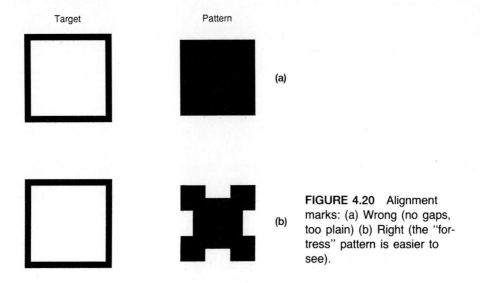

FIGURE 4.20 Alignment marks: (a) Wrong (no gaps, too plain) (b) Right (the "fortress" pattern is easier to see).

alignment. These marks are simply crosses in the corners that can be quickly aligned. Generally, fiducials are placed on the masks by the mask-making shop and do not have to be specified explicitly.

Yet another marking that process engineers need is called **critical dimensions**; these are used to check and adjust the quality of the fabrication. Critical dimensions are sets of minimum-width marks on each layer, which are checked under a microscope as the wafer is manufactured. When the process parameters are adjusted to obtain the proper critical dimensions, the rest of the chip is likely to turn out well.

Critical dimensions, like alignment marks, belong in a distinctive place on the chip that will be easy to find under a microscope. Rather than being hidden in some unused area on the chip, these marks belong in a corner, outside of the active circuitry. This extra chip space does not increase the probability of a defect that can damage circuit performance. In fact, having some extra space outside the active circuitry will help spread the individual chips on a wafer, which will make it easier to cut the wafer without damaging the chips.

Wafer cutting, sometimes called **scribing** or **sawing**, is the subject of another piece of mask graphics that a design system must handle. **Scribe lanes** are fairly wide lines that run horizontally and vertically across the wafer. They typically form a regular grid inside the round area of the wafer and do not extend to the edges (see Fig. 4.21). A CAD system that prepares wafer masks does not need to replicate the chip in each location on the wafer. Rather, it is sufficient to specify a complete chip with its scribe lanes and then indicate where on the wafer this chip will appear. Typically there are also test circuits that are inserted in a few of the chip locations to check different vicinities of the wafer.

One decision that must be made in placing scribe lanes is whether to put a full- or half-width lane around each chip. With a full-width lane, the chips will have to overlap on the wafer by one-half of the scribe-lane width in order to fit properly (see Fig. 4.22). This overlap is somewhat confusing to specify and costs extra mask-making time since all scribe lanes are then drawn twice. With a half-width scribe-lane border, each chip can abut its neighbors exactly and build a clean wafer. The disadvantage of half-width scribe lanes is that chips on the edge of the wafer will be damaged during scribing unless the other half of the outside scribe lane is specified explicitly. These edge chips are sometimes useless anyway since fabrication quality is best near the prime center area of the wafer. Another problem with half-width

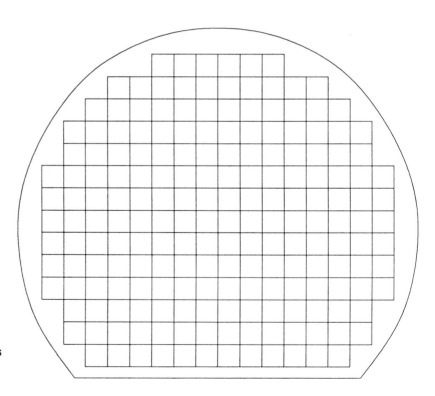

FIGURE 4.21   Scribe lanes on a wafer.

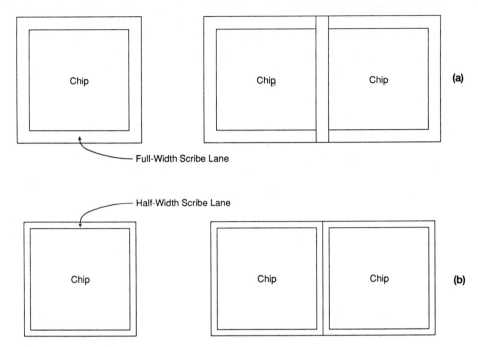

FIGURE 4.22   Full- and half-width scribe lanes: (a) Full-width scribe lanes overlap (b) Half-width scribe lanes abut.

scribe lanes is that the test chips may not have any scribe lanes, to allow them to fit in any wafer scheme. This means that the four chips adjoining each test chip will also lack full scribe edges.

The process engineer specifies exactly which layers should appear in the scribe lanes. In addition, a certain minimum spacing between the active circuitry and the scribe lane must be maintained. As a final consideration, the distance between scribe-lane centers must not be irregular because scribing machines are less accurate than are mask-making machines. For example, the scribe-lane spacing may have to be a multiple of 5 mils (five thousandths of an inch). All these details are process-dependent.

The last piece of graphics that a CAD system should be able to place on a layout is text. All designers document their circuitry by spelling out messages. This text will include the circuit name, different letters for each layer, and any other message of interest. This text not only is useful for documentation purposes but also serves to verify the correct side of a mask during fabrication. Instructions to the processing staff often include an indication as to whether the mask should be placed "right reading" or "wrong reading" according to how the text appears [Hon and Sequin]. This is easier to determine than some arbitrary symbol location or shape.

Mask graphics is a significant layout task that CAD systems often overlook; rather than providing these symbols, the designer is often forced to produce them manually. A good system not only will provide text and graphics but will help with the entire mask-preparation process by understanding what is necessary and where it should be placed.

4.7   Summary

This chapter has illustrated a number of techniques for automatic synthesis of VLSI circuitry. The list is not complete, since new methods are being proposed continuously. It is hoped that eventually the synthesis algorithms will be able to do the entire job of design. Until such time, however, humans will have to do much of the layout and will have to guide these synthesis tools carefully. Because this guidance is subject to errors, there will also have to be methods of analyzing a circuit for correctness. These analysis tools are the subject of the next two chapters.

Questions

1 How do PLAs and gate matrices differ in philosophy from gate-arrays?
2 Why do channel and switchbox routers produce better wiring patterns than do repeated applications of maze routers?
3 What is the most difficult aspect of global routing?
4 What other considerations should be addressed besides wire lengths when doing placement?
5 Why are rule-based systems attractive techniques for synthesis tools?
6 Why is there little hope that silicon compilers will ever be able to generate arbitrary and appropriate architectures?
7 In what sense are the results of simulated-annealing compaction incorrect?

References

Ackland, Bryan; Dickenson, Alex; Ensor, Robert; Gabbe, John; Kollaritsch, Paul; London, Tom; Poirier, Charles; Subrahmanyam, P.; and Watanabe, Hiroyuki, "CADRE—A System of Cooperating VLSI Design Experts," *Proceedings IEEE International Conference on Computer Design*, 99-104, October 1985.

Ackland, Bryan and Weste, Neil, "An Automatic Assembly Tool for Virtual Grid Symbolic Layout," *VLSI '83* (Anceau and Aas, eds), North Holland, Amsterdam, 457-466, August 1983.

Ayres, Ronald F., *VLSI Silicon Compilation and the Art of Automatic Microchip Design*, Prentice-Hall, Englewood Cliffs, New Jersey, 1983.

Baker, Clark M. and Terman, Chris, "Tools for Verifying Integrated Circuit Designs," *Lambda*, 1:3, 22-30, 4th Quarter 1980.

Balraj, T. S. and Foster, M. J., "Miss Manners: A Specialized Silicon Compiler for Synchronizers," *Proceedings 4th MIT Conference on Advanced Research in VLSI* (Leiserson, ed), 3-20, April 1986.

Beresford, Roderic, "Comparing Gate Arrays and Standard-Cell ICs," *VLSI Design*, IV:8, 30-36, December 1983.

Blank, Tom, "A Survey of Hardware Accelerators Used in Computer-Aided Design," *IEEE Design and Test*, 1:3, 21-39, August 1984.

Breuer, Melvin A., "A Class of Min-Cut Placement Algorithms," *Proceedings 14th Design Automation Conference*, 284-290, June 1977.

Buric, Misha R. and Matheson, Thomas G., "Silicon Compilation Environments," *Proceedings Custom Integrated Circuits Conference*, 208-212, May 1985.

Burstein, Michael; Hong, Se June; and Pelavin, Richard, "Hierarchical VLSI Layout: Simultaneous Placement and Wiring of Gate Arrays," *VLSI '83* (Anceau and Aas, eds), North Holland, Amsterdam, 45-60, August 1983.

Cheng, Chung-Kuan and Kuh, Ernest S., "Module Placement Based on Resistive Network Optimization," *IEEE Transactions on CAD*, 3:3, 218-225, July 1984

Deas, Alex R. and Nixon, Ian M., "Chromatic Idioms for Automated VLSI Floorplanning," *VLSI '85*, (Horbst, ed), 61-70, August 1985.

Denyer, Peter B.; Murray, Alan F.; and Renshaw, David, "FIRST—Prospect and Retrospect," *VLSI Signal Processing*, IEEE press, New York, 252-263, 1984.

Deutsch, David N., "A 'Dogleg' Channel Router," *Proceedings 13th Design Automation Conference*, 425-433, June 1976.

Do, James and Dawson, William M., "Spacer II: A Well-Behaved IC Layout Compactor," *VLSI '85*, (Horbst, ed), 283-291, August 1985.

Doreau, Michel T. and Koziol, Piotr, "TWIGY: A Topological Algorithm Based Routing System," *Proceedings 18th Design Automation Conference*, 746-755, June 1981.

Dunlop, A. E., "SLIM—The Translation of Symbolic Layouts into Mask Data," *Proceedings 17th Design Automation Conference*, 595-602, June 1980.

Entenman, George and Daniel, Stephen W., "A Fully Automatic Hierarchical Compactor," Proceedings 22nd Design Automation Conference, 69-75, June 1985.

Freeman, William J. III and Freund, Vincent J. Jr., "A History of Semicustom Design at IBM," *VLSI Systems Design*, Semicustom Design Guide, 14-22, Summer 1986.

Gajski, Daniel D., "ARSENIC Silicon Compiler," Proceedings International Symposium on Circuits and Systems, 399-402, June 1985.

Garey, Michael R. and Johnson, David S., *Computers and Intractability, A Guide to the Theory of NP-Completeness*, W.H. Freeman, San Francisco, 1979.

Goates, Gary B.; Harris, Thomas R.; Oettel, Richard E.; and Waldron, Harvey M. III, "Storage/Logic Array Design: Reducing Theory to Practice," *VLSI Design*, III:4, 56-62, 1982.

Hachtel, G. D.; Newton, A. R.; and Sangiovanni-Vincentelli, A. L., "Techniques for Programmable Logic Array Folding," Proceedings 19th Design Automation Conference, 147-155, June 1982.

Hamachi, Gordon T. and Ousterhout, John K., "A Switchbox Router with Obstacle Avoidance," Proceedings 21st Design Automation Conference, 173-179, June 1984.

Heller, W. R., "An Algorithm for Chip Planning," Caltech Silicon Structures Project file #2806, 1979.

Heller, William R.; Sorkin, G.; Maling, Klim, "The Planar Package Planner for System Designers," Proceedings 19th Design Automation Conference, 253-260, June 1982.

Hightower, D. W., "A Solution to Line-Routing Problems in the Continuous Plane," Proceedings 6th Design Automation Workshop, 1-24, June 1969.

Hon, Robert W. and Sequin, Carlo H., "A Guide to LSI Implementation," 2nd Edition, Xerox Palo Alto Research Center technical memo SSL-79-7, January 1980.

Hsueh, Min-Yu and Pederson, Donald O., "Computer-Aided Layout of LSI Circuit Building-Blocks," Proceedings International Symposium on Circuits and Systems, 474-477, July 1979.

Johannsen, D. L., "Bristle Blocks: A Silicon Compiler," Proceedings 16th Design Automation Conference, 310-313, June 1979.

Kahrs, Mark, "Silicon compilation of a very high level signal processing specification language," *VLSI Signal Processing*, IEEE press, New York, 228-238, 1984.

Keller, John, *Power and Ground Requirements for a High Speed 32 Bit Computer Chip Set*, Masters thesis, University of California at Berkeley, UCB/CSD 86/253, August 1985.

Kernighan, B. W.; Schweikert, D. G.; and Persky, G., "An Optimum Channel-Routing Algorithm for Polycell Layouts of Integrated Circuits," Proceedings 10th Design Automation Workshop, 50-59, June 1973.

Kim, Jin H.; McDermott, John; and Siewiorek, Daniel P., "Exploiting Domain Knowledge in IC Cell Layout," *IEEE Design and Test*, 1:3, 52-64, 1984.

Kirkpatrick, S.; Gelatt, C. D. Jr.; and Vecchi, M. P., "Optimization by Simulated Annealing," *Science*, 220:4598, 671-680, May 1983.

Kollaritsch, P. W. and Weste, N. H. E., "A Rule-Based Symbolic Layout Expert," *VLSI Design*, V:8, 62-66, August 1984.

Kowalski, T. J. and Thomas, D. E., "The VLSI Design Automation Assistant: Prototype System," Proceedings 20th Design Automation Conference, 479-483, June 1983.

Kozminski, Krzysztof and Kinnen, Edwin, "An Algorithm for Finding a Rectangular Dual of a Planar Graph for Use in Area Planning for VLSI Integrated Circuits," Proceedings 21st Design Automation Conference, 655-656, June 1984.

Lauther, Ulrich, "Channel Routing in a General Cell Environment," *VLSI '85*, (Horbst, ed), 393-403, August 1985.

Lee, C. Y., "An Algorithm for Path Connections and Its Applications," *IRE Transactions on Electronic Computers*, EC-10, 346-365, September 1961.

Liu, Erwin S. K., "A Silicon Logic Module Compiler," Project Report, University of Calgary Department of Computer Science, April 1984.

Lopez, Alexander D. and Law, Hung-Fai S., "A Dense Gate Matrix Layout Method for MOS VLSI," *IEEE Transactions on Electron Devices*, 27:8, 1671-1675, August 1980.

Luk, W. K., "A Greedy Switch-box Router," Carnegie-Mellon University Department of Computer Science VLSI Document V158, May 1984.

Maley, F. Miller, "Compaction with Automatic Jog Introduction," Chappel Hill Conference on VLSI (Fuchs, ed), 261-283, March 1985.

Mayo, Robert N., "Combining Graphics and Procedures in a VLSI Layout Tool: The Tpack System," University of California at Berkeley Computer Science Division technical report, January 1984.

McCreight, E.M., "Efficient Algorithms for Enumerating Intersecting Intervals and Rectangles," Xerox Palo Alto Research Center, CSL-80-9, 1980.

Metropolis, Nicholas; Rosenbluth, Arianna W.; Rosenbluth, Marshall N.; Teller, Augusta H.; and Teller, Edward, "Equation of State Calculations by Fast Computing Machines," *Journal of Chemical Physics*, 21:6, 1087-1092, June 1953.

Mitchell, Tom M.; Steinberg, Louis I.; and Shulman, Jeffrey S., "A Knowledge-Based Approach to Design," Proceedings IEEE Workshop on Principles of Knowledge-Based Systems, 27-34, December 1984.

Moore, E. F., "Shortest Path Through a Maze," Harvard University Press, Cambridge, Massachusetts, 285-292, 1959.

Mori, Hajimu, "Interactive Compaction Router for VLSI Layout," Proceedings 21st Design Automation Conference, 137-143, June 1984.

Mosteller, R. C., "REST—A Leaf Cell Design System," *VLSI '81* (Gray, ed), Academic Press, London, 163-172, August 1981.

Nogatch, John T. and Hedges, Tom, "Automated Design of CMOS Leaf Cells," *VLSI Systems Design*, VI:11, 66-78, November 1985.

Pope, Stephen; Rabaey, Jan; and Brodersen, Robert W., "Automated Design of Signal Processors Using Macrocells," *VLSI Signal Processing*, IEEE press, New York, 239-251, 1984.

Ramsay, Frank R., "Automation of Design for Uncommitted Logic Arrays," Proceedings 17th Design Automation Conference, 100-107, June 1980.

Rivest, Ronald L., "The 'PI' (Placement and Interconnect) System," Proceedings 19th Design Automation Conference, 475-481, June 1982.

Rivest, Ronald L. and Fiduccia, Charles M., "A 'Greedy' Channel Router," *Proceedings 19th Design Automation Conference*, 418-424, June 1982.

Schiele, W., "Design Rule Adaptation of Non-Orthogonal Layouts with Approximate Scaling," *VLSI '85*, (Horbst, ed), 273-282, August 1985.

SCI, *GENESIL System User's Manual*, Silicon Compilers, Incorporated publication 110016, November 1985.

Scott, Walter S. and Ousterhout, John K., "Plowing: Interactive Stretching and Compaction in Magic," *Proceedings 21st Design Automation Conference*, 166-172, June 1984.

Seattle Silicon, *The Mentor Idea/Concorde User's Manual*, Seattle Silicon Technologies, Incorporated, publication UMC Beta 300 Rev 1, March 1986.

Sechen, Carl and Sangiovanni-Vincentelli, Alberto, "The TimberWolf Placement and Routing Package," *Proceedings Custom Integrated Circuit Conference*, 522-527, May 1984.

Simoudis, Evangelos and Fickas, Stephen, "The Application of Knowledge-Based Design Techniques to Circuit Design," *ICCAD '85*, 213-215, November 1985.

Soukup, Jiri, "Circuit Layout," *Proceedings IEEE*, 69:10, 1281-1304, October 1981.

Southard, Jay R., "MacPitts: An Approach to Silicon Compilation," *IEEE Computer*, 74-82, December 1983.

Supowitz, Kenneth J. and Slutz, Eric A., "Placement Algorithms for Custom VLSI," *Proceedings 20th Design Automation Conference*, 164-170, June 1983.

Teig, Steven; Smith, Randall L.; and Seaton, John, "Timing-Driven Layout of Cell-Based ICs," *VLSI Systems Design*, VII:5, 63-73, May 1986.

Tompa, Martin, "An Optimal Solution to a Wire-Routing Problem," *Proceedings 12th Annual ACM Symposium on Theory of Computing*, 161-176, 1980.

VTI, *VLSI Design System*, VLSI Technologies Inc., 1983.

Wardle, C. L.; Watson, C. R.; Wilson, C. A.; Mudge, J. C.; and Nelson, B. J., "A Declarative Design Approach for Combining Macrocells by Directed Placement and Constructive Routing," *Proceedings 21st Design Automation Conference*, 594-601, June 1984.

Watanabe, Hiroyuki, *IC Layout Generation and Compaction Using Mathematical Optimization*, PhD dissertation, University of Rochester Computer Science Department, TR 128, 1984.

Weinberger, A., "Large Scale Integration of MOS Complex Logic: A Layout Method," *IEEE Journal of Solid State Circuits*, 2:4, 182-190, 1967.

Weste, Neil, "Virtual Grid Symbolic Layout," *Proceedings 18th Design Automation Conference*, 225-233, June 1981.

Weste, Neil and Eshraghian, Kamran, *Principles of CMOS VLSI Design*, Addison-Wesley, Reading, Massachusetts, 1985.

Williams, John D., "STICKS—A graphical compiler for high level LSI design," *Proceedings AFIPS Conference 47*, 289-295, June 1978.

Wilmore, James A., "Efficient Boolean Operations on IC Masks," *Proceedings 18th Design Automation Conference*, 571-579, June 1981.

Wing, Omar; Huang, Shuo; and Wang, Rui, "Gate Matrix Layout," *IEEE Transactions on CAD*, 4:3, 220-231, July 1985.

STATIC ANALYSIS TOOLS

5.1 Introduction

Given that it may take weeks or months to get an integrated circuit manufactured, it should not be surprising that designers often spend large amounts of time analyzing their layouts to avoid costly mistakes. It can be difficult, if not impossible, to cure design flaws, particularly on integrated circuits, so the old saw "an ounce of prevention . . ." is very true. This chapter describes some of the computer aids to circuit analysis that help to guarantee correctly manufactured parts.

A VLSI circuit is a complex maze of paths that combine to effect an overall function. There are a number of ways to analyze these paths to ensure that they are correct. One method is to place test data on certain probe points and to follow those data conceptually as they flow through the circuit. Output points can then be observed to determine whether the circuit has properly handled the data. The selection of input data is called **test-vector generation** and the observation of those data as they flow through the computer representation of a circuit is called **simulation**. The overall process is called **dynamic analysis**, because it examines an energized, time-functioning circuit. Both of these processes presume that the design is so complex that the only way to verify it is to try a sample of its capabilities rather than to analyze every part. Thus the proper selection of test vectors is crucial and the accuracy of the simulation can tell much about the circuit. These dynamic analysis tools will be discussed in Chapter 6.

Of interest in this chapter is the other kind of analysis, called **static analysis** because it examines a circuit that is devoid of input data and therefore does not change over time. There are two basic types of static analysis: rule checking and verification. **Rule checking** ensures that a circuit obeys the restrictions placed on it by the design environment and is further divided into geometric design rules, which are concerned with the physical placement of the layout, and electrical rules, which examine the interconnection of the layout. **Verification** ensures that a circuit obeys the restrictions placed on it by the designer so that the intended behavior agrees with the actual behavior. In functional verification, the operation of parts is aggregated into an overall function for the circuit. In timing verification, the delay of parts is aggregated into an overall speed for the circuit.

Computer analysis of VLSI circuits can be very time consuming because of the size of the designs and the required thoroughness of analysis. One way to reduce this time is to perform analysis as the circuit is being designed. Many small tasks can be done in the idle time between commands to a CAD system. In addition, simple bookkeeping done incrementally can save much time later. If error feedback can be supplied immediately, vast amounts of redesign time can be saved. The more responsive a CAD system is in its analysis, the better it will be for complex VLSI design.

5.2 Node Extraction

Before discussing static-analysis tools, it is useful to examine some operations that simplify the job. In many IC layout systems, the connectivity is not specified, but must be derived from the geometry. Since connectivity is crucial to most circuit-analysis tools, it must be obtained during or immediately after design. Ideally, network maintenance should be done during design as new geometry is placed [Kors and Israel], but the more common design system waits for a finished layout. The process of converting such a design from pure geometry to connected circuitry is called **node extraction**.

Node extraction of IC layout can be difficult and slow due to the complex and often nonobvious interaction between layers. In printed-circuit boards, there is only one type of wire and its interactions are much simpler. This allows PC node extraction to be easily combined with other analysis tools such as design-rule checking [Kaplan].

Integrated-circuit node extraction must recognize layer configurations for complex components. In MOS layout, for example, the recognition of transistors involves detection of the intersection of polysilicon and diffusion, with or without depletion and tub implants, but without contact cuts and buried implants. Rules for detecting such combinations are specially coded for each design environment and can be applied in two different ways: polygon-based or raster-based. Polygon-based node extraction uses the complex geometry that is produced by the designer, whereas raster-based node extraction reduces everything to a fine grid of points that is simpler to analyze.

5.2.1 Raster-Based Node Extraction

The raster method of node extraction views a layout as a unit grid of points, each of which is completely filled with zero or more layers [Baker and Terman]. Such a view is called a **raster image** since it changes the layout into a form that can be scanned in a regular and rectangular manner. Analysis is done in this raster scan order by

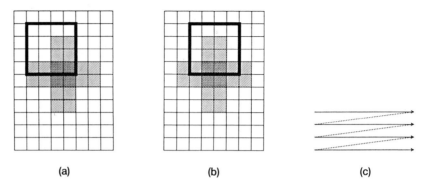

FIGURE 5.1 Raster-based circuit analysis: (a) First position of window (b) Second position of window (c) Raster order.

passing a window over the image and examining the window's contents (see Fig. 5.1). As the window is moved, the lower-right corner is always positioned over a new element of the design. This element is assigned a node number based on the contents and node numbers of the other elements in the window. In fact, since this method is valid only for Manhattan geometry, the window need be only 2 × 2 because there are only two other elements of importance in this window: the element above and the element to the left of the new point.

Rules for assigning node numbers are very simple (see Fig. 5.2). If the new point in the lower-right corner is not connected to its adjoining points, it is given a new node number because it is on the upper-left corner of a new net. If the new point connects to one of its neighbors, then it is a downward or rightward continuation of that net and is given the same node number. If both neighbors connect to the new point and have the same node number, then this is again a continuation of a path. However, if the new point connects to both neighbors, and they have different node numbers, then this point is connecting two nets. It must be assigned a new node number and all three nets must be combined. Node-number combination is accomplished by having a table of equivalences in which each entry contains a true node number. The table must be as large as the maximum number of different net pieces that will be encountered, which can be much larger than the actual number of nets in the layout. Special care must be taken to ensure that transistor configurations and other special layer combinations are handled correctly in terms of net change and connectivity.

When the entire layout has been scanned, an array of node numbers

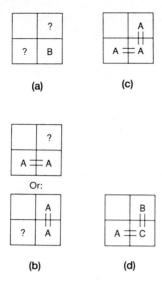

FIGURE 5.2   Raster-based node extraction: (a) Upper-right and lower-left quadrants have any node number, ?; lower-right not connected to either neighbor; lower-right assigned new node number, B (b) One corner (upper-right or lower-left) has node number, A, and is connected to lower-right corner; lower-right assigned same node number, A (c) Upper-right and lower-left have same node number, A; both corners connected to lower-right; lower-right assigned same node number, A (d) Upper-right and lower-left have different node numbers, A and B; both corners connected to lower-right; lower-right assigned new node number, C and adjoining nodes (A and B) are marked the same as C.

is available that can be used to tell which two points are connected. It is necessary to walk recursively through this table when determining true node numbers since a net may be equivalenced multiple times. Nevertheless, this information is easily converted to a netlist that shows components and their connections.

5.2.2   Polygon-Based Node Extraction

Network connectivity can also be determined from the adjacency and overlap of polygons that compose a design. Although many algorithms exist for this polygon evaluation, it helps to transform the polygons before analysis. In one system, all polygons are converted to strips that abut along only horizontal lines. This enables node extraction to proceed sequentially down the layout with simple comparison methods [McCormick].

Active components such as MOS transistors must be handled specially since they may contain single polygons that describe multiple nodes. The diffusion layer polygon must be split at the center of the transistor so that each half can be given a different node number.

In general, it is difficult to recognize all the special combinations of polygons that form different components. A common approach is to use libraries of component shapes to detect strange configurations without undue computation [McCormick; Chiba, Takashima, and Mitsuhashi; Takashima, *et al*.]. This works because CAD systems often use regular and predictable combinations of polygons when creating complex components.

5.2.3   Network Comparison

Once a network has been derived from a completed layout, it is useful to know whether this is the desired network. Although simulation and other network analyses can tell much about circuit correctness, only the designer knows the true topology, and would benefit from knowing whether the derived network is the same.

Often, an IC designer will first enter a circuit as a schematic and then produce the layout geometry to correspond. Since these operations are both done manually, the comparison of the original schematic with the node-extracted layout is informative. The original schematic network has known and correct topology, but the network derived from the IC layout is less certain. The goal is to compare these networks by associating the individual components and connections. If all of the parts associate, the networks are the same; if there are unassociated parts, these indicate how the networks differ.

Unfortunately, the comparison of two networks is a potentially impossible task when the networks are large. If the components of one network can be selected in order and given definite associations with the corresponding components in the other network, the entire comparison process is very fast. However, if uncertain choices are made and then rechosen during comparison as other dependent choices fail, this backtracking can take an unlimited amount of time to complete.

Many solutions to the problem make severe restrictions on the networks in order to function effectively. These restrictions attempt to reduce or eliminate the backtracking that occurs when an association between components is uncertain and must be tried and rejected. One set of restrictions organizes the networks in such a manner that they can both be traversed in a guaranteed path that must associate if the networks are equivalent [Baird and Cho]. The two restrictions that achieve this organization are a limit on the fan-out or fan-in, and a forced correspondence in the order of pin assignment on each component. The first restriction requires that every net have exactly one component that produces output to it. This associates components and nets with a one-to-one mapping, which makes network traversal easy. The use of wired-and or wired-or is disallowed and must be implemented with explicit components. Because every net is guaranteed to have exactly one source, it is possible to traverse the network, starting from the outputs, and reach every component in order.

In Fig. 5.3, the two output pins correspond and are each driven by exactly one component (inverters B and D). With these inverters associated, it is next necessary to examine their inputs and determine the sources. In this example, each inverter has one input that is driven by the NAND gates A and C. Finally, the inputs of the NAND gates must be associated: One is from the inverter and the other is from the network input source. Note that this fan-in restriction of one output per net could be a fan-out restriction of one input per net if the algorithm worked instead from the inputs. All that matters is that every net have exactly one component that can be associated.

The other necessary restriction to linearize search through a network is that the component's input connections are always assigned in the same order. Thus, for the previous example, each NAND gate's top input connection must be on the In signal and the bottom connection must feed back to the inverter. If one of the NAND gates has reversed inputs, then the traversal algorithm will fail. This restriction allows the nets to be associated correctly with no guesswork. However, it is a severe restriction that cannot always be guaranteed. One way of achieving this ordering is to "canonicalize" the input connections so that they are always ordered according to some network characteristic such as fan-out, wire length, or alphabetized net names. This may not work in all cases, so users must be allowed to correct the connection ordering if the network comparison has trouble. Of course, not all

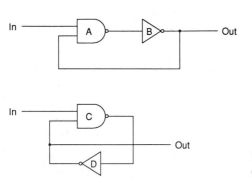

FIGURE 5.3 Network comparison.

components have interchangeable inputs; so in most cases they will be properly ordered by design.

This network comparison scheme works well much of the time, but does have its shortcomings. For instance, certain pathological situations with strange connectivity may occur that cause the algorithm to loop. This can be overcome by marking components and nets as they are associated so that looping is detected. Also, bidirectional components such as MOS transistors could present a problem. Nevertheless, the comparison of networks is a necessary task that, even in its most restricted form, is useful in design.

When the restrictions on the networks described here are impossible to meet, other comparison schemes that take only slightly more time can be used. The method of **graph isomorphism** has been used successfully for network comparison [Ebeling and Zajicek; Spickelmier and Newton]. What graph isomorphism does is to classify every component according to as many characteristics as possible. Classifications include the component type, its connectivity (number of inputs and outputs), the type of its neighbors, and so on. After classification, the two networks are examined to find those classes that have only one component in each. These components are associated and the remaining components are reclassified. The purpose of reclassification is to take advantage of the information about previously associated components. Thus a new classification category is "number of neighbors already associated" and this information helps to aggragate additional associations about the initially distinct components.

Actual implementation of the graph-isomorphism process uses a hashing function to combine the component attributes into a single integer. This integer value is easier to tabulate. In addition, classification values from neighboring components can be combined into a component to refine its uniqueness further. The graph isomorphism technique of network comparison consumes time that is almost linear with the network size and is therefore competitive with the previously described method that restricts the network. Also, like the previous method, it is mostly unable to help identify problems when two networks differ.

5.3 Geometrical Design-Rule Checkers

Geometrical design rules, or just **design rules** as they are known more commonly, are layout restrictions that ensure that the manufactured circuit will connect as desired with no short-circuits or open paths. The rules are based on known parameters of the manufacturing process, to ensure a margin of safety and to allow for process errors. For example, some nMOS processes require metal wires to be no less than 6 microns wide, as narrower wires could break when manufactured, leaving an open circuit. Similarly, the minimum distance between two metal wires is 6 microns, to prevent fabrication inaccuracies from causing the wires to touch. In some processes the design rules are very complex, involving combinations of layers in specific configurations. **Design-rule checking (DRC)** is the process of checking these dimensions.

Recently, it was proposed that all design rules could be simplified so that checking would be easier, and uniform scaling of layout would work [Mead and Conway]. Rather than design rules being specified in absolute distances such as microns, they would be given in terms of a relative and scalable unit called **lambda** (λ). For nMOS design, a metal wire must be 3λ wide, and unconnected wires must be kept 3λ apart. When λ is 2 microns, this is equivalent to the previous rule. Since integer multiples of λ are easiest to use [Lyon], many real design-rule sets are approximated when converted to λ rules. The advantage of this relaxation of the rules is that, when process fabrication allows smaller geometries, the design can easily scale simply by changing the value of λ. Although layer sizes do not actually scale linearly, the design rules contain tolerances to allow for many scales. This is because the rules are based on practical relationships between process layers that hold true at most scales.

5.3.1 Polygon-Based Design-Rule Checking

The earliest attempts at design-rule checking were based on an algebra of polygons [Baird; Crawford]. This algebra consists of polygon-size adjustment and set operations on the polygon boundaries. For example, the six operations *merge*, *bloat*, *and*, *or*, *xor*, and *touch* can be used to find design-rule violations [Crawford]. *Merging* combines overlapping polygons on the same layer into one that contains the union of their areas. *Bloating* expands all polygons on a layer by a specified distance from their center. *And*, *or*, and *xor* are interlayer operations that find the intersection, union, and exclusive areas of two different layers and produce resulting polygons on a third layer. *Touch* finds all polygons on one layer that adjoin or intersect polygons on another layer. Figure 5.4 shows an example of using this method to detect a spacing error. The *merged* polygons are *bloated* by one-half of the minimum spacing and then *anded* with each other to find any overlap. Algebraically, the error polygon is derived as:

$$error = bloat\left(poly1, \frac{minspace}{2}\right) \wedge bloat\left(poly2, \frac{minspace}{2}\right)$$

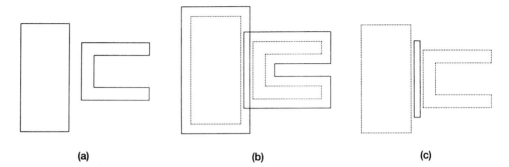

(a) (b) (c)

FIGURE 5.4 Polygon operations for design-rule checking: (a) Original polygons (b) Bloated polygons (originals are dotted) (c) Intersection of bloated polygons (violation).

Implementation of polygon-based design-rule checking requires the availability of intermediate layers that can hold temporary results of the algebra. When run incrementally to check new polygons, these operations can run without consuming excessive time or space. However, when an entire layout is checked at once, the intermediate layers can use a large amount of storage. In addition, polygon intersection can be very time consuming when all the polygons on a layer must be checked. For this reason, many techniques exist to speed the polygon comparison process.

Bounding-box values, stored with each polygon, can be used to eliminate quickly those polygons that should not be considered. However, even with this information it is necessary to compare all polygons when searching for overlap. A better method is one that breaks the layout into a grid of polygon subsets [Crawford; Bentley, Haken and Hon]. Each polygon need be checked against only those in its immediate grid area. As the number of grid lines increases, the number of polygons in a grid subset decreases and the neighborhood of a polygon becomes faster to check. However, polygons that cross or are near grid lines must be included in multiple grid subsets and this redundancy becomes more expensive as the grid becomes finer. At the limit, the grid lines are as fine as the minimum unit of design and the data structure becomes a bit-map of the layout. Since the extremes of this grid scheme are not effective, it is important to find a good value of grid spacing that optimizes time and space. Bentley, Haken, and Hon found that for designs that have n boxes, a grid size of $\sqrt{n} \times \sqrt{n}$ is best.

Another method of comparing neighboring polygons sorts the edges along one axis. If, for example, all polygons are sorted by their left edge, then the layout can be effectively scanned from left to right.

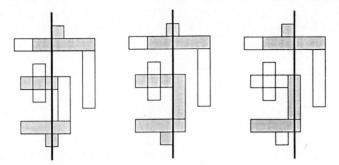

FIGURE 5.5 Working-set polygon comparison. The vertical line defines a close subset of polygons (shaded).

This is done by moving a vertical scan line horizontally across the layout and maintaining a working set of polygons that intersect this scan line (see the shaded polygons in Fig. 5.5). As the scan line advances, new polygons are added to the working set and old ones are removed. If the polygons are presorted, they can be efficiently paged from disk and large designs can be represented with a small amount of memory. When the set is initially formed, all pairs of polygons must be compared, but as the scan line advances, each new polygon need be compared only to the others already in the set. To enhance efficiency further, the scan line can be made to advance a variable distance, always stopping at polygon edges. This is possible because no polygons will be added to or deleted from the working set if the scan line moves to a position that does not border a polygon. The sorted list of polygons can act as a source of coordinates for the advancement of the scan line.

Another approach to polygonal design-rule checking is to simplify the geometry so that the distance calculations can be quickly done. For example, if the corners of all polygons are rounded, then non-Manhattan distances are easier to calculate. Even simpler is the reduction of a design to a collection of variable-width lines [Piscatelli and Tingleff]. This notion is especially useful for printed-circuit board design-rule checking since the boards have only wires and no components. A three-part algorithm determines first whether two lines are at all close, then whether they intersect, and finally whether lines are too close but not connected. In this algorithm, a line from (x_0, y_0) to (x_1, y_1) is represented parametrically as:

$$x = U_1 t + x_0 \qquad y = U_2 t + y_0$$

where:

$$0 \leq t \leq L$$
$$L = (x_1 - x_0)^2 + (y_1 - y_0)^2$$

and:

$$U_1 = (x_1 - x_0)/L$$
$$U_2 = (y_1 - y_0)/L$$

The first test of line proximity is to find, for both endpoints of one line, whether they are close to the other line, regardless of the other line's endpoints. Thus this test is inaccurate because two line segments may be very far apart but will appear close if the projected extension of one line is close to an endpoint of the other. This coarse distance measure is:

$$D = U_1 (y_p - y_0) - U_2 (x_p - x_0)$$

Where (x_p, y_p) is an endpoint of one line and (x_0, y_0), U_1, and U_2 are from the equation of the other line. The distance factor D can include line widths and minimum-spacing information to determine a cutoff beyond which the lines are safely distant. This distance measure is signed to indicate which side of the line contains the test point. The sign information is used in the second part of the algorithm to determine whether the lines intersect. If both lines have endpoints with different signs relative to the other line, then they do intersect. The third part of the algorithm, necessary only if the the lines are close but do not intersect, determines exact line-segment distance using standard geometric methods.

One observation that is often made about design-rule checking is that its process is similar to that of node extraction [Losleben and Thompson]. For example, the three-part algorithm can also do node extraction by following intersections of wires. When wires are too close, they are considered to connect. If two different signals of a layout end up connected, then an electrical-rule violation has occurred. On printed-circuit boards, complete node extraction is done by using this information in combination with the interlayer drill holes to propagate network paths to all layers [Kaplan].

5.3.2 Raster-Based Design-Rule Checking

An alternative to using polygon algebra is the raster method of design-rule checking [Baker and Terman]. This is similar to the raster method of node extraction, in which a small window is used to examine the design uniformly. Every position of the window examines a different configuration and matches it against a set of error templates. If, for example, metal wires must be 3λ apart, then a 4 × 4 window that matches either of the templates in Fig. 5.6 will indicate a design-rule violation. In these templates, the "?" character indicates that any combination of layers may be present and the "M" and " " (blank) require metal and empty layers. Although these templates do not cover all configurations of the error, they will be passed over every location in the layout and will therefore eventually find the errors. Two templates are needed to handle the horizontal and vertical orientations of the geometry. Notice that this method works best with manhattan geometry because non-Manhattan polygons must be converted to stair-stepped grid points, which requires more complex error templates. For 45-degree angles, there is only a modest increase in the number of templates [Seiler], but for arbitrary angles, the problem becomes very difficult.

There are a number of problems with the raster method of design-rule checking. The time spent comparing error templates with windows can be excessive, especially if the templates are numerous and the window is large. It is only by using simplified design-rules, in which the largest spacing is 3λ and all distances are integers, that a window as small and manageable as 4 × 4 can be used. With a window this small, all templates for one layer can be referenced directly through a table that is $2^{4 \times 4} = 65,534$ entries long [Baker and Terman]. Another problem with this raster-based analysis is that it suffers from tunnel vision and can miss information that helps to clarify a particular configuration. This causes extraneous errors to be reported. For example, since the window cannot see global connectivity, it cannot understand that two points may approach each other if they connect elsewhere. When faced with the choice of reporting too many errors or too few, most design-rule checkers fail on the side of inclusion and let the designer ignore the spurious complaints.

The advantages of raster-based design-rule checking are that it is a simple operation and that it takes little space. The simplicity of this method has led to hardware implementations [Seiler]. The saving of

?	M	?	?
			?
			?
?	M	?	?

?			?
M			M
?			?
?	?	?	?

FIGURE 5.6 Templates for metal spacing errors. The "M" matches metal, the " " (blank) matches no metal, and the "?" matches any layer.

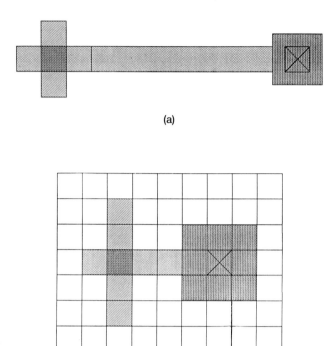

(a)

(b)

FIGURE 5.7   Essential bit-maps: (a) Original polygons (b) Essential bit-map.

space is based on the fact that raster data can be produced from polygons in scan-line order and then the data can be destroyed when the window has passed [Wilmore]. Thus, for a 4 × 4 window, only four scan lines need to be kept in memory. In addition, the representation can be further compacted by having the unit squares of the raster be of variable size according to the spacing of polygon vertices in the original layout [Dobes and Byrd]. These **essential bit-maps** (see Fig. 5.7) have more complex templates with rules for surrounding spacing but they save much memory because they represent the layout in a virtual grid style.

5.3.3   Syntactic Design-Rule Checking

The polygon approach to design-rule checking takes too much space and the raster approach has too narrow a view to be effective, but there are other ways to check geometric spacing. The Lyra design-rule checker [Arnold and Ousterhout] checks the corners and inter-

Corners and Intersections

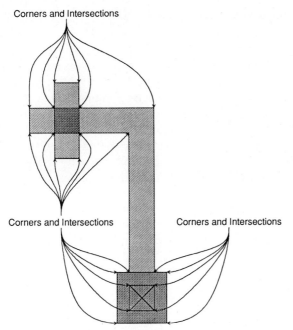

Corners and Intersections

Corners and Intersections

FIGURE 5.8 The corners and intersections of a layout.

sections of polygons against arbitrary spacing rules (see Fig. 5.8). Since checking is done only where the geometry changes, the scope of a corner is equivalent to a 2×2 window on an essential bit-map. However, the spacing rules that are applied to each corner can examine arbitrarily large areas that would exceed a fixed-window capability, even when using essential bit maps.

Figure 5.9 illustrates the Lyra design rules. Once again, the rule is that no two metal wires may be closer than 3λ apart. A **template** is compared to any corner that has the specified configuration and, if it matches, Lyra checks the associated **rule**, which tests the vicinity of the corner. In some situations, the matching of a template indicates

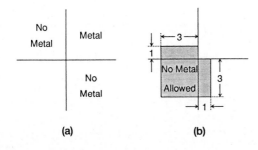

No Metal   Metal

No Metal

No Metal Allowed

(a)   (b)

FIGURE 5.9 Corner-based design-rule checking: (a) Template (b) Rule.

a design-rule violation and no rule need be applied. Null rules then suffice to flag the error.

When applying the templates, Lyra uses special code for efficiency. Lyra uses a tabular approach in which the corner configuration is indexed into a set of dispatch tables to determine which rules apply. In addition, all rotations of the template must be checked. A simple compiler produces the templates and rules from higher-level design-rule descriptions and automatically handles different template orientations. Although Lyra uses only Manhattan geometry, it would be feasible to extend this corner approach to non-Manhattan checking by including corner-angle ranges in the template.

If design-rule checking can be done with syntactic methods, then it is not unreasonable to apply traditional syntax-checking techniques used in compilers. In particular, it is possible to do design-rule checking by following state diagrams [Eustace and Mukhopadhyay]. Since compiler-state machines work on one-dimensional data, a design-rule checking-state machine must be modified to have two entries for state selection so that it can look in two orthogonal directions. Thus, for a 13-state machine, there are 13^2 table entries to account for all combinations of the states in two orthogonal directions. Figure 5.10 shows a state table for the 3λ metal-spacing rule. Starting in state 0, the vertical (V) and horizontal (H) neighbor states are examined along with the presence or absence of metal. This determines the state for the current point and flags errors (E).

Producing state diagrams for design rules can become very complex. What is needed is a compiler that produces the state tables from higher-level-rule specifications. With such a facility, design-rule checking can be done in such a simple manner that it can be easily implemented in hardware. Augmentation of the state table to include memory would reduce table size with only a modest increase in state-machine complexity. This totally state-oriented approach is a generalization of the Lyra technique in that both methods use rules for handling layer-change points. However, with enough states, the state-oriented method can also handle global information between change points.

5.3.4   Hierarchical Design-Rule Checking

Hierarchy is an important factor in simplifying design and is therefore useful in simplifying design analysis. Besides speeding the process, the use of hierarchy can make error messages more informative by retaining knowledge of the circuit structure. If two instances of a cell are created in similar environments, then it should not be necessary to check the contents of both of them. Hierarchical design-rule checking

V	H	Metal	No Metal	V	H	Metal	No Metal	V	H	Metal	No Metal	V	H	Metal	No Metal
0	0	0	1	3	0	0	3	6	0	0	3	9	0	0	3
0	1	11/E	4	3	1	-	-	6	1	-	-	9	1	-	-
0	2	-	-	3	2	-	-	6	2	-	-	9	2	-	-
0	3	-	-	3	3	-	-	6	3	-	6	9	3	-	6
0	4	11/E	7	3	4	-	-	6	4	-	-	9	4	-	-
0	5	-	-	3	5	-	-	6	5	-	-	9	5	-	-
0	6	-	-	3	6	-	-	6	6	-	-	9	6	-	9
0	7	0	7	3	7	-	-	6	7	-	-	9	7	-	-
0	8	0	7	3	8	-	-	6	8	-	-	9	8	-	-
0	9	0	7	3	9	-	-	6	9	-	-	9	9	0	9
0	10	-	-	3	10	-	-	6	10	-	-	9	10	-	-
0	11	0	-	3	11	-	-	6	11	-	-	9	11	-	-
0	12	0	-	3	12	-	-	6	12	-	-	9	12	-	-
1	0	11/E	2	4	0	-	-	7	0	-	2/E	10	0	-	-
1	1	-	10	4	1	-	-	7	1	-	-	10	1	-	-
1	2	-	-	4	2	11/E	5	7	2	-	5	10	2	-	2
1	3	-	-	4	3	-	-	7	3	-	6/E	10	3	-	-
1	4	-	-	4	4	-	7	7	4	-	7	10	4	-	-
1	5	-	-	4	5	-	-	7	5	-	8	10	5	-	2
1	6	-	-	4	6	-	-	7	6	-	9	10	6	-	-
1	7	-	10	4	7	-	-	7	7	-	8	10	7	-	-
1	8	-	-	4	8	-	-	7	8	-	8	10	8	-	2
1	9	-	-	4	9	-	-	7	9	-	8	10	9	-	-
1	10	-	-	4	10	-	4	7	10	-	-	10	10	-	4
1	11	-	-	4	11	-	12	7	11	-	-	10	11	-	-
1	12	-	-	4	12	-	-	7	12	-	12	10	12	-	-
2	0	11/E	3	5	0	-	-	8	0	-	-	11	0	0	1
2	1	-	-	5	1	-	-	8	1	-	-	11	1	-	-
2	2	-	-	5	2	-	-	8	2	-	-	11	2	-	-
2	3	-	6	5	3	-	6	8	3	-	9/E	11	3	-	12
2	4	-	10	5	4	-	-	8	4	-	7	11	4	-	-
2	5	-	-	5	5	-	-	8	5	-	-	11	5	-	12
2	6	-	3	5	6	-	9	8	6	11/E	9	11	6	-	-
2	7	-	-	5	7	-	-	8	7	-	-	11	7	-	-
2	8	-	-	5	8	-	-	8	8	-	9	11	8	-	-
2	9	-	9	5	9	-	9	8	9	-	9	11	9	-	-
2	10	-	-	5	10	-	4	8	10	-	-	11	10	-	-
2	11	-	-	5	11	-	12	8	11	-	-	11	11	-	12
2	12	-	-	5	12	-	-	8	12	-	-	11	12	-	-
12	0	0	-	12	4	-	7	12	8	-	-	12	12	0	12
12	1	-	4	12	5	-	-	12	9	-	-				
12	2	-	-	12	6	-	-	12	10	-	-				
12	3	-	-	12	7	-	-	12	11	-	-				

FIGURE 5.10 State-based design-rule checking.

has been used to eliminate those polygons that are redundant in a layout [Whitney]. A precheck filter removes all but one instance of a cell and leaves only those combinations of cell intersections that are unique. This same technique has been used in node extraction by reducing a cell to its external connection geometry and examining the contents only once to assign network labels [Chao, Huang, and Yam]. The extracted network can be hierarchically structured, which may be useful to other tools [Newell and Fitzpatrick]. In fact, many analysis tasks can be improved by hierarchically examining a circuit so that only one instance of a cell and its overlap is considered [Hon].

Care must be taken to ensure that instances of a cell eliminated from the check have identical environments of geometry at higher levels of the hierarchy. For example, a piece of layout placed arbitrarily across a cell may have varying effects on the design rules of identical subcells farther down the hierarchy. In regular and well-composed hierarchies, a hierarchical design-rule checking filter eliminates much of the layout and vastly speeds checking. Restricted design styles, such as those that do not allow cell overlap, will make hierarchical checking most efficient [Scheffer]. However, in poorly-composed hierarchies, the filter can spend more time looking for unique instance configurations than would be spent in a nonhierarchical check.

A simpler use of hierarchy in speeding design-rule checking is the use of cell location to restrict the search for a polygon's neighborhood [Johnson; Shand]. Since cell instances provide bounding boxes for their contents, this information can be used to eliminate entire cells and their subcells from consideration. Where cells overlap, checking recurses through their contents, using boundary information to find neighboring polygons quickly. Shand also uses this scheme for node extraction. Thus it can be seen that hierarchy, in combination with good search methods, speeds design-rule checking and other analysis tasks.

5.4   Electrical-Rule Checkers

Geometrical design rules ensure that the circuit will be manufactured correctly by checking the relative position, or syntax, of the final layout. However, there is nothing to ensure that this circuit will work. Correct functionality is left to the simulators and verifiers that manipulate circuit activity and behavior. Nevertheless, there is a middle ground between simple layout syntax and complex behavioral analysis, and it is the domain of **electrical-rule checkers**, or **ERC**.

Electrical rules are those properties of a circuit that can be determined from the geometry and connectivity without understanding the behavior. For example, the estimated power consumption of a circuit can be determined by evaluating the requirements of each device and trying to figure out how many of the devices will be active at one time. From this information, the power-carrying lines can be checked to see whether they have adequate capacity. In addition to **power estimation**, there are electrical rules to detect incorrect transistor ratios, short-circuits, and isolated or badly connected parts of a circuit. All these checks examine the network and look for inconsistencies. Thus, whereas design-rule checking does syntax analysis on the layout, electrical-rule checking does syntax analysis on the network.

5.4.1 Power Estimation

There are several reasons to watch the power consumption of a circuit, such as limited power-carrying capacity on wires, total power limits in a single package, and cost of system manufacturing. As integrated circuits grow in size and complexity, there will be an increasing need to watch for power problems.

Metal wires that are required to carry too much power suffer **metal migration**, in which the atoms move within the wire, leaving a break in the conductor. In order to prevent this phenomenon, the wire must be widened to reduce its current density. Typical process rules specify the number of amps per unit width of wire. Another limit is the total amount of power, and therefore heat, that a single packaged design can dissipate before it destroys itself and its package. Standard IC packages can handle dissipations up to a few watts. However, new packaging designs employ heat fins and other cooling methods to allow greater power dissipation.

Power estimation is dependent on the fabrication process. In nMOS, every depletion transistor that appears in a pullup configuration consumes power when pulled down to ground (see Fig. 5.11). Estimating the

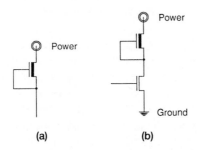

FIGURE 5.11 nMOS power consumption: (a) nMOS depletion transistor in pullup configuration (b) Pullup and pulldown.

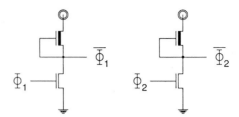

FIGURE 5.12   Clocking considerations in power estimation.

number of these pullups that may be pulled down allows an overall static power consumption to be derived. This estimation can use other information in the network; for example, the operation of clock signals and the interconnection of circuitry that can be determined to mutually exclude certain pulldowns. If a clock with known phases appears on the pulldown gate or on some control of that gate then it can be determined which pullups will be active in nonoverlapping times. Figure 5.12 illustrates two pullups that will not be active at the same time and can therefore be estimated to use the power of only one pullup. Figure 5.13 illustrates two coupled pullups that can also be guaranteed never to be active at the same time. When all of these heuristics have been exhausted, it must be assumed that the remaining pullups are active since their state is data-dependent. This analysis assumes static (or low-speed) power consumption. For high-speed operation, a CMOS type of analysis must also be considered.

Power estimation for CMOS is different because there is usually no dissipation during stable states. Rather, CMOS circuits consume power when they switch, with the power consumption being proportional to the total load capacitance and the switching frequency. Power estimation is therefore a matter of determining the capacitance of nodes within a circuit and the rate at which the potential on these nodes changes. Power estimation for CMOS is often much more difficult than that for static nMOS.

On printed-circuit boards it is also important to determine power consumption; however, all the power-use information is described in the specification of the individual IC packages. These packages have minimum and maximum power consumption, so it is safe to sum the maximum values to determine the maximum board usage. Cooling considerations are difficult to predict. Many methods have been employed, ranging from simply allowing enough air space above the board to using freon cooling pipes running near the surface. What good is the most sophisticated electronic circuit if it burns up when plugged in?

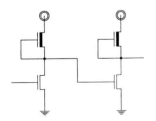

FIGURE 5.13   Coupling considerations in power estimation.

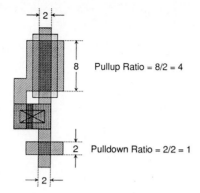

FIGURE 5.14 nMOS transistor ratios. The ratio is the length (polysilicon size) divided by the width (diffusion size).

5.4.2 Transistor-Ratio Checkers

When doing VLSI layout, it is necessary to scale transistors so that they function effectively, relative to each other. In order to achieve proper logic transition, a threshold voltage must be crossed and the speed at which this happens is determined by the relative scales or ratios of the driving components. The problem is particularly important in nMOS design, because an imbalance exists between rising and falling transition times. To ensure that such a circuit will operate correctly, an analysis of the relative transistor ratios must be done.

An isolated nMOS inverter requires a pullup-to-pulldown size ratio of at least 4 : 1 which means that the length-to-width scale of the pullup must be four or more times that of the pulldown (see Fig. 5.14). When the pulldown is a series of components, the ratios of the latter are additive. When parallel transistors form a pulldown, however, each individual transistor must completely account for a proper ratio (see Fig. 5.15). This is because of the simple rule that every path from power to ground must have correct ratios of pullup and pulldown.

As a final rule in nMOS transistor-ratio checking, the nature of the pulldown gate must be considered. Gates that are driven directly from a pullup are considered to have a **restored** signal, and follow the

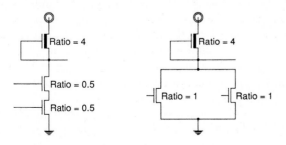

FIGURE 5.15 Series and parallel transistor ratios.

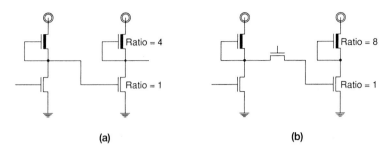

FIGURE 5.16 Restored and unrestored transistor ratios: (a) Restored (b) Unrestored.

(a)

(b)

rules for sizing in the previous paragraph. However, gates that are driven through one or more pass transistors have **unrestored** signals, and the ratio rule must be doubled [Mead and Conway]. Figure 5.16 illustrates this requirement, showing that the pullup on the right of Fig. 5.16(b) must change its ratio due to the nature of the pulldown gate signal. These rules are fairly straightforward to implement by traversing the circuit network and evaluating the appropriate properties. It is necessary to know the nature of all external signals to determine the restorative nature of each component. For safety, they should be assumed to be unrestored, although proper modular design dictates that all generated values should be restored.

5.4.3   Short-Circuits and Isolated Circuits

A short-circuit is a path between any two unconnected nets such as from power to ground. Detection of direct paths is simply a process of ensuring that the nets are distinct. Often this is done during node extraction, when a simple test of the net numbers will tell the truth. Also important is a check to ensure that no transistor gates power and ground signals. Although this configuration is not a static short-circuit, it does not make electrical sense and is easy to detect (see Fig. 5.17).

Another static analysis is the detection of circuitry that is not properly connected. For example, when the input to a component is not connected to other parts of the circuit, it is said to be a **floating input**. This condition can be detected by propagating external inputs through the circuit, marking all reachable elements. When an input signal connects to the gate of a transistor, it indicates that the source and drain of the transistor are properly controlled. If other active signals appear on the source or drain end of that transistor, then the other end will be active too. By propagating the power, ground, input,

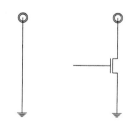

FIGURE 5.17 Short-circuit configurations of power to ground.

and clock signals through the circuit, the machine can check for incorrect output drive, inconsistencies of signal specifications, unconnected circuit elements, and much more. Exceptions must be allowed for artwork, which is not intended to be part of the circuit, and should not be flagged as isolated layout.

5.4.4   Miscellaneous Electrical Rules

There is an unending number of specialized electrical rules. When two output signals connect such that their values combine, they are called **tied outputs**. This may be purposeful, but it may also be an error. Microwave circuit boards require that the wires be tapered to prevent incorrect impedance conditions. Some design environments limit the number of components that an output signal can drive (**fan-out** limitation) or the number of input lines that a component can accept (**fan-in** limitation).

All the electrical rules mentioned here can be checked at one time as the circuit network is traversed. Thus a single electrical-rule checking program is often used to check these and other circuit properties [Baker and Terman]. Such a tool can take a unified approach to checking by reducing everything to constraints on the circuit network [Karplus]. Since electrical rules tend to be specific to a particular design environment, they should be easily tailorable. The use of rule-based systems allows electrical rules to be updated easily for any given design environment and even allows the rule checker to focus its attention on the sensitive parts of a circuit [De Man *et al.*].

5.5   Verification

The essential difference between verifiers and rule checkers is that verifiers are given operating requirements to check for each different circuit. Thus any of the previously described rule checkers could be called a verifier if the conditions being checked were explicitly specified as part of the design. However, such conditions do not usually vary between designs because checkers are verifying low-level correctness conditions that vary only with the environment. The verifiers discussed here do the same kind of rule checking, but work with individualized rules for each circuit.

Two types of verifiers will be discussed: timing and functional. **Timing verifiers** determine the longest delay path in a circuit to optimize performance and to make sure that the clock cycles are correct. **Functional verifiers** compare symbolic descriptions of circuit functionality with the derived behavior of the individual parts of the circuit, also described symbolically. In each case a specification is given (time delays or behavior) and the checker ensures that the rules are met.

5.5.1   Timing Verification

As circuits become complex, it is increasingly difficult to analyze them accurately and completely for precise timing characteristics. Rather than run a circuit simulator such as SPICE [Nagel] that may take weeks to finish, a timing verifier can approximate delays quickly and report much of the same information [Harrison and Olson]. Timing verification is more like electrical-rule checking because its essential function is to traverse the circuit network. For every input and output signal, there are many possible paths through the circuit. Each path consists of a set of network nodes that connect the output of one component to the input of another. If the delay of each node is first determined and stored, then the verifier's job consists of recursively finding the worst-case path to every output:

OverallDelay = **max**(WorstDelay(EachOutputNode))

WorstDelay(Node) =

    if Node is INPUT **then** WorstDelay = 0 **else**

        WorstDelay = ComponentDelay(InputComponent(Node)) +

            WireDelay(Node) +

            max(WorstDelay(EachInputNode(InputComponent(Node))))

The delay of nodes and components can be determined in a number of ways. Printed-circuit board components have documented delays that must be combined with wire delays and signal conditions to obtain overall timing ranges [McWilliams]. Integrated-circuit components vary widely in delay and are even data-dependent in some environments such as nMOS. An **RC tree**, which models the delay of circuit elements with resistance and capacitance, can quickly approximate most timing [Penfield and Rubenstein; Horowitz; Lin and Mead]. RC trees are described in Chapter 6. To determine a path delay precisely, many timing verifiers extract the details of the longest path and use circuit simulation to evaluate the timing fully [Ousterhout; Jouppi].

There are a number of additional considerations that must be addressed to verify accurately the delays in a circuit. As was mentioned before, nMOS components are data-dependent, and therefore a chain of inverters that feed each other should have their delay times redefined, allowing for the fact that every other one will switch much more quickly. Another problem that arises with MOS transistors is that they can function bidirectionally, which means that complex arrays of these components can present an exponential number of paths to the timing verifier. The Crystal system uses user-specified directionality information to reduce the possible paths through a circuit [Ousterhout]. Crystal also has special code to detect buses so that massively connected precharging schemes do not appear to have infinite paths onto and off of the bus. The TV system has complex interconnect rules to determine signal flow directions automatically [Jouppi].

Knowledge of fixed signals can help to limit the search possibilities. For example, clock lines create firewalls throughout the network by removing clocked components from the analysis, since it is known when these components will operate. A difficult problem is the detection of delays that continue to propagate on one clock phase while another is active. These cross-phase signals are common, and should be allowed to continue propagating during the other clock cycle provided that they quiesce before they are used.

One major problem with timing analyzers is their user interface. Since the operation focuses on some subset of the circuit, much help is needed to make it focus correctly. Although intelligent analysis techniques help some, only the user knows what is correct. Thus it is necessary to allow user-specified information such as wire direction, initial conditions, and elimination of circuitry from analysis. The timing analyzer should also be able to indicate paths graphically and to explain why they are critical.

In addition to finding critical-path delays, timing verifiers can be used to do miscellaneous static analysis. For example, there are timing-verification systems that find high-speed components off the critical path that can be slowed down to save power [Jouppi; Trimberger]. One system couples timing analysis with transistor sizing so that it can optimize either time or space [Fishburn and Dunlop]. Once again it is seen that much static analysis can conveniently be done at one time, because the methods overlap. This is also good because the more an analysis tool looks for, the more it knows and is able to analyze.

5.5.2 Functional Verification

The last static analysis technique to be discussed is functional veri-fication. Each primitive component can be described behaviorally in terms of its inputs, outputs, and internal state. By combining these component descriptions, an overall circuit behavior is produced that symbolically represents the circuit's function. Verification consists of comparing this derived behavior with a designer-specified behavior. Although not identical, the two descriptions should be mathematically equivalent, in which case the circuit is successfully verified. The three difficult aspects of this process are the selection of a behavioral representation, the aggregation of individual behavioral descriptions to form overall circuit specifications, and the comparison of derived functionality with specified functionality. It is also much harder to verify asynchronous circuits than it is to verify those that are synchronous.

One useful model of circuit behavior consists of equations for the internal state and the component's output [Gordon 81]. These equations use the values of component input and any previous state so that a notion of time is available, much like the temporal logic described in Chapter 2. For unclocked and unbuffered components, there is no state, so these equations are simply formulas that give the output in terms of the input. For example, a NAND gate has no state and is described as:

$$output \equiv input1 \;\overline{\wedge}\; input2$$

Conditional expressions are also available. For example, a multiplexor that chooses between two inputs according to a third selector input is described as:

$$output \equiv \textbf{if } selector \textbf{ then } input1 \textbf{ else } input2$$

Registers that store data must be described with both output equations and state equations. A buffer is:

$$output \equiv \textbf{last } memory \qquad memory \equiv input$$

Gordon has shown how this scheme can be used in both logic gate and nMOS transistor descriptions. However, the MOS transistor cannot be modeled bidirectionally and so is described as:

$$drain \equiv \textbf{if } gate = 1 \vee \textbf{last } memory = 1 \textbf{ then } source$$
$$\textbf{else } \text{UNDEFINED}$$
$$memory \equiv gate$$

This retains the state of the output that is held on a wire even after the gate current is cut off.

Combining state and output equations is a simple matter of substitution. The resulting equations are long and complex, but they can be reduced with standard techniques. Special induction methods can be employed at this stage to aggregate multiple parallel bits of processing into multibit values [German and Wang]. When components are connected in a loop, it is the use of clocked registers that helps to break the infinite recursion. This is because clocked components are typically found inside of such loops, which breaks them down to one state and one output specification for the entire loop. It is for this reason that synchronous circuits are easier to verify than are asynchronous ones. Also, during this behavioral-aggregation phase, certain electrical-rule checks can be performed to ensure that the components connect correctly [Barrow].

The final stage of verification compares the aggregated behavior with the user-specified behavior. This involves showing equivalence between two sets of equations, and it can be done in a number of ways [Barrow]. Pure identity is rare but should be checked first. After that, the equations are reduced to canonical form so that they can again be compared. This reduction involves the elimination of constants ($A \wedge 1$ becomes A) and the rearrangement of clauses into normal conjunctive form (($A \vee (B \wedge C)$) becomes ($A \vee B$) $\wedge$ ($A \vee C$)). Another way to test for equivalence is to enumerate all values of variables and compare the results. This works only when the number of inputs and state variables is small.

Besides fully automatic verification techniques, there are methods that involve the user in the proof process. Whereas an automatic verifier finds the equivalence between two sets of equations, a manual verifier checks a user-proposed proof of equivalence [Ketonen and Weening]. In between the two are semiautomatic systems that do what they can, and then refer to the user when they encounter difficulty [Gordon 85].

Although not widely used, functional verification is a significant analysis technique. The mere specification of behavior is a valuable exercise for a designer, and the rigorous analysis of such specifications is very informative. Verification may not become a common design practice, but the use of formal methods is very important and may find acceptance in the design of standard cell libraries, silicon compilers, and other design environments that must combine correctly without constant planning and analysis.

5.6 Summary

This chapter has illustrated a number of techniques for the data-free examination of VLSI circuits. Such static analyses require a full topological description of the circuit that is often a product of node extraction. Geometric design rules check the layout of a circuit, and electrical rules check the topology. Verification tools check individual specifications to ensure that the design will perform as intended. The next chapter deals with actual performance considerations that can be analyzed by treating a circuit dynamically.

Questions

1 What is the problem with raster-based design-rule checking that does not occur with raster-based node extraction?

2 How can network-comparison methods fully describe the differences between dissimilar networks?

3 How would you program a preprocessor to convert design rules into state-based design-rule checking tables?

4 Why is timing verification more user-intensive than is design-rule checking?

5 Why is the functional verification of asynchronous circuits so difficult?

6 Why are rule-based systems an ideal choice for implementation of electrical-rule checkers?

7 What is the drawback of hierarchical design-rule checking?

References

Arnold, Michael H. and Ousterhout, John K., "Lyra: A New Approach to Geometric Layout Rule Checking," Proceedings 19th Design Automation Conference, 530-536, June 1982.

Baird, Henry S., "Fast Algorithms for LSI Artwork Analysis", Proceedings 14th Design Automation Conference, 303-311, June 1977.

Baird, H. S. and Cho, Y. E., "An Artwork Design Verification System," Proceedings 12th Design Automation Conference, 414-420, June 1975.

Baker, Clark M. and Terman, Chris, "Tools for Verifying Integrated Circuit Designs," Lambda, 1:3, 22-30, 4th Quarter 1980.

Barrow, Harry G., "VERIFY: A Program for Proving Correctness of Digital Hardware Designs," Artificial Intelligence, 24:1-3, 437-491, December 1984.

Bentley, Jon Louis; Haken, Dorthea; and Hon, Robert W., "Fast Geometric Algorithms for VLSI Tasks," Proceedings 20th IEEE Compcon, 88-92, February 1980.

Chao, Shiu-Ping; Huang, Yen-Son; and Yam, Lap Man, "A Hierarchical Approach for Layout Versus Circuit Consistency Check," Proceedings 17th Design Automation Conference, 270-276, June 1980.

Chiba, Toshiaki; Takashima, Makoto; and Mitsuhashi, Takashi, "A Mask Artwork Analysis System for Bipolar Integrated Circuits," Proceedings 21st IEEE Compcon, 175-183, September 1981.

Crawford, B. J., "Design Rules Checking for Integrated Circuits Using Graphical Operators," *Computer Graphics*, 9:1, 168-176, 1975.

De Man, Hugo J.; Bolsens, I.; Meersch, Erik Vanden; and Cleynenbreugel, Johan Van, "DIALOG: An Expert Debugging System for MOSVLSI Design," *IEEE Transactions on CAD*, 4:3, 303-311, July 1985.

Dobes, Ivan and Byrd, Ron, "The Automatic Recognition of Silicon Gate Transistor Geometries: An LSI Design Aid Program," Proceedings 13th Design Automation Conference, 327-335, June 1976.

Ebeling, Carl and Zajicek, Ofer, "Validating VLSI Circuit Layout by Wirelist Comparison," *ICCAD '83*, 172-173, September 1983.

Eustace, R. Alan and Mukhopadhyay, Amar, "A Deterministic Finite Automaton Approach to Design Rule Checking For VLSI," Proceedings 19th Design Automation Conference, 712-717, June 1982.

Fishburn, J. P. and Dunlop, A. E., "TILOS: A Posynomial Programming Approach to Transistor Sizing," *ICCAD '85*, 326-328, November 1985.

German, Steven M. and Wang, Yu, "Formal Verification of Parameterized Hardware Designs," Proceedings IEEE International Conference on Computer Design, 549-552, October 1985.

Gordon, M., "A Very Simple Model of Sequential Behaviour of nMOS," *VLSI '81* (Gray, ed), Academic Press, London, 85-94, August 1981.

Gordon, Mike, "HOL—A Machine Oriented Formulation of Higher Order Logic," University of Cambridge Computer Laboratory, technical report 68, July 1985.

Harrison, Richard A. and Olson, Daniel J., "Race Analysis of Digital Systems Without Logic Simulation," Proceedings 8th Design Automation Workshop, 82-94, June 1971.

Hon, Robert W., *The Hierarchical Analysis of VLSI Designs*, PhD dissertation, Carnegie-Mellon University Computer Science Department, CMU-CS-83-170, December 1983.

Horowitz, Mark, "Timing Models for MOS Pass Networks," Proceedings International Symposium on Circuits and Systems, 198-201, May 1983.

Johnson, Stephen C., "Hierarchical Design Validation Based on Rectangles," Proceedings MIT Conference on Advanced Research in VLSI (Penfield, ed), 97-100, January 1982.

Jouppi, Norman P., "TV: An nMOS Timing Analyzer," Proceedings 3rd Caltech Conference on VLSI (Bryant, ed), Computer Science Press, 71-85, March 1983.

Kaplan, David, "A 'Non-Restrictive' Artwork Verification Program for Printed Circuit Boards," Proceedings 19th Design Automation Conference, 551-558, June 1982.

Karplus, Kevin, "Exclusion Constraints, a new application of Graph Algorithms to VLSI Design," Proceedings 4th MIT Conference on Advanced Research in VLSI (Leiserson, ed), 123-139, April 1986.

Ketonen, Jussi and Weening, Joseph S., "EKL—An Interactive Proof Checker User's Reference Manual," Stanford University Department of Computer Science, report STAN-CS-84-1006, June 1984.

Kors, J. L. and Israel, M., "An Interactive Electrical Graph Extractor," Proceedings 21st Design Automation Conference, 624-628, June 1984.

Lin, Tzu-Mu and Mead, Carver A., "Signal Delay in General RC Networks with Application to Timing Simulation of Digital Integrated Circuits,'' Proceedings MIT Conference on Advanced Research in VLSI (Penfield, ed), 93-99, January 1984.

Losleben, Paul and Thompson, Kathryn, ''Topological Analysis for VLSI Circuits,'' Proceedings 16th Design Automation Conference, 461-473, June 1979.

Lyon, Richard F., ''Simplified Design Rules for VLSI Layouts,'' *Lambda*, 2:1, 54-59, 1st Quarter 1981.

McCormick, Steven P., ''EXCL: A Circuit Extractor for IC Designs,'' Proceedings 21st Design Automation Conference, 616-623, June 1984.

McWilliams, Thomas M., ''Verification of Timing Constraints on Large Digital Systems,'' Proceedings 17th Design Automation Conference, 139-147, June 1980.

Mead, C. and Conway, L., *Introduction to VLSI Systems*, Addison-Wesley, Reading, Massachusetts, 1980.

Nagel, L. W., ''Spice2: A Computer Program to Simulate Semiconductor Circuits,'' University of California at Berkeley, ERL-M520, May 1975.

Newell, Martin E. and Fitzpatrick, Daniel T., ''Exploiting Structure in Integrated Circuit Design Analysis,'' Proceedings MIT Conference on Advanced Research in VLSI (Penfield, ed), 84-92, January 1982.

Ousterhout, John K., ''Crystal: A Timing Analyzer for nMOS VLSI Circuits,'' Proceedings 3rd Caltech Conference on VLSI (Bryant, ed), Computer Science Press, 57-69, March 1983.

Penfield, Paul Jr. and Rubenstein, Jorge, ''Signal Delay in RC Tree Networks,'' Proceedings 18th Design Automation Conference, 613-617, June 1981.

Piscatelli, R. N. and Tingleff, P., ''A Solution To Closeness Checking of Non-Orthogonal Printed Circuit Board Wiring,'' Proceedings 13th Design Automation Conference, 172-178, June 1976.

Scheffer, Louis K., ''A Methodology for Improved Verification of VLSI Designs Without Loss of Area,'' Proceedings 2nd Caltech Conference on VLSI (Seitz, ed), 299-309, January 1981.

Seiler, Larry, ''A Hardware Assisted Design Rule Check Architecture,'' Proceedings 19th Design Automation Conference, 232-238, June 1982.

Shand, Mark A., ''Hierarchical VLSI Artwork Analysis,'' *VLSI '85*, (Horbst, ed), 419-428, August 1985.

Spickelmier, Rick L. and Newton, A. Richard, ''Wombat: A New Netlist Comparison Program,'' *ICCAD '83*, 170-171, September 1983.

Takashima, Makoto; Mitsuhashi, Takashi; Chiba, Toshiaki; and Yoshida, Kenji, ''Programs for Verifying Circuit Connectivity of MOS/LSI Mask Artwork,'' Proceedings 19th Design Automation Conference, 544-550, June 1982.

Trimberger, Stephen, ''Automated Performance Optimization of Custom Integrated Circuits,'' *VLSI '83* (Anceau and Aas, eds), North Holland, Amsterdam, 99-108, August 1983.

Whitney, Telle, ''A Hierarchical Design-Rule Checking Algorithm,'' *Lambda*, 2:1, 40-43, 1st Quarter 1981.

Wilmore, James A., ''Efficient Boolean Operations on IC Masks,'' Proceedings 18th Design Automation Conference, 571-579, June 1981.

DYNAMIC
ANALYSIS TOOLS*

* This chapter was contributed by Robert W. Hon, Schlumberger Palo Alto Research.

6.1   Introduction

Dynamic analysis of designs is the study of a circuit's behavior as a function of time. One might, for example, be interested in the logic values on a set of outputs some time t after certain logic values become present on the circuit's inputs. Or, it might be desirable to determine the expected delay from a change in an input until the output settles. In general, it is computationally impractical to determine a circuit's response to all possible inputs as a function of time; that is, it is not possible to solve in closed form the equations that represent the behavior of the circuit.

For this reason, most dynamic analysis of designs involves an empirical rather than an analytical approach to determining behavior. **Simulation** is the process of empirically determining a circuit's response to a particular set of inputs called a **test vector**. The simulation techniques that are examined here all require some **model** of the computational elements of the circuit and some way of handling **time**.

In spite of the fact that simulation is used in preference to analytical techniques for computational reasons, simulation still requires a considerable amount of computation time for circuits of only modest size. Ideally, one would prefer to simulate a design in detail. At present, this means approximating the analog-waveform behavior of the design. Simulators that operate at this level of detail are called circuit-level simulators. Circuit-level simulators give the waveform tracings for selected circuit nodes as a function of time (see Fig. 6.1). Unfortunately, circuit-level simulation is a computationally costly process and is therefore impractical for circuits of more than a few thousand transistors. Current VLSI designs consisting of a few hundred thousand transistors are well beyond the reach of circuit-level simulators.

Since it is impossible to simulate entire complex designs at the circuit level, two alternatives exist for analyzing large designs. The first is to break the design into a number of smaller pieces and to simulate each piece at the circuit level. This technique is widely used in practice and is particularly useful when some parts of the design are used in several places (for example, an output buffer, a slice of an arithmetic-logical unit). Verification that the subpieces interact correctly can be achieved by dividing the design and simulating at the interfaces, or by other means, including hand simulation. The second method is to simulate a larger portion of the design at a coarser level of detail; for example, accepting only the logical levels 0, 1, and X (undefined) instead of full analog waveforms. As described in subsequent sections, simulators exist that view circuits as composed of switches, gates, or even more complex units. Generally, coarser approximations

```
---------------     0.000d+00     1.250d+00     2.500d+00     3.750d+00  5.000d+00
                - - - - - - - - - - - - - - - - - - - - - - - - - - - - - - - - - -
1.200d-08   4.837d-05  *              .              .              .          X
1.250d-08   4.141d-05  *              .              .              .          X
1.300d-08   3.240d-05  *              .              .              .          X
1.350d-08   2.169d-05  *              .              .              .          X
1.400d-08   1.706d-05  *              .              .              .          X
1.450d-08   1.680d-05  *              .              .              .          X
1.500d-08   1.185d-05  *              .              .              .          X
1.550d-08  -1.266d-02  *              .              .              .      =    +
1.600d-08  -2.354d-02  *              .              .              .  =        +
1.650d-08  -2.832d-02  *              .              .          =   .          +.
1.700d-08  -1.283d-02  *              .              .      =       .          +.
1.750d-08   3.128d-02  *              .          =                  .          + .
1.800d-08   1.096d-01  .*             .      =       .              .        +  .
1.850d-08   2.285d-01  .    *         .  =           .              .        +
1.900d-08   3.996d-01  .      *     =                .              .     +
1.950d-08   6.369d-01  .       =*                    .              .    +
2.000d-08   9.273d-01  =           *                 .              .  +
2.050d-08   1.249d+00  =             *               .              .+
2.100d-08   1.560d+00  =             .   *           .              +
2.150d-08   1.859d+00  =             .       *       .           +.
2.200d-08   2.146d+00  =             .          *    .          +  .
2.250d-08   2.414d+00  =             .             *.          +
2.300d-08   2.661d+00  =             .             .   *     +
2.350d-08   2.887d+00  =             .             .     * +
2.400d-08   3.093d+00  =             .             .       *+
2.450d-08   3.278d+00  =             .             .       + *
2.500d-08   3.444d+00  =             .             .       +   *  .
2.550d-08   3.615d+00  .      =      .             .       +    *  .
2.600d-08   3.762d+00  .         =   .             .       +       *
2.650d-08   3.885d+00  .             .   =         .       +        *
2.700d-08   3.979d+00  .             .       =     .          +     *
2.750d-08   4.062d+00  .             .             =         +      *
2.800d-08   4.132d+00  .             .             .      +=        *
2.850d-08   4.187d+00  .             .             .   +      =     *
2.900d-08   4.229d+00  .             .             .+              = *
2.950d-08   4.275d+00  .             .         +   .              * =  .
3.000d-08   4.329d+00  .             .     +       .              *    =
3.050d-08   4.381d+00  .             .  +          .              *     =
3.100d-08   4.439d+00  .             .+            .             *      =
3.150d-08   4.498d+00  .           +.              .             *      =
3.200d-08   4.557d+00  .       +    .              .            *       =
3.250d-08   4.614d+00  .     +      .              .           *     =
3.300d-08   4.666d+00  .    +       .              .           *      =
3.350d-08   4.714d+00  .    +       .              .          *       =
3.400d-08   4.756d+00  .   +        .              .          *       =
3.450d-08   4.794d+00  .  +         .              .          *     =
3.500d-08   4.827d+00  .  +         .              .          *     =
                - - - - - - - - - - - - - - - - - - - - - - - - - - - - - - - - - -
```

FIGURE 6.1 Circuit-level simulation output.

of behavior can be achieved in less computation time than is used for more detailed approximations, given a circuit of the same complexity.

The following sections give brief overviews of each of the main types of simulators, followed by a closer look at some issues of interest in building simulators.

6.2   Circuit-Level Simulators

Circuit-level simulators determine the analog waveforms at particular nodes in the design [Nagel]. Circuit elements are modeled as transistors, resistors and wires with propagation delays determined by their geometric structure and the underlying technology. These simulators rely on basic physical principles and thus can be highly accurate and general.

Circuit-level simulators provide fine-grain detail about the waveforms at nodes in a design, at the expense of being slow and therefore unable to process very large designs in a reasonable amount of time. Generally, circuit-level simulation is used to check critical parts of a design, whereas overall simulation is left to a higher-level simulator. Circuit-level simulators provide detailed timing information (for example, worst-case input-to-output delays) and are profoundly affected by the design-implementation technology.

Circuit-level simulation actually begins in the node-extraction phase of static analysis. The node extractor must provide the simulator with the capacitance and resistance of the wires, transistors, and resistors that the design comprises so that the delays can be accurately determined (see the later section, Delay Modeling). The values of these parameters are determined by the technology and the geometric properties of the structures. Once the connectivity, resistance, and capacitance parameters are determined, the circuit-level simulator combines this information with built-in or user-supplied **device models**.

A device model is a functional approximation of the behavior of an actual device. For example, Fig. 6.2 shows a typical model for a MOS transistor as used in the SHIELD simulator [Grundmann].

All this information is combined to produce a system of coupled, nonlinear differential equations that must be solved. Often, a so-called **direct method** is used, which is based on Newton's method, sparse-matrix techniques, and numerical integration [Nagel; IBM]. At each time step an initial guess at the node values is made based on the values at the previous time step. This guess is improved by iteration until some predetermined error tolerance is met, at which point the system records the values and moves on to the next time step. The direct method has the advantage that it can be used to simulate any

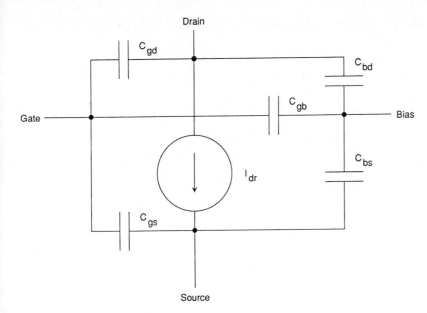

Drain

Gate

Bias

Source

FIGURE 6.2   MOS transistor model.

electronic circuit. Its primary disadvantage is that the computation time is long for large circuits.

Other methods have been developed to decrease the computation time at the expense of generality. One key observation is that a circuit often may be partitioned into several sections that do not interact very much, or that interact in a limited fashion. Exploiting this observation and properties of particular technologies has led to the development of a class of circuit simulators based on **relaxation techniques** instead of on Newton's method. Relaxation methods have the advantage that they generally require less computation than does the direct method, but they are of use only for MOS circuits and can be significantly slower than the direct method where there is feedback in the design [Saleh, Kleckner, and Newton; Newton and Sangiovanni-Vincentelli; Hennion and Senn].

Regardless of method chosen, this relatively expensive computation is performed for each node in the system at each time step. This accounts for the relatively long running times of circuit-level simulators.

6.3   Logic-Level Simulators

Logic-level simulators replace the analog waveforms of circuit-level simulators with the three values 0, 1, and X (undefined). Circuit elements are modeled as simple switches or gates connected by wires the propagation delay of which is typically zero. Logic-level simulators provide the ability to simulate larger designs than circuit-level simulators can, at the expense of detail. Logic-level simulators may simulate at the device level or at the gate level.

6.3.1   Switch-Level Simulators

Switch-level simulators [Bryant] model circuits as a collection of transistors and wires, and must be provided with connectivity information. This information consists of a list of transistors and wires, which provide the equivalent of a schematic of the design. Transistors are modeled as simple switches and wires are modeled as idealized, zero-delay conductors. Usually a simple delay model is incorporated into the transistor model; for example, unit-delay in switching. Other delay information, such as transistor delays derived in a previous circuit simulation, may be incorporated if available.

Processing is a matter of solving simplified equations based on approximate circuit theory. Rudimentary timing information, within the limits of the simplified model, is available for the nodes being monitored. In spite of the approximations involved, the timing information may be sufficient to determine overall system parameters such as worst-case delays and maximum clock rates.

6.3.2   Gate-Level Simulators

Gate-level simulators use the same logic values as switch-level simulators use, but circuit elements are modeled at the gate rather than the transistor level. Gate-level simulators can be used to simulate still larger designs in terms of logic values. However, since they are based on a model of gates, they cannot easily model some common VLSI structures (for example, pass transistors, tri-state bus structures).

Input to gate-level simulators is a schematic of the circuit in terms of logic gates (for example, NAND, OR, D-flip flop) and wires. Some delay is associated with each gate. Usually the schematic is directly available from the schematic-entry system used to draw the design. Truth tables, extended to include the undefined logic value, can be used to model each gate. Figure 6.3 shows the truth table for a two-input AND gate.

Processing is similar to that in switch-level simulators, although larger designs can often be simulated because the computational elements are more complex. In addition, the lack of certain constructs available in a transistor-based model reduces the complexity of the computation involved.

In1	In2	Out
0	0	0
0	1	0
0	X	0
1	0	0
1	1	1
1	X	X
X	0	0
X	1	X
X	X	X

FIGURE 6.3 Truth table for two-input AND gate.

6.3.3 Multiple-State Simulators

In order to obtain more complete information than is provided by the values 0, 1, and X, and yet retain the advantage of being able to handle complex designs, some logic-level simulators are augmented by a larger set of logic levels [VLSI]. It is not uncommon to find commercially offered simulators with a dozen or more states. States are added to provide finer resolution of conflicts at nodes; for example, when a resistive pullup is in opposition to an output that is driving the same output low or when a tri-state bus is in a high-impedance state. Component behavior is specified by large truth tables, as in the simpler three-state simulators.

6.4 Functional- and Behavioral-Level Simulators

Functional- and behavioral-level simulators allow the use of sophisticated data and control representations. This class offers the highest performance but the connection to the implementation that ultimately emerges is not always obvious and is even more rarely automatically generated. Functional-level simulators may use the same logic values as logic-level simulators, or may allow more sophisticated data representations (such as integers). Circuit elements are modeled as functional blocks that correspond to the architecture's hardware functional blocks. Behavioral-level simulators allow sophisticated data representations and model only the behavior of a design. The partitioning of the behavior into pieces may not directly map to isolatable hardware functional units.

Functional- and behavioral-level simulators are similar, differing chiefly in that functional-level simulators map functional units onto actual hardware units [Insinga; Lathrop and Kirk; Frey; Lightner *et al.*]. In both approaches the design consists of a number of units that have behavior that is specified by the designer.

In a functional-level simulator, the structure of the behavior (what units are connected to what other units, the signals entering and leaving a unit) of each unit duplicates the intended hardware structure. Behavioral-level simulators seek to mimic the behavior of the intended design, but not necessarily in precisely the same way that the hardware will implement it. Functional-level simulators might be thought of as the generalization of gate-level simulators, where the design can contain user-defined gates (the function boxes). Sometimes high-level data representations are allowed; for example, ones that show integers as single entities rather than as a collection of bits.

Often the behaviors of computational units in this class of simulator are specified in a simulation language that is similar to a modern programming language. This technique offers good performance and ease of experimentation, but is also far from the actual hardware implementation and therefore cannot catch many types of low-level errors.

Simulating at a higher level of abstraction usually allows the inclusion of more of the design for a given amount of computation time. This gives the designer confidence in the correctness of the design as a whole. The drawback is that subtle errors can be missed because of the lack of detail; for example, race conditions may well be detectable only by a circuit-level simulation. This is a consequence of the fact that the link to the underlying implementation technology becomes progressively weaker as the simulation moves to higher levels of abstraction.

It is useful to be able to simulate part or all of a design at varying levels of abstraction. For this reason, several **multilevel** simulators have been developed that allow different parts of the design to be simulated at different levels [Insinga]. Sometimes the level of simulation can intermixed in the course of the same simulation run; for example, 49 of the instances of a buffer might be simulated at the switch level, whereas the fiftieth is simulated at the circuit level. Presumably, all instances would produce the same behavior so that the expense of simulating 49 of the instances in detail is avoided.

Other simulators mix two or more computational techniques at the same level of abstraction [Chen *et al.*; Lathrop and Kirk]. These **mixed-mode** simulators achieve improved performance by the dynamic, judicious choice of algorithms.

The impossibility of exhaustively simulating a design means that the quality of the verification process is ultimately limited by the designer's patience and the available computer time.

6.5   Simulation Issues

Regardless of the level of abstraction chosen for simulation, a number of key issues must be addressed. Some form of input is required, and the content may change with level of abstraction. Input includes a specification of the circuit or system to be simulated, information about the implementation technology (for example, characteristics of the particular CMOS process used), and logic or voltage values to apply to inputs. The basic simulation process uses this information to determine the values on a set of outputs after some amount of simulated time has passed. Finally, some means of displaying the results of the simulation must be incorporated in the simulator.

6.5.1   Input Requirements

One of the tasks described in Chapter 5 was **node extraction**, which is the process of identifying circuit elements and their connectivity from a geometric representation of the layout. This information, along with some underlying information about the technology used to implement the circuit, is used by circuit- and logic-level simulators to determine a circuit's behavior. Circuit-level simulators require additional information regarding the capacitance of structures so that delays may be accurately calculated. It is the node extractor's job to calculate this additional information from the circuit layout (see the later section, Delay Modeling).

Gate-level simulators require a schematic of the design, expressed as standard logic gates connected by wires. Functional- and behavioral-level simulators operate on input that often closely resembles standard programming languages, with the behavior of a particular box given in a subroutinelike construct. In functional-level simulators, connectivity may be explicitly shown by wires carrying signals between functional units, or implicitly shown by functional units calling (in the manner of subroutines) other functional units with parameters. In behavioral-level simulators, connectivity is usually shown implicitly by call structure.

As each type of simulator requires a different type of design specification, maintaining consistency between the several representations of the design is a major problem in design-aid systems. Most systems leave the problem of maintaining consistent representations (for example, making sure that the geometric representation of a circuit does indeed match the functional description) up to the human user. Others attempt to maintain consistency automatically. Automatic schemes have been the most successful at the lower levels of abstraction; for example, automatically compiling geometric layout from a gate-level schematic.

Test Inputs Simulation results can be useful in verifying the correctness of a design before implementation, as well as for diagnosing implementations of a design. Most designs are sufficiently complex that it is not possible to try all combinations of inputs systematically and to verify that the resulting behavior is correct. A much narrower range of tests needs to be provided as input to the simulator. **Test inputs** are used to specify the set of values that should be applied to a set of inputs at certain times during the simulation. Test inputs can be represented by a list of three-tuples, where each tuple consists of (*input value time*). *Input* is the location in the circuit that will be set to the logic value or voltage given by *value* at the time specified.

It is a nontrivial task to determine a set of test inputs that will sufficiently test a design. Most often test inputs are generated by hand by someone who is familiar with the design. Hand generation of test inputs has the advantage of exploiting a human's knowledge of the tricky or critical sections of the design. Of course, it is quite common for a human to miss or to test incompletely part of a design, particularly when the human is the same person who created the design.

So that human bias and error can be minimized in such situations, a good deal of research has gone into finding algorithms that can be systematically applied to a design in order to find errors. One example is the **D-algorithm** [Roth]. The D-algorithm was originally designed to test gate-level combinational logic designs, but other people have extended the algorithm (see, for example, [Jain and Agrawal] for an application to switch-level simulations).

Given a device that contains combinational logic with some set of inputs and some set of outputs, the goal of the D-algorithm is to find an assignment of input values that will allow the detection of a particular internal fault by examining the output values. In the D-algorithm, the existence of two machines—the "good machine" and the "faulty machine"—is hypothesized. The good machine works perfectly, whereas the faulty machine has some specific internal input stuck at a particular logic value (say, an input to an AND gate stuck at zero). The existence of the error causes a discrepancy between the behaviors of the good machine and the faulty machine for some values of inputs. The D-algorithm provides a means of systematically assigning input values so that this discrepancy is driven to an output, where it may be observed and thus detected. In other words, the D-algorithm provides a test input that is applied to a design. The output values are compared to the expected output values and any discrepancy indicates the presence of the particular internal fault that the test input was specifically designed to find (see [Miczo] for a good tutorial introduction to the D-algorithm as well as an overview of several other commonly used testing techniques).

6.5.2   Time

The method by which a simulator handles time can intimately affect its performance. There are two primary methods of handling time; sometimes hybrid techniques that combine the two methods are used.

The first method runs the simulation as a series of fixed-length time steps. These **incremental-time** simulators advance the simulation clock a fixed amount of time, then calculate the values for all of the nodes in the design at the end of the time step. The time step is usually controllable by the user and is selected so that the finest-grain events of interest are not missed. Sometimes, circuit-level simulators use a variable time step that is controlled by the convergence properties of the equations.

The second method views the simulation as a series of **events**, where an event is a change to a node. A time-ordered priority queue of events is kept; when the simulator is running it processes events in the following way:

1 Remove from the queue all events that occur at the next time t. Each event is expressed as a change to an input at time t.

2 Note each input change at the appropriate node.

3 Calculate a new value for the output of each affected node, based on the new inputs.

4 Create an event for each node that is connected to an output node that changed in the previous step. Insert it into the priority queue at time $t + delay$, where $delay$ is determined by the propagation time through the node. Calculation of $delay$ may be based on a physical model, or may be as simple as assuming a unit delay per node.

The simulator repeats this loop until a user-specified time is reached, or until the event queue is empty.

In both methods the calculation of a new value for a node can be quite expensive, depending on the accuracy desired. Typically, the calculation involves iteration until the nodes of interest do not change between successive iterations.

Each method has advantages and disadvantages. Incremental-time algorithms are relatively simple to implement and offer explicit control of the time grain of the simulation. Unfortunately, there may be long periods in the simulation during which few nodes change, yet the incremental-time method continues the expensive process of checking many nodes. In its simplest form, the incremental-time method often processes nodes that do not or cannot change from the previous time step to the current one. Various schemes have been developed to identify nodes or subcircuits that are inactive and therefore do not need to be recalculated. Event-driven algorithms attempt to avoid this problem by doing work only when something actually changes; their chief disadvantage is that they inherently have a variable time step and therefore may process many fine-grain events that are of no importance.

It is worth noting that most simulators have no built-in notion of time. The essential property is that time is a nondecreasing sequence of numbers. The relation of simulation-time units to real-time units is implicitly determined by the user when the component behaviors are specified. For example, in a gate-level simulation, it is convenient to give delays in units of nanoseconds rather than 0.000000001 seconds. In a switch-level simulation, units of basic inverter delay might be appropriate.

Delay Modeling   There are primarily two types of delay that are of concern to circuit designers: delay through "active components" and delay through wires. The exact definition of active component varies with the level of simulation—it might be a single transistor for a switch-level simulation, or it might be an entire ALU in a functional simulation.

Circuit-level simulators inherently calculate active-component and wire delay, since they are normally taken into account in the equations that are solved. For efficiency, logic-level simulators often assume a simplified model of delay, perhaps using a unit delay through active components and assuming zero delay through wires.

Many variations are possible. For example a functional simulator might use a table of delays for various functional blocks. The table could be based on actual observed delays for off-the-shelf components, if appropriate. Another class of simulator, called **timing** simulators [Terman; Chawla, Gummel, and Kozak], fits just above circuit-level simulators. Timing simulators use relatively complex (compared to most logic-level simulators) delay models combined with the simplification provided in assuming a fixed set of logic levels. For example,

RSIM [Terman] uses a linear model of transistors to determine delay and final logic values for each node. The model takes into account the capacitance at each node and the drive capabilities, and is carefully tuned for each technology so that accurate results are obtained.

In cases in which a simple delay model, such as a unit-delay or table-driven approach, is used, the simulator needs only the connectivity of the circuit and the type of each node. However, much of the value of simulation lies in the ability to determine circuit delays accurately. When this detail is required, a more complex delay model is used.

These delay models approximate the fine-grain behavior of nodes and wires. The time that it takes a wire to change state, say from near 0 volts to near 5 volts, is determined by the resistance R and capacitance C of the wire. The delay depends exponentially on the product RC (for a more detailed explanation, see [Mead and Conway]). Thus, the simulator must be given the resistance and capacitance of each node and wire, in addition to device type and connectivity information.

The RC information is calculated in the node-extraction phase of design analysis. Consider a wire made of polysilicon in a MOS technology. The resistance of the wire is determined by its volume ($= length \times width \times depth$) and the resistance of the polysilicon. Since the resistance of the material and the depth of the wires is fixed for a particular implementation process, the node extractor need only calculate the length and width of the wire and multiply by a constant in order to determine its resistance. Similarly, the capacitance of the wire depends on its area, its separation from the underlying substrate, and the permittivity of the insulator that separates it from the substrate. The separation distance and the permittivity are fixed for the implementation process, and therefore the capacitance can be determined from the length and width of the wire [$C = (permittivity \times length \times width) / separation$].

Often this simple delay model suffices for approximating timing behavior. In cases for which it does not, more accurate calculation of resistance and capacitance can be performed and more complex delay equations used [Glasser and Dobberpuhl]. Nonplanar technologies such as bipolar may also require complex calculations to determine the resistance and capacitance of nodes and wires. As usual, the cost is time.

6.5.3 Device Model

The choice of device model affects the accuracy and the speed of the simulation. The most accurate device models, often used by circuit-level simulators, are analytical and require an iterated solution to a set of equations. This type of model provides a close approximation to the analog behavior of the device, but requires a great deal of computation time.

Logic-level simulators often use simple table-driven models; for example, the behavior of a gate can be stored in a simple truth table and its output can be looked up once the inputs are known. Device delay can be assumed to be a single time unit, or can be calculated from a delay model that takes the actual device geometry into account.

The efficiency of table-driven models can be used in simulators that use multiple logic states by augmenting the truth tables with terms for each undefined or transition state.

Power Consumption As advancing technology allows more devices to be placed on a chip, it becomes increasingly difficult to dissipate the heat generated. In high-density, high-speed technologies, the exotic physical packaging required to remove heat can easily double the price of a system and can lead to attendant problems, for example in connecting to circuit boards for debugging. Thus, designers are often interested in the amount of power that their chips consume.

In nMOS designs, most of the power is consumed in depletion loads that are pulled low. As outlined in Chapter 5, a static estimation of power consumption can be made by determining how many loads can be pulled down. This may be needlessly pessimistic, since it is unlikely that all of the loads are pulled down simultaneously. Dynamic analysis can be used to provide a more accurate estimate of power dissipation, and can also take into account the actual switching frequencies.

In CMOS designs, depletion loads are not needed, since both n- and p-type transistors are available. As shown in the CMOS inverter of Fig. 6.4, when the input is zero the n-type transistor is off, and it can be easily seen that when the input is one the p-type transistor is off. This means that when the circuit is stable with input at zero or one, there is always one transistor that is off and hence no connection between V_{dd} and V_{ss}. The only power dissipated is from leakage currents, which are typically small.

When the input makes a relatively rapid transition from zero to one, the p-type transistor turns off while the n-type transistor turns on. Until they finish switching, V_{ss} is connected to V_{dd} through a changing resistance (the two transistors in series), and so some power is dissipated. In addition, power is consumed in charging whatever capacitance is connected to the output. As shown in [Weste and Eshraghian], the power (P) consumed during these transitions is given by:

$$P = C_{load} \times V_{dd}^2 \times f$$

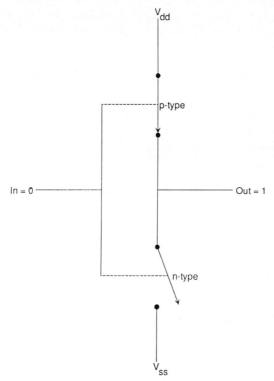

FIGURE 6.4 Idealized CMOS inverter.

where C_{load} is load capacitance and f is the switching frequency.

The total power is the sum of the power consumed while stable and the power consumed while switching. Once again, dynamic analysis can provide a more accurate estimate than can static analysis since the actual switching frequencies are known.

6.5.4 Output

The output of simulators is at least a textual listing containing the values present on the monitored nodes as a function of time. This type of output has several significant disadvantages:

- Separating errors from correct behavior is difficult because of the sheer bulk of information presented
- Spotting trends in the data is difficult because of the discontinuous nature of the textual presentation
- Relating an error to the place in the design where it occurred often is inconvenient

The next step up in output is exemplified in Fig. 6.1, where a rough graphical approximation to the waveforms is provided. The recent proliferation of graphics workstations enables simulation output to be presented in near-analog form, which makes comprehending the results much easier. Integration of the simulator results with the design-entry system further allows nodes to be graphically pinpointed.

6.6   Event-Driven Simulation

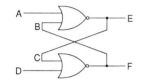

FIGURE 6.5   Cross coupled NOR gates.

This section gives a brief example of how a gate-level event-driven simulation might operate on the cross-coupled NOR gates of Fig. 6.5.

6.6.1   Example

In order to simulate the behavior of the circuit, a model of the behavior of each gate is needed. In simulating at the gate level, a truth table as shown in Fig. 6.6 is appropriate. Given the circuit and its functional behavior, only a test vector and a delay model are required to begin the simulation. Reasonable assumptions are that each gate requires a delay of one time unit, and that wires have a delay of zero time units. At time 0, all inputs and outputs are at logic value X. At some time t, input A is changed to logic value 1 and input D to 0.

The simulator then begins the algorithm given previously, which is repeated here:

1 Remove all events from the queue that occur at the next time t.

2 Note each input change at the appropriate node.

3 Calculate a new value for the output of each affected node.

In1	In2	Out
0	0	1
0	1	0
0	X	X
1	0	0
1	1	0
1	X	0
X	0	X
X	1	0
X	X	X

FIGURE 6.6   Truth table for NOR gate.

4 Create an event for each node that is connected to the output of the affected node and insert it into the queue at time $t + delay$, where *delay* is determined by the propagation time through the node.

Applying the test vector places two events on the queue:

$$(A\ 1\ t)\ (D\ 0\ t)$$

In executing the algorithm, at time t the inputs to the top gate are 1X, which causes the creation of event (E 0 t + 1). Since input C is connected to output E, the event (C 0 t + 1) is also created. The inputs to the bottom gate are X0, which causes the creation of events (F X t + 1) and (B X t + 1). At the end of the first iteration of the algorithm, representing the circuit at time t, the event queue contains:

$$(E\ 0\ t + 1)\ (C\ 0\ t + 1)\ (F\ X\ t + 1)\ (B\ X\ t + 1)$$

Continuing at step 1, event (E 0 t + 1) is removed. This change to output E requires no action, other than reporting the value by printing or plotting. In considering the event (B X t + 1), it is observed that the event does not change the inputs to the top gate. Therefore no new events are created for that gate. (The simulator could have chosen to not check the previous value of the B input when considering the event [B X t + 1], and entered the resulting events [E 0 t + 2] [C 0 t + 2] on the queue. This would not have affected the correctness of the simulation, but would have caused a considerable amount of needless computation.) The inputs to the bottom gate are now 00, so the events (F 1 t + 2) (B 1 t + 2) are added to the queue. At the end of time t + 1, the queue contains:

$$(F\ 1\ t + 2)\ (B\ 1\ t + 2)$$

At time t + 2, the inputs to the top gate are 11, so the events (E 0 t + 3) (C 0 t + 3) are added. At the end of time t + 2, the queue contains:

$$(E\ 0\ t + 3)\ (C\ 0\ t + 3)$$

In the processing of these events no inputs are changed, so no new events need to be added to the queue, which is empty. At this point the algorithm terminates.

6.6.2 Discussion

This example focuses several important issues in simulation. A medium level of abstraction was chosen for this simulation. What are the implications of that choice? In simulating this circuit, about 10 events were processed. Even so, care was taken to avoid processing several events that had no affect on the circuit. In a circuit with several thousand gates, many hundreds of thousands of events might be processed. So it is important that the simulator is implemented efficiently, and that unnecessary events are not processed.

One way to avoid unnecessary event processing is to simulate with more abstract circuit elements. For example, we could replace the cross-coupled NOR gates with a two-input, two-output logic block the behavior of which is specified with a truth table. One line in the table indicates that inputs of 10 gives outputs of 01 after a delay of two time units. Simulating with this new logic block would result in only four events being processed (two for the test vector and two output events). This gives a faster simulation, but some information has been lost.

The NOR-gate simulation showed that the outputs actually settled independently: E changed to 0 at time $t + 1$, whereas F did not change to 1 until time $t + 2$. This information might have important implications, depending on the rest of the circuit, yet was not available when simulating with the more abstract circuit. The independent settling of the outputs is caused by the inclusion of **feedback** in the circuit, which is a common circuit-design technique. Feedback can cause many subtle timing problems, including race conditions and oscillations. As just demonstrated, the improper or unlucky choice of abstraction level can mask such problems.

Similarly, the NOR-gate simulation cannot model some types of behavior. It is well known that the simultaneous changing of gate inputs can lead to metastable behavior. The simple models used in this simulation cannot detect such behavior. A more detailed timing simulation or circuit-level simulation would be likely to find this type of problem, at the expense of time.

6.7   Hardware and Simulation

Actual hardware fits into the dynamic analysis of designs in two ways: first as the computing engine on which the simulator of choice runs, and second as the goal of the design, verification, and construction process.

6.7.1   Simulation Machines

Several examples of hardware solutions to the computational burden imposed by simulators have been described [Denneau; Wang; Abramovici, Levendel, and Menon].

The Yorktown Simulation Engine [Denneau] is a special-purpose architecture designed to simulate up to two million gates at a rate exceeding three billion gate computations per second. The machine consists of 256 logic processors interconnected by a switch. Each has its own instruction and data memory and a functional unit capable of evaluating logic functions. In use, a gate-level simulation is divided into several networks of gates. Each network is loaded onto one logic processor. Internetwork communication is achieved through the interprocessor switch. In this way, parallel hardware is used to simulate the proposed design rapidly. There is a clear mapping between the intended hardware and the simulation.

Wang proposes a different solution: parallel hardware support for solving systems of nonlinear differential equations using the direct method [Wang]. This would provide rapid circuit-level simulation of designs larger than those that currently can be processed.

Other, intermediate solutions are also possible. For example, the simple addition of a floating-point coprocessor or a vector processor can speed the inner loops of the simulation calculations.

6.7.2   Testing Prototype Hardware

A side effect of a careful simulation effort is a set of test vectors that are used to debug the design. Those vectors can be used to test the prototype hardware, and simulation results can be compared to the results obtained from the real part.

The ultimate goal of the design, verification, and construction process is working hardware. The advent of multilevel simulation systems brings about the possibility of using hardware as part of the simulation of a large system. For example, if an ALU chip became available it might be used to replace the simulation of the ALU in the overall system. This would serve both as a testbed for the new hardware part and as a means of improving the performance of the simulation as a whole. Several commercial vendors market products that allow the intermixing of simulator and prototype hardware [VLSI].

6.8   Summary

The state of the art demands that designers make compromises in size of design versus level of detail in simulation. The best performing simulators can simulate large designs, but not in sufficient detail to catch subtle errors of the type likely to occur in a process that still involves a great deal of low-level design. Circuit-level simulators can provide all the detail that a designer is likely to need, but cannot process large designs.

A particularly effective approach is to use detailed (circuit-level) simulation to check critical functional or timing paths, and less-detailed simulation for the remainder of the design. The critical sections can be determined using human insight, or through the use of higher-level simulations that identify potential problems in the design.

Although thorough static-analysis techniques proliferate, they still do not replace the actual use of data in debugging a design. Simulation is therefore of continued importance in a CAD system. Simulation tools will have to keep up with the rapidly changing needs of design environments, both in their capabilities and in their style of use.

Questions

1 Invent a more abstract logic block that could replace the cross-coupled NOR gates of the example in the section, Event-Driven Simulation. Give its extended truth table.

2 Show the event queue that results from simulating your logic block under the same assumptions as before (unit gate delay, zero wire delay).

3 Build a truth table for an *n*-type MOS transistor that includes the four states: 0, 1, X, and Z (resistive-pullup high). Consider the gate and source to be inputs and the drain to be an output.

4 What problems occur when simulating bidirectional MOS transistors? How can these be overcome?

5 How can more accurate models of time improve the quality of simulation that is concerned only with logic values at nodes (and not time)?

6 How can hierarchy aid in simulation?

References

Abramovici, M.; Levendel, Y.H.; and Menon, P.R., "A Logic Simulation Machine," Proceedings 9th Symposium on Computer Architecture, SI-GArch Newsletter, 10:3, 148-157, April 1982.

Bryant, Randal Everitt, *A Switch-Level Simulation Model for Integrated Logic Circuits*, PhD dissertation, Massachusetts Institute of Technology Laboratory for Computer Science, report MIT/LCS/TR-259, March 1981.

Chawla, Basant R.; Gummel, Hermann K.; and Kozak, Paul, "MOTIS—An MOS Timing Simulator," *IEEE Transactions on Circuits and Systems*, CAS-22:12, 901-910, December 1975.

Chen, C.F.; Lo, C-Y.; Nham, H.N.; and Subramaniam, Prasad, "The Second Generation MOTIS Mixed-Mode Simulator," Proceedings 21st Design Automation Conference, 10-17, June 1984.

Denneau, Monty M., "The Yorktown Simulation Engine," Proceedings 19th Design Automation Conference, 55-59, June 1982.

Frey, Ernest J., "ESIM: A Functional Level Simulation Tool," *ICCAD '84*, 48-50, November 1984.

Glasser, Lance A. and Dobberpuhl, Daniel W., *The Design and Analysis of VLSI Circuits*, Addison-Wesley, Reading, Massachusetts, 1985.

Grundmann, John W., "Event-Driven MOS Timing Simulator," *ICCAD '83*, 141-142, September 1983.

Hennion, B. and Senn, P., "A New Algorithm for Third Generation Circuit Simulators: The One-Step Relaxation Method," Proceedings 22nd Design Automation Conference, 137-143, June 1985.

IBM, *Advanced Statistical Analysis Program (ASTAP)*, IBM Corporation Data Products Division, Publication SH20-1118-0, White Plains, New York.

Insinga, Aron K., "Behavioral Modeling in a Structural Logic Simulator," *ICCAD '84*, 42-44, November 1984.

Jain, Sunil K. and Agrawal, Vishwani D., "Modeling and Test Generation Algorithms for MOS Circuits," *IEEE Transactions on Computers*, C-34:5, 426-433, May 1985.

Lathrop, Richard H. and Kirk, Robert S., "An Extensible Object-Oriented Mixed-Mode Functional Simulation System," Proceedings 22nd Design Automation Conference, 630-636, June 1985.

Lightner, M.R.; Moceyunas, P.H.; Mueller, H.P.; Vellandi, B.L.; and Vellandi, H.P., "CSIM: The Evolution of a Behavioral Level Simulator from a Functional Simulator: Implementation Issues and Performance Measurements," *ICCAD '85*, 350-352, November 1985.

Mead, C. and Conway, L., *Introduction to VLSI Systems*, Addison-Wesley, Reading, Massachusetts, 1980.

Miczo, Alexander, *Digital Logic Testing and Simulation*, Chapter 2: "Combinational Logic Test," Harper and Row, New York, 1986.

Nagel, L. W., "Spice2: A Computer Program to Simulate Semiconductor Circuits," University of California at Berkeley, ERL-M520, May 1975.

Newton, Arthur Richard and Sangiovanni-Vincentelli, Alberto L., "Relaxation-Based Electrical Simulation," *IEEE Transactions on CAD*, CAD-3:4, 308-331, October 1984.

Roth, J.P., "Diagnosis of Automata Failures: A Calculus and a Method," *IBM Journal of Research and Development*, 10:4, 278-291, July 1966.

Saleh, Resve A.; Kleckner, James E.; and Newton, A. Richard, "Iterated Timing Analysis in Splice1," *ICCAD '83*, 139-140, September 1983.

Terman, Christopher J. "RSIM—A Logic-level Timing Simulator," Proceedings IEEE International Conference on Computer Design, 437-440, October 1983.

VLSI Systems Design Staff, "1986 Survey of Logic Simulators," *VLSI Systems Design*, VII:2, 32-40, February 1986.

Wang, Paul K.U., "Approaches to Hardware Acceleration of Circuit Simulation," Proceedings IEEE International Conference on Computer Design, 724-726, October 1985.

THE OUTPUT OF
DESIGN AIDS

7.1 Introduction

Computer-aided design systems must be able to communicate with the machines that manufacture and test circuits. This may seem obvious and is often taken for granted; without a good connection to such devices, however, the entire design system becomes a useless toy. The contents of the display screen are not acceptable input to manufacturing, nor are wallpaper-scale plots. These are strictly meant to help the designer visualize a circuit. Manufacturing and testing machines have their own descriptive formats, which all CAD systems must be able to generate.

There are many different manufacturing machines, most of which use vastly different input formats. There are several reasons for these different formats. First, specification formats tend to be optimized for the type of artifact being manufactured, ranging from integrated-circuit masks to printed-circuit and wire-wrap boards. Second, patent protection and other marketing factors often make standardization undesirable and impractical. Finally, technology is evolving at such a rate that machines and their interfaces are rapidly made obsolete.

Testers have traditionally been programmed independently of design systems, with little thought given to integration. Although some testers are designed to mimic simulators, the input formats are rarely standardized or properly connected to the design process.

Besides having the ability to communicate in all the necessary manufacturing and test formats, a good CAD system should be able to understand the many **interchange** formats that allow it to exchange designs with other CAD systems. These interchange formats not only allow free flow of design information, but also enable obscure manufacturing styles, understandable by a subset of CAD systems, to be accessible from other systems.

One common characteristic of manufacturing, testing, and interchange formats is that they are fundamentally unreadable to humans. These specifications are occasionally designed as binary bit streams, but more commonly as highly abbreviated text. It is unreasonable to expect designers to be able to manipulate this sort of text because it contains way too much detail about the exact geometry, topology, and functionality of the circuit. This overabundance of numbers is exactly the thing that design systems seek to avoid. There would be no need for these systems if designers were content to specify circuits in the manners described in this chapter. However, since CAD systems abound, these formats must not be intended for humans and remain to be used for only intermachine communication.

This chapter will discuss interfacing for the manufacturing of different types of electronic circuits. The first section on circuit boards discusses formats for wire-wrap, printed circuitry, and board drilling machines. The Integrated Circuit section covers interchange formats and manufacturing formats. The last section of the chapter describes implementation issues for VLSI circuits, including tester formats and a look at the MOSIS implementation service. Full details of the formats mentioned in this chapter can be found in Appendixes A through E.

Figure 7.1 is a table that summarizes the capabilities of the formats described in this chapter. Some formats are meant as interchange between CAD systems, some are for testing, and some are intended to be read directly by manufacturing machines. It is interesting to note that the interchange formats all support hierarchical description, whereas none of the manufacturing formats do. The table shows that some formats are in human readable text and others are purely binary. Also shown is the nature of the represented data—the extent to which they support topology, geometry, or behavior. Finally, the allowable types of geometry are shown—their ability to handle lines, polygons, circles, arbitrary curves, or text. The following sections provide detail for the table.

	Wire-Wrap	Gerber	N.C. Drill	CIF	GDS II	EBES	EDIF	SDIF	CADDIF
Type	Man	Man	Man	Int	Int	Man	Int	Test	Test
Hierarchy	N	N	N	Y	Y	N	Y	Y	N
Human-readable	Y	Y	Y	Y	N	N	Y	Y	N
Contents:									
Topology	Y	N	N	N	Y	N	Y	N	N
Geometry	N	Y	Y	Y	Y	Y	Y	N	N
Behavior	N	N	N	N	N	N	Y	Y	Y
Geometry:									
Lines	N	Y	N	N	Y	N	Y	N	N
Polygons	N	Y	N	Y	Y	Y	Y	N	N
Circles	N	Y	Y	Y	N	N	Y	N	N
Curves	N	Y	N	N	N	N	Y	N	N
Text	N	Y	N	N	Y	N	Y	N	N

FIGURE 7.1 Summary of output formats.

7.2   Circuit Boards

7.2.1   Format of the Wire-Wrap Board

Before manufacturing a printed-circuit board, many designers build a prototype on a wire-wrap board. Wire-wrapping is the technique by which wires are attached to pins by being tightly wrapped around them. The electrical connection is good and yet changes are easily made.

Wire-wrap boards have a mesh of **pins** on one side connected to **sockets** on the other. IC packages are plugged into the sockets and interconnection is done on the pin side. Since the leads on ICs are typically 0.1 inch apart and since IC packages are often 0.3 inch wide or multiples thereof, the pins and sockets on a wire-wrap board are usually spaced 0.1 inch apart in columns spaced 0.3 inch apart. Additional pins and sockets are often found near the edge of the board to enable wiring of off-board connectors.

Since the wrapped wires are insulated, they can cross each other arbitrarily. Thus there is no routing problem, and the specific geometry of wire paths is not needed. All that matters is the location of the packages and the endpoints of connections. Essentially it is a netlist such as is given to switch-level simulators.

Although there are no rigid standards for this information, certain conventions should be followed. Wire-wrap boards typically identify their pins with alphabetic columns and numeric rows. The upper-left pin is A01 and the pin to its right is B01. Since the columns are spaced wider apart than are the rows, there are fewer columns and single letters often suffice to enumerate them. However, after column Z comes column AA, BB, and so on. To list the components on the board it is necessary to give the following information:

> Package name
> Package orientation
> Location of pin 1

For example, a line of text that reads 7404 1 M15 will place a 74-04 chip in a vertical orientation (orientation number 1) with the first pin of the chip at location M15 of the board. As a double check, additional information may also be requested:

> Number of pins on the package
> Package width
> Location of the power pin
> Location of the ground pin

Wiring of the packages is then done by listing coordinate pairs to be connected. For example, the line `C14 D25` will place a wire that runs from column C, row 14 to column D, row 25.

Some wire-wrap formats allow definition of signal names so that symbolic pin locations can be given and color-coordinated wires can be used. Other options include the specification of twisted-pair wiring that requires two source and two destination pins.

Care should be taken when specifying the order of wires. Since wire-wrapping machines must make a physical movement when traveling from pin to pin, it is best to keep the start of one wire close to the end of the previous wire. This speeds wrapping time and saves money. It is tempting to use **daisy chains**, in which the next wire begins on the ending pin of the previous wire, since these are most efficient when specified in sequence. However, for ease of debugging, no wire should be covered by other wires that are further covered, because that will require too many removals to make a correction. The best order is to place every second wire in a run, and then to wrap the odd ones on top.

To generate the most useful wire-wrap boards, consider their purpose for existing. Wire-wrapping is a prototyping stage, which means that the designer will want to make changes to the circuit. Each change requires removal and rewrapping of wires. If the most volatile wires are wrapped last, they will be on top, which will make them easier to change. Power and ground wires rarely change so they should be specified first. Long wires should be placed later so they do not become trapped by short wires, which are more easily removed. User indication of volatility should even be allowed as a consideration when producing a wire-wrap specification.

7.2.2   Format of the Printed-Circuit Board

A printed-circuit board is a sheet of nonconducting material, usually fiberglass, that has flat wires laid onto it. These wires run between holes in the board into which components will be attached. Most printed-circuit boards are multilayer boards that have wires on both sides or even on layers sandwiched inside the board.

Each layer of a printed-circuit board is manufactured in a photographic process that uses a sheet of film to mask the location of wires. This film can be produced by hand or automatically from CAD system output. One popular automatic technique is the use of a **photoplotter**, which takes the output of a CAD system and produces film to be used in printed-circuit manufacturing (see Fig. 7.2).

The most popular format for photoplotter control is called **Gerber**

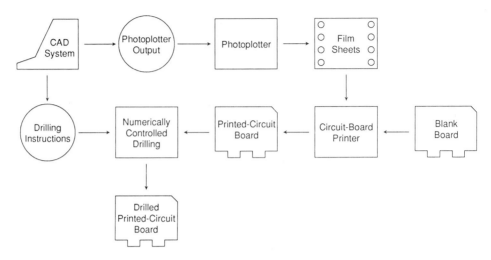

FIGURE 7.2   The manufacturing steps of printed-circuit boards.

format, created for Gerber Scientific Instrument Company machines [Gerber]. Today other companies' machines accept Gerber format, making this an industry standard.

Appendix A gives the details of Gerber format. Essentially, a Gerber file is a collection of ASCII commands and parameter values. The commands control a virtual pen that moves about the film drawing points, lines, curves, and text. Mode commands set state, and coordinate commands control the actual drawing. For example, once line-drawing mode is set, subsequent coordinate commands draw lines; when cubic-spline mode is set, subsequent coordinate commands are taken as inflection points for a curve. The virtual pen can be raised or lowered to switch between drawing and moving.

There is no hierarchy in Gerber format; the drawing commands are listed in the order to be performed. Thus an important consideration in producing these files is locality. Photoplotters take time to move the pen position, so the order of plotting should follow a minimal spatial path through the design to keep costs down. Some manufacturing companies even have programs to reorder plotting files so that the machine time is reduced. However, a good CAD system does things right the first time.

In addition to specifications for the wiring of a printed-circuit board, two other types of information should be available: the graphic (nonelectronic) **artwork** and the **drilling instructions**. Graphic artwork is an optional step that puts nonconducting ink on the printed-circuit board. This graphics contains helpful messages such as part numbers and connection names, which aid in the final production and debugging of the board. The methods for specifying and manufacturing this artwork are the same as for printed wiring: a Gerber file. Drilling instructions are necessary because they make the holes for the chip leads. These specifications are very different from wiring and artwork specifications, as the next section will illustrate.

7.2.3 Drilling Machines

As Fig. 7.2 shows, a printed-circuit board must not only be covered with wires, but also be drilled. The holes are used either for component connections or as **through holes** to connect different layers of the board. Drilling instructions are separate from those for the wiring and graphic artwork because different machines do the drilling. Typically these machines are **numerically controlled** (**NC**) and they work from instructions on punched paper tape.

Numerically controlled drilling machines are very simple devices. They have a switchable drilling head that can automatically select from a set of tools, and they have a movable table that can place any part of the board under the head. Thus there are only two commands to an NC drilling machine:

$$T dd$$

switches to tool *dd*, a number such as 01. This tool switching takes time and so it should be done infrequently. The second command makes the tool function by giving a coordinate in the form:

$$X ddddd Y ddddd$$

where the *ddddd* values are five-digit coordinates, usually measured in inches, with an implied decimal point after the first two digits (*dd.ddd*). Since it is not known where the board will be placed on the drilling machine, the origin of the coordinate system should be adjustable so that any point on the board can be set to (0, 0) and the other coordinates will slide along. With NC drilling, as well as all formats, it is best to ask a manufacturing company to specify the precise style that it expects.

7.3 Integrated Circuits

Integrated circuits are manufactured on large disks of silicon called **wafers**. A group of about 20 of these wafers is called a **boatload** and a large collection of boatloads manufactured together is called a **run** or a **batch**. A wafer is round, from 3 to 6 inches in diameter, and has one side flattened (see Fig. 4.21). Although it can contain hundreds of different IC **dies**, there is typically only one type of chip on a wafer and it is replicated in all but a few of the die locations. Some prototyping operations combine multiple chip dies on a single wafer, or place multiple IC designs into a single die. Also, there are occasional test dies inserted at various locations on the wafer. The placement of a different die on an otherwise uniform mask is called a **drop-in**. The entire production process is illustrated in Fig. 7.3.

A wafer consists of multiple layers of conducting and semiconducting material. Each layer is specified with a **mask** that controls the deposition or etching of material during fabrication. Masks are produced either optically or by electron-beam techniques, and always under computer

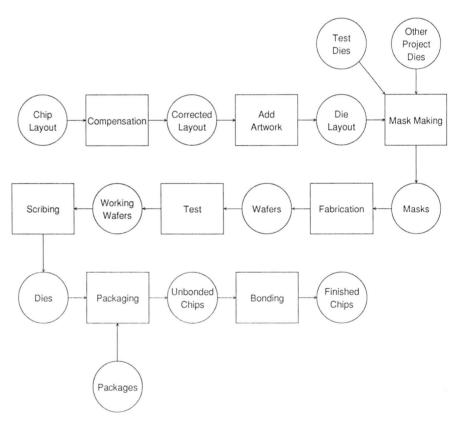

FIGURE 7.3 Chip manufacturing steps.

control. In some prototyping situations, the mask is not used and an electron-beam device writes directly on the wafer. Regardless of the masking technology, the controlling formats are the same.

Although there are many mask-manufacturing formats and interchange formats targeted at mask making, there are a few standards that will be discussed here. Caltech Intermediate Format (CIF) and Calma GDS II Stream Format are well-known interchange standards, whereas Electronic Design Interchange Format (EDIF) is newer and less popular, but its vast extent may carry it to a wide acceptance. The Electron Beam Exposure System (EBES) format is an actual machine specification, like the Gerber format.

7.3.1   Caltech Intermediate Format

Caltech Intermediate Format (CIF) is a simple method for describing integrated circuits [Hon and Sequin]. The goals of CIF are to be concise, to be readable by humans and machines, and most of all to remain constant so that it can be relied on as a standard. Its authors have allowed "user extensions," but permit no change to the basic commands. Appendix B describes CIF in detail.

A CIF file is a set of textual commands to set layers and draw on them. CIF describes only the geometry that is found on IC masks: rectangles, polygons, wires, and circular pads. There are no higher-level graphics operations such as text or curves. There is hierarchy, however, which makes CIF easy to generate and can be used to build complex graphics. In addition, the language is structured to allow multiple CIF files to easily be aggregated into a single larger file so that multiproject chips can be specified. The only drawback of CIF is that it is not readable by any mask-making machines and thus must be further processed before it can be used in fabrication.

7.3.2   Calma GDS II Stream Format

Calma GDS II Stream Format is older than CIF and is more widely used [Calma]. Its early entry in the market caused subsequent vendors to accept it for compatibility reasons; thus it became a standard. In addition to representing mask geometry, it also has facilities for topological information and arbitrary other attributes. The reason for this is that GDS II is the complete database representation for Calma CAD systems, and not just the output specification for mask-making. Thus it includes text, arrays, structures, and hierarchy (see Appendix C).

GDS II has changed over the years as new constructs have been introduced. Nevertheless, backward compatibility has been maintained, which has allowed the format to remain the industry standard. Like CIF, this is an intermediate format that must be converted before it can be made into a mask. The other drawback of GDS II is that it is binary, which makes it not readable by humans. Some would argue that these formats are not meant for humans and that binary provides more compaction of data. However, the CAD programmer does appreciate a readable format and, as far as compaction is concerned, a concise text language such as CIF does not use much more space.

7.3.3   Electronic Design Interchange Format

The most recent VLSI interchange format is the **Electronic Design Interchange Format (EDIF)**, which was proposed by a consortium of electronics companies [EDIF Committee]. There is hope that this format will replace others as the standard, because it is so powerful, easy to parse, and still open to the future needs of CAD systems. Additionally, many CAD vendors have promised EDIF compatibility in their products [VLSI Design]. The format is described in Appendix D.

EDIF is a textual format that resembles the LISP programming language [McCarthy *et al.*]. Each statement is enclosed in parentheses and has a keyword first, followed by its parameters. Also, the entire EDIF file is structured so that all statements are parameters to other statements. This means that an EDIF file has only one statement:

(edif *ediffile*)

where *ediffile* is a list of other statements such as:

(library *name libcontents*)

and *libcontents* include cell statements such as

(cell *name cellcontents*)

and so on. The nesting of parenthetical statements makes parsing easy and allows arbitrary complexity to be included in the EDIF language. Needless to say, EDIF files can be hierarchical and they can contain any form of information including topology, geometry, and behavior. The geometry can describe circles, text, and much more.

Although EDIF ambitiously attempts to include everything of interest to a CAD system, the specification is imperfect in a number of areas, as even the EDIF committee acknowledges. Behavior and constraints are poorly understood, so the specification merely mentions the need for this information without suggesting how it can be done. Also, there is a definite problem with wordiness and overspecification. Some of the ordering of the syntax is set so that identifiers are used before they are first declared (for example, the `viewmap` section, which associates different views of a cell, must come before the views). This makes parsing more difficult. Even the so-called human readability of EDIF could be improved if some of the long keywords (for example, `portimplementation`) had dashes or hyphens to help separate the individual words that compose them.

A cleaner and more complete scheme for this kind of information is SHIFT [Liblong], which was proposed earlier but was not selected as a standard. This interchange format has well-specified behavior and constraints as well as the capability of describing topology and geometry. It is remarkable that SHIFT is so similar to EDIF in structure and intent. It even has LISPlike syntax.

EDIF does provide nonstandard escape mechanisms so that additional information can be included. This could be useful, for example, when specifying non-Manhattan orientations of cell instances that are not yet supported in EDIF. Since the standard openly admits that it is incomplete, these extensions will soon be accepted. However, care must be exercised by the CAD system designer to ensure that these extensions are properly coordinated and well thought out, or else EDIF will diverge and fail to be a standard.

7.3.4 Electron Beam Exposure System Format

Sooner or later the CAD system must produce a chip description that can be read by a mask-making machine. With today's smaller IC features, it is becoming clear that the electron-beam lithography methods are preferable to optical techniques. Within the domain of electron-beam devices, a very popular format is **EBES**, the **Electron Beam Exposure System** developed at Bell Laboratories [Gross, Raamot, and Watkins].

EBES is readable directly by a number of different mask-making machines and this explains why it allows no hierarchy. Although these machines have computers in them, they spend all their time driving the electron beam and cannot accept a complex input format. In fact, there are only three different figures that can be drawn: rectangles, parallelograms, and trapezoids. Everything else must be composed from these. Although some machines have allowed extensions to EBES such as arrays and extended addressing, the original format is the only guaranteed standard.

The major drawback of EBES format is that it is binary and that the binary word size is restrictively small for modern tasks. With only 16-bit words, a complete chip must be described in a square of only 65,536 units on a side. This used to be a large field but it is easily exceeded today, forcing chips to be described as multiple abutting dies. Another problem with EBES, as Appendix E shows, is that the geometry must be sorted so that it is presented to the electron-beam machine in spatial order. This geometry must be clipped into stripes and then all the pieces must be sorted by y coordinate. Rounded figures must be approximated by polygons and complex polygons must be broken into four-sided figures. Combined with a lack of hierarchy, these problems cause EBES files to be large and slow to produce.

7.3.5 Bonding Machines

In this discussion of IC manufacturing specification, it is appropriate to discuss the final steps: packaging and bonding. Packaging is the selection and attachment of the IC die to a plastic or ceramic chip body that has metal leads [Johnson and Lipman]. Bonding is the step that runs wires from the chip-body's leads to the IC pads on the die. Both of these steps belong in a CAD system although they are more commonly done by hand.

Packaging options for ICs change very often so it would be helpful if the CAD system maintained a database of packages showing the number of pins, the external pin arrangement, the internal cavity size for the die, and other physical package properties. The system could then help to select the most appropriate package for the task at hand.

Programmable bonding machines do exist to wire integrated-circuit pads to the chip package leads. However, they typically work in an interactive-learning mode that requires an operator to bond one chip completely before the machine can do more. Subsequent chips are then automatically inspected to locate special fiducial marks so that variances in orientation can be handled. It would certainly be reasonable to link the CAD system with this machine so that the latter knows the relative placement of all of the pads and needs no human instruction.

7.4 Implementation Issues

There are many other considerations in the manufacturing of a circuit besides its mere specification. Data conversion may be required to translate between interchange formats. Testing must be accommodated to verify a circuit and to weed out badly manufactured devices. Also, there are many standard procedures that may be available to speed the design and manufacturing process. This section discusses these side issues that relate to the implementation of VLSI systems.

7.4.1   Data Conversion

As this chapter mentioned earlier, interchange formats exist not only to exchange circuits between CAD systems, but also as a path to manufacturing, by converting from one format to another. Therefore one important implementation facility is the ability to translate between formats. When doing so for the purpose of manufacturing, only geometric information needs to be converted, because topology and behavior is not used in fabrication. Nevertheless, a complete conversion system should attempt to retain as much information as possible, in case those data are needed later to reconstruct the original file.

Most geometric format conversion is straightforward and can be accomplished on a figure-by-figure basis. CIF specifies a box by giving its center point, size, and rotation; GDS II describes it with four corner coordinates. Such conversion demands the simplest of algorithms.

A somewhat more difficult conversion occurs when there is no equivalent figure in the other format. Thus the Gerber TEXT commands have no equivalent in CIF and may require a full typeface description so that appropriate geometry can be produced. The same problem arises when converting curved objects to rectilinear ones.

A useful technique in format conversion is to use a third representation as an intermediate. This allows the input file to be read completely before any output is generated, so that all the data can be considered before processing. For example, conversion to CIF needs to examine all coordinate values before choosing scale factors for the CIF cells. Another advantage of using an intermediate representation is that it can be tailored to the needs of the output format. Thus when generating EBES files, an intermediate raster representation helps immensely in handling nonrectilinear shapes such as curves or text. These figures are approximated in the raster array and any overlap is automatically merged. Also, the resulting EBES files need to be sorted and that is much easier when the data are already in a matrix.

Format conversion may deteriorate the data or even fail to work at all. If the unit sizes in the two files are sufficiently incompatible, then the data may have to be approximated. For example, an EDIF file with 200 units to the micron will have to be rounded by one bit when converted to CIF, in which the smallest unit is one-hundredth of a micron. An even more severe problem, however, arises when a CIF cell instance is rotated by 45 degrees. If this CIF file is converted to EDIF, there will be no way to place the instance correctly, since EDIF cells must appear at Manhattan orientations. The conversion program can hierarchically flatten the cell instance and describe all its enclosing geometry, or it can simply give up. When conversion degrades or fails, the user should be told of the problem and shown exactly where it has occurred. Conversion must then continue, however, so that unimportant problems do not block the overall process.

7.4.2   Testing Formats

Testing is the same as simulating, only it uses real parts rather than software emulation. Some simulators allow actual components to be used for parts of the process, and some testers make use of software models for part of their operation. Therefore it is not surprising that tester formats are often the same as simulator formats. In fact, most testing-machine manufacturers design input specifications that mimic known simulators.

Unfortunately, there are no standards either in simulator commands or in tester specifications. EDIF considers the possibility of tester description but it does not have a complete language for that purpose. A follow-on to EDIF is the **Stimulus Data Interchange Format (SDIF)**, which has similar syntax and attempts to fill the gap in EDIF [Pieper]. Another possible tester standard is CADDIF [Factron], which has been released to the public in an attempt to receive wide acceptance. CADDIF is a binary representation with no control structures: It merely lists the stimulus values to be applied and the expected values to watch. SDIF, however, has the full expressibility of EDIF and LISP, so it is able to describe loops and other concise testing control. SDIF's textual appearance is more flexible than CADDIF's binary structure, which sometimes limits fields to 8 bits and thus cannot express large values.

Tester-interface formats are also variable because of the wide range of testing that can be done to a circuit. Both digital and analog devices must be handled, with tolerances available to capture all the necessary values. For digital circuits, the voltages must be classified according to discrete values of high, low, threshold, floating, and other notions that are digitally described but continuously analyzed. The tester language must therefore speak to all designers in a suitable style. Some tester languages are even able to accept waveforms from the simulator as stimulus and response data.

As the processors in testers grow more sophisticated, they will be able to accept a wider range of test specifications. This will open up the possibility of standard tester formats that cut across manufacturer's limitations.

7.4.3   Standards on the VLSI Chip

Another implementation issue that is worthy of consideration is the use of standardized structures on the VLSI chip to ease interface with the outside world. The most common standard structures are the bonding pads, which are provided for each design environment. Use of such pads does more than relieve the designer from bonding considerations such as pad layers and spacing. These standard pads often have necessary circuitry for static protection, TTL driving, and all the analog considerations that are not needed elsewhere on the chip. Thus the digital designer can view VLSI systems in a purely logical sense by using standard parts for the nondigital interface.

One can go even further in providing on-chip standards. For example, it is possible to provide the chip designer with a standard frame of pads, already placed and connected to power and ground. The VLSI designer can then use these pads without worrying about die size or proper pad organization. For maximum flexibility, the pad frame can be a simple template on which real pads are placed. This allows the input, output, and bidirectional pads to appear in appropriate number and location. Power and ground pads can still be fixed as part of the frame, since their locations are often critical and their rearrangement is not necessary to the chip design.

The ultimate on-chip standard is the **design frame** that includes pads, clock lines, and fully interfaced I/O [Borriello *et al*.]. The designer is presented with an inner chip area and a set of connection points along the frame. Besides a standard frame on the chip, there is also a standard circuit board outside the chip that correctly interfaces to the bus of a computer. A scheme such as this is excellent for instructional purposes and also provides rapid turnaround for the manufacture and test of prototype circuits. A library of different design frames would be a valuable addition to any CAD system.

7.4.4   Standards of Fabrication: MOSIS

The final implementation issue to be covered here is the use of standardized services for the manufacturing of VLSI systems. Earlier sections of this book noted that, in many situations, the manufacturer must be asked to supply the proper mask graphics, compensation, and other specifications for circuit production. However, thanks to our highly specialized and service-oriented society, there are "brokers" that can take care of these details. One such broker is the **MOS Implementation Service (MOSIS)** which does rapid prototyping of integrated circuits and printed-circuit boards [MOSIS], based on methods developed at Xerox Palo Alto Research Center [Conway, Bell, and Newell].

MOSIS is a service funded by the U.S. government that is available to those people who have received grants from the National Science Foundation (NSF) or the Defense Advanced Research Projects Agency

(DARPA). This service is widely used in American universities for implementation of student VLSI projects and faculty research projects. It is also available to other clients who are willing to pay. There are similar services active in other countries, particularly Canada [CMC] and Australia [Hellestrand *et al.*]. A number of firms offer portions of these services, calling themselves **silicon foundries**.

Input to MOSIS is either a CIF description or a GDS II tape. CIF is the preferred method of submitting chips because it can be done via electronic mail. In fact, all communication with MOSIS is best done by computer mail. On the ARPAnet, use the address "MOSIS@MOSIS.EDU". On the TeleMail net, use "[MOSIS/-USCISI]TELEMAIL/USA".

Fabricated chips from MOSIS will be placed into 28-, 40-, or 64-pin DIPs, or 84-pin grid arrays, depending on the pad requirements. MOSIS encourages the use of standard pad frames to make packaging easier. This service includes mask making, fabricating, packaging, bonding, and delivery of chips along with SPICE parameters for the run. The reason that the SPICE parameters are reported with each run is simply that MOSIS is a broker for many fabrication firms and does not know in advance where a chip will be built. Users are assured of high fabrication standards because prototype chips cannot tolerate a low yield. In addition to chips, printed-circuit boards can be manufactured from CIF specifications.

CIF files submitted to MOSIS must meet certain restrictions to enable smooth fabrication. The Definition Delete (DD) statement may not be used. Fortunately, this is usually safe because the statement exists only for implementors such as MOSIS and is almost never generated by a CAD system. Polygons must have at least three points, and round flashes must have a nonzero diameter. Needless to say, pads must be present in the CIF file, because their location and number helps to classify the chip automatically. As an aid to implementation, it is suggested that the "XP" layer be used to duplicate pad geometry.

GDS II tapes submitted for fabrication must include the name of the top-level cell, since that information is not part of the data. Users should be cautioned that NODEs, BOXs, and PLEXs will be ignored as will the information in ELFLAGS and ELKEY records. GDS II files must be mailed on a magnetic tape since the format is tape-oriented and its binary data cannot travel well by network.

As it was mentioned before, most conversation with MOSIS can take place electronically. Automatic mail readers know how to respond to messages in the right format, so many requests can be handled without human intervention. Users can obtain documentation, pad-frame descriptions, and even cell libraries containing pads, PLA elements, and so on. It is also possible to submit a chip for fabrication merely by talking to the automatic mail reader at MOSIS. Of course, proper accounting must be established before services can be rendered.

The format of an electronic message to MOSIS is as follows:

> REQUEST: *keyword*
> *parameters*
> REQUEST: END

Each REQUEST controls actions to be taken and the message terminates with an END action. Messages that have incorrect format will evoke a response that explains the correct format and gives some initial help. Thus first-time users of MOSIS can get started simply by sending a message with any random content.

Requests to MOSIS can be to ask for INFORMATION, to SUBMIT chips, to demand human ATTENTION, and more. The options are always growing as the service expands, so consult the current MOSIS user's manual for details.

7.5   Summary

This chapter has described the necessary steps for the fabrication of VLSI circuits. The CAD system must be able to generate manufacturing specifications in a number of different formats, depending on the particular environment of design. Also, interchange formats are a necessary part of a CAD system, to let it communicate with the other systems in the world. This chapter also covered specific implementation issues such as format conversion, testing interface, layout standards, and manufacturing standards. CAD systems and their users should understand these issues in order to complete their task and realize their designs.

Questions

1 Why is there no hierarchy in descriptions of wire-wrap or printed-circuit boards?

2 List three disadvantages of binary file formats.

3 What is the most important feature of any standard output format?

4 List two reasons to convert between output formats.

5 What is the programming-language equivalent of cell libraries? Of design frames?

6 What is the main difficulty that may prevent EDIF from becoming a standard?

7 What technological advance will remove the distinction between manufacturing formats and interchange formats?

References

Borriello, Gaetano; Katz, Randy H.; Bell, Alan G.; and Conway, Lynn, "VLSI System Design by the Numbers," *IEEE Spectrum*, 22:2, 44-50, February 1985.

Calma Corporation, *GDS II Stream Format*, July 1984.

CMC, *Guide to the Integrated Circuit Implementation Services of the Canadian Microelectronics Corporation*, version 2:0, Kingston Ontario, January 1986.

Conway, Lynn; Bell, Alan; and Newell, Martin E., "MPC79: The Demonstration-Operation of a Prototype Remote-Entry Fast-Turnaround VLSI Implementation System," Proceedings MIT Conference on Advanced Research in Integrated Circuits, January 1980 (also reprinted in *Lambda*, 1:2, 10-19, 2nd Quarter 1980).

Electronic Design Interface Format Steering Committee, *EDIF—Electronic Design Interchange Format Version 1 0 0*, Texas Instruments, Dallas, Texas, 1985.

Factron, "CADDIF Version 2.0 Engineering Specifications," Schlumberger Factron, October 1985.

Gerber Corporation, "Gerber Format," Gerber Scientific Instrument Company document number 40101-S00-066A, July 1983.

Gross, A. G.; Raamot, J.; and Watkins, S. B., "Computer Systems for Pattern Generator Control," *Bell Systems Technical Journal*, 49:9, 2011-2029, November 1970.

Hellestrand, G. R.; Tan, C. H.; Yong, F. N.; and Forster, R. L., "Australian Multi-Project Chip Activities, 1982-1986," Joint Microelectronics Research Centre, University of New South Wales, October 1986.

Hon, Robert W. and Sequin, Carlo H., "A Guide to LSI Implementation," 2nd Edition, Xerox Palo Alto Research Center technical memo SSL-79-7, January 1980.

Johnson, Dean P. and Lipman, Jim, "IC Packaging: An Introduction For the VLSI Designer," *VLSI Systems Design, VII:6, 108-116, June 1986.*

Liblong, Breen M., *SHIFT—A Structured Hierarchical Intermediate Form for VLSI Design Tools*, Masters Thesis, University of Calgary Department of Computer Science, September 1984.

McCarthy, John; Abrahams, Paul W.; Edwards, Daniel J.; Hart, Timothy P.; and Levin, Michael I., *LISP 1.5 Programmer's Manual*, MIT Press, Cambridge, Massachusetts, 1962.

MOSIS, *MOSIS User's Manual*, University of Southern California Information Sciences Institute, 1986.

Pieper, Chris, "Stimulus Data Interchange Format," *VLSI Systems Design*, Part I: VII:7, 76-81, July 1986; Part II: VII:8, 56-60, August 1986.

VLSI Design Staff, "A Perspective On CAE Workstations," *VLSI Design*, IV:4, 52-74, April 1985.

PROGRAMMABILITY

8.1   Introduction

There are a number of factors that determine the size and complexity of circuits that can be designed. One major factor is the quality and power of the design tools. Generators can produce large arrays, routers and compacters can manipulate large layouts in special-purpose ways, and silicon compilers can produce great amounts of layout from small, high-level specifications. Another factor in the production of large circuits is the use of good design techniques such as hierarchy and modularity. However, there comes a point in many large design efforts where something more is needed to help with special-purpose tasks.

To go beyond the limitations of standard circuit production facilities, it is necessary to develop custom tools and techniques. Occasionally, this can be done by combining existing facilities, for example by using array constructs to aid in repetitive tasks. More often it can be accomplished through command macros that allow complex operations to be built from simpler commands. This still leaves several unresolved issues. For example, how can an ordinary set of editing commands be used to design correctly a cell with proper pullup and pulldown ratios? How can they be used to adjust the spacing of an entire hierarchy when a low-level cell changes size? These issues and many others can be resolved only when the available commands provide the expressiveness and power of a programming language—when they are able to examine circuits and to react appropriately. Since the parallels between VLSI design and computer programming are strong, it should come as no surprise that programming is actually a design technique.

8.1.1   Dichotomy of Programmability

Programmability can be found in two forms in a design system: imperative or declarative. **Imperative programming** is the usual form: an explicit set of commands that are sequentially executed when invoked by the designer. **Declarative programming** is the implicit execution of commands that result as a side effect of other design activity. Dataflow programming is declarative, because execution is determined by the data and not by the code order. Declarative programming in a design environment is often viewed as **constraints** or **daemons** because the commands reside on pieces of the circuit. Figure 8.1 illustrates the imperative and declarative ways of expressing a relationship between two values.

In addition to there being two forms of programming in a design system, there are two ways that these programs can be expressed.

Imperative:
    if pullup-ratio $\times$ 4 pulldown-ratio then
        pullup-ratio = pulldown-ratio/4

Declarative:
    pullup-ratio $\equiv$ pulldown-ratio/4
    pulldown-ratio $\equiv$ pullup-ratio $\times$ 4

FIGURE 8.1 Imperative versus declarative code.

Some languages are purely **textual**, such that a page of code defines a circuit. Others are **graphic** in nature, using an interactive display to build the circuit. Textual design languages tend to be primarily imperative in their sequencing, with a few declarative options, whereas graphic languages reverse this tendency, being mostly declarative with occasional imperative sections. Figure 8.2 shows graphic ways of expressing the relationships in Fig. 8.1.

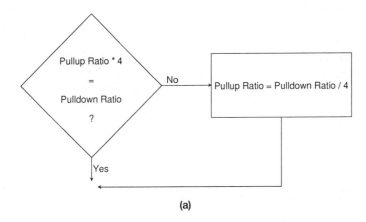

(a)

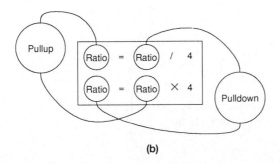

(b)

FIGURE 8.2 Graphical imperative and declarative code: (a) Imperative (b) Declarative.

8.1.2   Adding Programmability

The unfortunate aspect of computer programming is that it is often so complicated that designers do not learn to do it. This is especially true in graphic design systems, which allow the designer to build an entire circuit without needing to write any code. The solution to this problem is to present programming in a restricted sense that is tailored to the design task and thus is easier to learn. For example, if the command interface allows macros, variables, algebraic expressions, conditionals, and other language features, then it is powerful enough to express any algorithm. Although the use of these features constitutes textual imperative programming, it is merely an extension of commands that are already familiar to the designer. This makes design programming easier to learn because it can be done only when it is needed.

Another way to add programmability is to provide constraining structures as part of the layout. A simple example is the constraint that a wire must remain connected to its components, and not change length. Constraints react to changes in the circuit by invoking other changes. In this example, a change to one component fires a constraint that changes the wire, which then fires a reverse constraint that affects the other component. This declarative programming is more intuitive than are conventional languages because the commands fit into the circuit being designed and are a natural extension of the layout.

The most obvious method of adding programmability is to provide a full programming language. Many systems begin with known languages and add macros or subroutines to make them handle circuitry [Batali and Hartheimer; Weste]. The resulting programming languages are called **embedded languages** because they have added design constructs. There are also design languages that have been created exclusively for the design task but still resemble full programming languages [Sastry and Klein]. Embedded languages have the advantage that there is already a body of knowledge and experience about the language in which the new constructs are embedded. Although the original language is usually imperative, the added design constructs can be declarative, with constraint statements that affect the circuit structure.

8.1.3   Hierarchy

Another issue that must be addressed in programmability is the use of hierarchy. Textual programming languages can be confusing because they may have a code hierarchy that is different from the design hierarchy. A clean design language will equate subroutines with cells so that each code routine generates exactly one layout routine. Of course, exceptions must be made for auxiliary code routines that have no corresponding cells.

The nature of information flow through the design hierarchy is a major issue that must be addressed when providing programmability. If the programming of each cell is dependent on its surrounding circuitry where it is instantiated higher in the hierarchy, then the programming works in a standard subroutine fashion: Parameters passed at the call (the cell instance) affect that particular version of the subroutine (the cell definition). Just as each different call to a subroutine results in a different execution sequence, so each different instance of a cell will yield a different layout at that point in the circuit. This facility, called **parameterized cells**, results in **top-down** programmability because information is passed down the design hierarchy to affect all lower levels. It is commonly found in textual design systems.

The opposite type of design programmability is **bottom-up**, in which a change to a cell definition affects all the instances and thus causes specification information to be passed up the hierarchy. In such a language, cells are not parameterizable and all instances of a cell are identical. This kind of hierarchy management is found in graphic systems and in those systems without programmability in which changes to a cell definition merely result in equivalent changes to every instance. Bottom-up schemes can be made programmable by having cell-definition changes react properly and make correct alterations to the circuitry surrounding their instances at higher levels of the hierarchy.

The rest of this chapter discusses the three issues of programmability introduced here: imperative programming, declarative programming, and hierarchy. Although many VLSI designers have been trained first as programmers, there will always be those who dislike programming and do not want to deal with it explicitly. Nevertheless, programming tasks exist to a certain degree in all design systems and most people find this programming very useful. The enhancement of programming provides powerful possibilities for productivity.

8.2   Imperative Programming

Imperative programs are the kind that most programmers know best: algorithms that execute sequentially from one statement to the next. Imperative languages are used for three classes of activity: database creation, database reformatting, and design-system control. The first activity is what is commonly thought of as programmed design: issuing statements such as `new-transistor(10, 50)` to place and connect components. Besides creating layout, these **hardware-description languages** can be used to produce different views of a circuit simply by switching primitive libraries [Holt and Sapiro]. The second activity done by imperative languages is the examination and reformatting of a database. These **formatting languages** are necessary because there are so many different circuit-description formats for tool interfacing and manufacturing. The third programmable activity in a design system is the execution and manipulation of design-system commands. These **command languages** can range in power from the simplest of control to the most complete of programmability.

To provide an imperative-programming capability the design system must support a language that can interact with the circuit. The first choice of a language is often the design system's own command interpreter that can already interact with the circuit. However, such a language is rarely powerful enough. An alternative is to start with a known language and extend it for design work. The final option is to create a new programming language, devised specifically for circuitry, that does not depend on the command interpreter for compatibility and can be tailored for the particular design needs.

8.2.1   Enhancing the Command Interface

One common way to get full programmability for all the necessary design activities is to enhance the command language of the design system. It has been shown that the two characteristics necessary for a programming language to be able to express any algorithm are a vast amount of memory and a conditional expression [Turing]. Thus all that is needed to convert a command interface into a programming language is the addition of variables and a conditional change of control. However, much more is needed to make such a language convenient to use.

The problem with Turing's programs is that they grow to unmanageable size and consume excessive storage. The reason for this is the lack of structure. In order to develop a complex program it is important to view that program hierarchically (so the lesson of VLSI design comes full circle and teaches about programming!). Hierarchy is just one way of structuring programming activity. It is found in subroutine or macro structures that share common code, and record or cell structures that share related data. Figure 8.3 lists other structuring

Structuring Method	Program Code	Program Data	Circuit Design
Hierarchy	Subroutines	Records	Cells
Iteration	Loops	Arrays	Arrays
Modularity	Parameters	Locals	Ports
Abstraction	Macros	Defines	Multiple views

FIGURE 8.3   Structuring methods.

methods and shows how they apply to programming and circuit design. More advanced structured-programming techniques demand specialized constructs to encourage more readable code and permit easier debugging. Clearly, a full set of language constructs provides the best environment for writing programs.

Unfortunately, many CAD systems add programmability as an afterthought. This means that the user interface contains two distinct sets of commands that do not fit consistently: design commands to create the data structures and programming commands to manipulate the design commands. For example, some early systems had variables that were totally separate from the design database. These "command-interpreter" variables were unable to interact with the actual design structure and could be used only to control other interactive commands for manipulating the design. In one system, searching the database could be done only by the use of repeated "select" statements, probing different physical locations with an imaginary cursor.

To improve this ad hoc programming interface, macro packages are often written to provide a third level of command interface that sits on top of, and replaces, the original commands. The result is that only "wizards," who have studied all of the details, can understand what is happening and maintain these packages. This is because the CAD system has defined a new and inconsistent programming language that only a few people can use.

Extending a known command interface is generally a bad idea because there are too many compatibility requirements. If a programming facility is desired, it should be planned with the language at the center and editing commands at the periphery.

8.2.2   Extending Conventional Languages

When programmability is built into the design interface from the start, the result is almost always the same: A known language grows into a hardware-description language. When extensions are added to existing programming languages, the general syntax is widely known and the extensions are easier to learn. Typically, the extensions will fit the style of the language so that they make sense syntactically.

Of course, even the use of known programming languages can suffer from database detachment if the languages are not well integrated into the design system. Two examples of languages that cannot access the database are ICPL [Computervision], a BASIC extension, and IAGL [Applicon], a PL/I extension. Although GPL [Calma] can access the database, its APL flavor and the particular database interface make programming difficult. Silicon Design Labs provides the L language, modeled after C, which can access the database to create layout [Buric and Matheson]. CAE Systems provides three languages, all modeled after Ada, each tailored for a different function [CAE]. One, the command language, is detached from the database; another, the hardware-description language, is integrated with the database for circuit creation; the third, the formatting language, is used to interface other tools.

Design languages can interact better if they are interpreted by the CAD system, rather than being compiled. This is obvious from the fact that interpreters are more interactive than compilers are. Interpretive programming environments can be found in LISP systems such as DPL [Batali and Hartheimer] and NS [Cherry *et al.*]. These systems are additionally convenient because a uniform language interface controls the editing process, database access, tool internals, and even the operating system.

When the design language is not normally interpreted, it can present difficulties to the user. There is a time cost that is incurred whenever changes are made and the code must be recompiled, and there is a quality cost that arises from the decreased ability to enforce structured subroutine use. The time cost can be lessened through the use of language systems that allow new object modules to be dynamically linked with the running program [Wilcox, Dageforde, and Jirak]. This enables hardware-description modules to be merged into the design system for immediate execution. Without this ability, the design language code will have to execute independently, communicating with the design system via common disk files.

8.2.3   Creating New Languages

The final option in selecting a design language is to invent one from scratch. In most textual languages that are designed to create graphic objects, there is a heavy use of declarative constructs to link the objects (see Fig. 1.4). However, the graphic objects can be linked in other ways that are more appropriate to imperative programming styles.

One textual imperative language that is able to express graphical relations does so functionally [Henderson]. The basic object is a **picture** that is manipulated with functions that create other pictures. Composition functions such as `beside` and `above` build up and pick apart the pictures, thus expressing graphical relationships. Other functions flip and rotate pictures, or create them with explicit data. This language is powerful enough to describe any geometry in an elegant style.

8.2.4   Imperative Code and Graphics

In a graphic design system, imperative programming can be used to control layout sequences so that the designer is freed from repetitious tasks. However, such code is rarely integrated with the circuit, but instead exists separately such that a change to either one will not affect the other. In most systems, for example, modified hardware-description code must be reexecuted to re-create the graphic layout, and a subsequent change to the layout will not be reflected in the code [Rosenberg and Weste; Batali and Hartheimer]. The Tpack system partially integrates code and graphics by allowing some specification to be done graphically [Mayo].

The SAM system is unusual in its ability to link code and graphics such that a change to either results in a change to both [Trimberger]. The two issues addressed by SAM are parameterized components and loops. When a component on the screen is described algorithmically, and the user modifies the algorithm, it is easy to recompute the graphics. However, if the user alters the graphics, it may not be easy to adjust the algorithm. Trimberger identifies three possible ways to keep the code consistent in such a situation: (1) replace the algorithm with a constant, (2) append an appropriate constant to the algorithm, or (3) rewrite the algorithm completely (see Fig. 8.4). The first choice is too destructive of code and the last choice is too hard to implement, so SAM uses the second method, which turns out to be what is needed in many circumstances.

Another issue in linking code with graphics is the handling of code loops that produce many components. When one such component is altered graphically, SAM chooses to alter every one that was produced

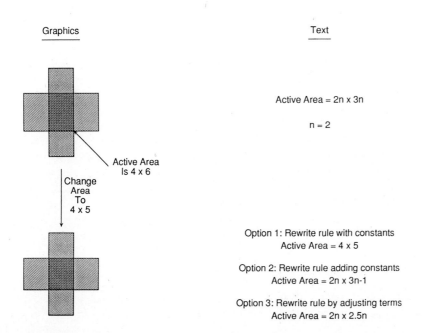

Graphics

Text

Active Area = 2n x 3n

n = 2

Active Area
Is 4 x 6

Change
Area
To
4 x 5

Option 1: Rewrite rule with constants
Active Area = 4 x 5

Option 2: Rewrite rule adding constants
Active Area = 2n x 3n-1

Option 3: Rewrite rule by adjusting terms
Active Area = 2n x 2.5n

FIGURE 8.4 Associating text and graphics.

by the loop, to preserve the code structure. This is not necessarily the right thing to do and shows that the text-to-graphics linkage is necessarily ad hoc. The SAM language is also not fully expressible and is admittedly a toy system for experimentation only. Its author tried to perform a very difficult task and the results show that much work is needed before the problem can be solved.

An elegant way to handle graphical code loops is found in the Escher system [Clarke and Feng]. Rather than having a graphic array of components declared by a textual loop, Escher allows the loop to be declared graphically. A separate "specification" cell has three components in which are placed the starting index, increment, and ending index of the loop (see Fig. 8.5). This cell is then converted into the actual array. Explicit placement of the first and last loop elements allows boundary conditions to be expressed if the structure needs to be different at the ends of the array. Escher also allows looping by recursion, wherein a specification cell contains an instance of itself, with an algorithmic modification of its called value. Although recursion is never allowed in an actual graphical design, it can appear in a graphic specification that, when executed, generates a standard, recursion-free circuit.

Of course, the most obvious way to combine text and graphics is to tie pieces of imperative code to parts of an interactive design. When this happens, however the code is merely an imperative procedure for implementing a constraint and is therefore part of a larger declarative program. Such programs are the subject of the next section.

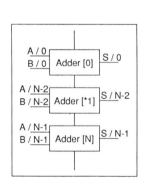

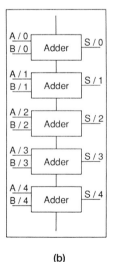

FIGURE 8.5 Graphical loop specification in Escher:
(a) Declared as Adder(N)
(b) Instantiated as Adder(4).

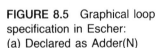

(a)   (b)

8.3   Declarative Programming

Declarative programming is the use of unsequenced statements that are performed in response to external stimuli. In circuit design, this style of programming usually appears as constraints that are attached to a design. There is no linear flow of control because the sequencing of these constraints is dependent on the structure of the circuit. Nevertheless, such code is powerful enough to perform any function. Many declarative languages are textual [Kingsley; Johnson; Lipton *et al.*; Davis and Clark], but these languages place their constraints on spatially located objects, thus functioning in a similar way to graphical languages.

Constraints can be used for a number of purposes. In layout systems, they commonly hold together pieces of the circuit so that the geometry will be correct. Textual constraint languages, which typically are evaluated in a batch mode, consider all constraints at once and produce a completed physical layout. Graphical constraint languages are more interactive, allowing continuous changes to the constraints and the geometry. Such systems demand incremental evaluation of declarative code so that, as changes are made, the constraints adjust appropriate components and wires. This is the situation in the Electric design system which has constraining wires that keep all components connected (see Chapter 11).

Although constraints are typically applied to geometry, they can also be used for other purposes. In VLSI design, many relevant pieces of circuit information can be coded as constraints. The Palladio system [Brown, Tong, and Foyster] has constraints for restricted interconnect so that components will be properly placed with respect to power, ground, input, output, clocks, restored signals, and so on. Other systems make use of timing constraints [Arnold; Borriello] or arbitrary code constraints that can respond to any need [Zippel].

In totally unrestricted constraint systems, users can write their own code to manipulate the circuit. These systems allow constraints on any value in the database so that code can be executed whenever necessary. This is implemented by having each attribute contain an optional code slot that executes when the attribute changes. Such systems can do arbitrary functions, such as hierarchical simulation [Saito, Uehara, and Kawato].

8.3.1   Constraint Sequencing

One difference between declarative and imperative programs is that there is no clear sequence of execution in a declarative constraint system. This means that the constraint-solving method is crucial to proper execution, and different solution methods will produce different results. Consider the four constraints in Fig. 8.6. If the constraint solver begins with object A, it has a choice of satisfying two constraints. Choosing the *above* constraint will place object B below object A. Now there is a choice between finishing the other constraint on A and proceeding with the constraint on B. If the solver chooses to go on to B, it will place object C to the right of B by three units. However, if it continues in this order, and places object D according to the constraint on C, it will be unable to satisfy the second constraint on object A because D will be too close.

This example illustrates many issues in constraint satisfaction. The basic problem in the example is that there is a **constraint loop**. Although this particular loop is easily solved in a number of ways, not all loops can be solved. In fact, **overconstrained** situations occur in which loops cannot be resolved because the constraints contradict each other. Detection of these loops requires some form of constraint marking to prevent infinite reevaluation. Once it detects these loops, the system must be prepared to deal with them in an appropriate manner. For VLSI design, in which each connection is a signal path, it is better to change the constraints than to change the topology.

The basic method of solving constraint loops, called **backtracking**, requires reversal of constraint execution until a point is reached at which an alternative choice will work. A stack must be kept with all the premises for each constraint decision so that it is possible to know how far to back up [Steele; Stallman and Sussman]. It may take many steps of backtracking to find the correct solution, and even simple backtracking may not be possible if unforeseen constraints continue to affect the search (see Fig. 8.7). Also, backtracking is time consuming.

In some situations it is possible to remove loops by precomputing their characteristics and reducing them to simpler constraints [Gosling].

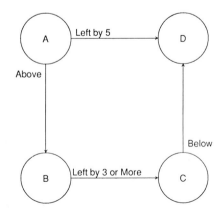

FIGURE 8.6 Example of constraint sequencing.

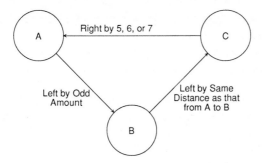

FIGURE 8.7 Simple backtracking fails for complex constraints. Solution steps: (1) solve constraint on A: set A left of B by 1; (2) solve constraint on B: set B left of C by 1; (3) solve constraint on C: CANNOT, must change A; (4) backtrack: remove constraint on B; (5) backtrack: remove constraint on A; (6) re-solve constraint on A with consideration for C: set A left of B by 5; (7) re-solve constraint on B: set B left of C by 5; (8) re-solve constraint on C: STILL CANNOT.

Such loop transformation involves the creation of a new constraint that acts like the old one but has the loop conflict resolved. This is done by algebraically manipulating the expressions that compose the original constraints to eliminate references to internal loop state and produce a set of equations that fully describe their characteristics.

As an alternative to computationally expensive solution methods, there are some simple techniques that can make constraints easier to solve. For example, it may be possible to prioritize the constraints according to their effects on other constraints, and to use this to obtain improved sequencing. In the example of Fig. 8.6, the constraint *left by 5* is more restrictive than is *below* and should be solved first to ensure consistency. Another feature that may help solve constraints is the ability to work them from any known point to the other, unknown points. This allows the solver to execute all constraints at any point, without having to work around a loop. In constraints that involve more than two objects, it should be possible to derive any value, given the others. If this is not possible, the user should be encouraged to add new constraints that are equivalent to the existing set but provide new opportunities for solution [Borning; Steele]. An alternative to adding new constraints is to have multiple views of a circuit with totally different ways of evaluating constraint systems [Sussman].

Another issue in sequencing constraints is the distinction between **breadth-first** and **depth-first** search. The example of Fig. 8.6 chose depth-first sequencing when it evaluated the constraint on object B before finishing the constraints on object A. Thus it went deeper into the constraint tree before finishing the constraints at the current level. Depth-first sequencing is used when the consequences of any one constraint must be fully known before other constraints can be evaluated. Breadth-first is used when the environment of a particular object is more important than that of others, and all constraints on that object must be evaluated first. It is not unreasonable to switch between the two types of search depending on the nature of the constraints found, using depth for important constraints and then finishing the less important ones by breadth. Care must be taken to ensure that all constraints get their turn.

8.3.2   Solution to Simple Geometric Constraints

As an example of implementing a constraint system, a simple geometric-constraint solver will be described. This particular technique, based on **linear inequalities**, is very popular and can be found in the textual design languages Plates [Sastry and Klein], Earl [Kingsley], ALI [Lipton *et al.*], LAVA [Mathews, Newkirk, and Eichenberger], SILT [Davis and Clark], I [Johnson], and many more. The constraint solver is essentially the same in all these systems because it allows only Manhattan constraints and solves those in *x* independently from those in *y*, thereby reducing the problem to one dimension. Also, these systems always restrict their constraints to simple inequalities between only two component coordinates. Thus the constraints *left by 3 or more* and *above by 5* can be expressed, but complex expressions such as *right by the same distance as that from A to B* cannot be handled.

With all of these restrictions, solution is simple: The constraints for a given axis are sorted by dependency, and values are substituted in order. If the constraints *A left of B by 5 or more* and *B left of C by 6* are given, then C is the least constrained and the first to be given a value. Subsequent values are set to be as small as possible, so that, once a position is established for C, B will be to C's left by 6 and A will be to C's left by 11. It is possible to rewrite all constraints in terms of C (as EARL does) or simply to solve the system by working from C to B and then to A. When two constraints appear for the same variable, the larger value is used to keep all constraints valid. So, for example, if the additional constraint *A left of C by 12* is given, then it will override the spacing of 11 from the other two constraints. Of course, constraints may conflict, which will cause unknown results. EARL can detect this situation because it rewrites the constraints before solving them, and then arrives at a nonsensical constraint of the form *A left of A by X*. Other systems will merely place objects incorrectly.

Although these constraints appear to be simple, they do form a powerful basis for complex operations. ALI, for example, has the primitive relationships *inclusion*, *minimum size*, *separation*, and *attachment*, which are all built on simpler inequalities [Lipton *et al.*]. It is also possible to use this kind of constraint system to compact a design to minimal design-rule spacing [Mathews, Newkirk, and Eichenberger; Hsueh and Pederson; Williams; Mosteller; Sastry and Klein]. All that is needed is to incorporate the design rules as implicit constraints between the edges of objects. Many of these systems also allow additional user constraints.

8.3.3 Parallel Sequencing

The preceding discussion presumes that constraints must be sequentially executed in order to be solved. Sequential execution techniques are called **propagation**. There are, however, a number of satisfaction techniques that work in parallel. The Juno system has a simple enough set of constraints that each one can be expressed algebraically and the entire collection can be placed into a system of simultaneous equations that is solved with Newton-Raphson iteration [Nelson].

More complex systems of constraints can be solved in parallel by using a method called **relaxation**. In this method, each constraint has an error function that indicates the degree to which the constraint is satisfied. Changes are made to every object in an attempt to achieve a minimum total constraint error. However, since it may not be possible to satisfy all constraints completely, the error functions may still be nonzero. Thus relaxation iterates between evaluating the error function and adjusting the constrained objects. Termination occurs when all error functions are zero or when the values cease to improve. As an example of this, the constraint system of Fig. 8.7 can be solved with the error and adjusting functions in Fig. 8.8. The very first graphical constraint system, Sketchpad, used relaxation to solve its constraints [Sutherland].

The problem with relaxation is that it is slow. This is because of its iteration and the fact that every object in the system must be checked, regardless of whether it is affected by constraint change.

Constraint	Error Function	Adjustment
A	([left amount] - 1) mod 2	Alternate: move B left by 1 / move B right by 1
B	| [left amount] - [A's left amount] |	Move C to 2B - A
C	min (| [right amount] - 6 |, | [right amount] - 5 |, | [right amount] - 7 |)	Alternate: move A to C-5 / move A to C-6 / move A to C-7

FIGURE 8.8 Relaxation solves complex constraints. Solution steps: (1) Initialize: A-to-B = 1 [error = 0], B-to-C = 1 [error = 0], C-to-A = 2 [error = 3]. (2) Adjust constraint C: move A left by 3. Now: A-to-B = 4 [error = 1], B-to-C = 1 [error = 3], C-to-A = 5 [error = 0]. (3) Adjust constraints A and B: move B left by 1; move C right by 1. Now: A-to-B = 3 [error = 0], B-to-C = 3 [error = 0], C-to-A = 6 [error = 0].

The ThingLab system attempts to improve response time by dynamically compiling the constraints [Borning]. Nevertheless, relaxation and all other parallel constraint-satisfaction methods suffer from the need to examine every object. Propagation methods work more efficiently because they do not need to spread to every object if the nature of the constraint application is localized to a small area of the design. This is important in VLSI design systems, which often have very large collections of objects to be constrained.

8.4   Hierarchy

The presence of hierarchy adds a new dimension to programmability. No longer can code simply scan from object to object; it must now consider subobjects and superobjects in the hierarchy. Two major issues are the equating of code hierarchies with design hierarchies, and the direction of information flow. The problem of equating hierarchical code with circuit structures is simply one of good design practice: Layout-generating programs will be easier to develop if their code structure is similar to the intended circuit structure. Both structures define a hierarchy, but the two may not necessarily correspond, which makes debugging difficult. Imperative programs are especially prone to this disparity since they have a sequential structure rather than a spatial feel.

A classic example of poor structure matching occurs in embedded languages that use subroutine calls to generate layout. Most subroutine libraries do not distinguish between the call that creates a new cell and the call that creates a new component in a cell. This means that a single code subroutine can create an arbitrary number of cells. Users of such a language should be encouraged to write more structured code that has a one-to-one correspondence with the generated design. Each code subroutine should begin with a cell-creation call followed by the placement of components in that cell.

The best way to foster structured use of imperative design languages is to restrict the nature of the code that can be written. Design languages that provide new statements have a better chance of guiding the designer in this way. For example, DPL has `defun` to define a code subroutine and `deflayout` to define a hierarchical circuit level [Batali and Hartheimer]. Only the `deflayout` subroutine can produce a cell, and it can create only one. Thus the hierarchical structure of the circuit is immediately obvious from the equivalent hierarchy in the code.

Relating code and circuitry in a graphic environment is a much less severe problem because there is only one hierarchy, defined by the circuit structure. However, it is still possible to create a separate hierarchy of graphic specifications for manipulating that circuit, and this opens up the difficulty of maintaining relations between the two. The CAD system must be able to display both specification and circuitry, and be able to identify related parts of each.

8.4.1 Hierarchical Information Passing

The main issue in hierarchical programmability is how to pass information among levels. Computer programs have subroutine parameters to pass information down the hierarchy and returned values to pass information back up. There are also global variables to communicate arbitrarily. Textual design languages make use of the same facilities, but graphic languages generally cannot. For graphic systems, it may be necessary to use textual descriptions to enable certain information to be communicated.

In practice, much of the commonly used information that is passed among hierarchical levels can be described graphically. A typical value that is passed across hierarchical levels is the location of connection points. The port location inside the cell defines the location of connecting wires on instances of the cell. Conversely, the location of wires on an instance can affect the port location in the cell definition. Systems typically choose between these two alternatives and thus define a direction of information flow in the hierarchy. Some systems, however, pass this port-alignment information in both directions across hierarchical levels in order to get the positions to line up [North; Kingsley]. These systems begin by building the cells from the bottom, declaring port positions up the hierarchy, and then pitch matching cells from the top, altering the port positions accordingly.

An important issue in design languages is the hierarchical order of execution. Do the top levels of hierarchy get evaluated first or last? In **bottom-up** programmability, information is passed up the hierarchy. Each cell definition is evaluated completely and the resulting circuit is used to guide higher levels that use instances of the cell. Every instance is identical and the static cell definitions are referred to as **database cells**. Database cells also occur in nonprogrammable environments, when the complete contents must be known before a higher-level floor-plan can be produced. The difference between nonprogrammability and bottom-up programmability is that changes made to nonprogrammable structures will not be noticed by the higher levels of the hierarchy and may therefore make the circuit inconsistent. Graphical systems tend to be bottom-up or nonprogrammable because each cell is designed explicitly and cannot automatically change.

In **top-down** programmability, the higher levels are evaluated first and the resulting information guides the lower-level cells that appear as instances. Top-down programmability is more like standard program execution because circuit creation begins at the outermost block and proceeds down the hierarchical organization. Cells can be parameterized, and individual instances can be differentiated according to their surroundings. The cell definitions in such an environment are called **procedural cells**. Textual design languages tend to be top-down because the order of execution runs with the text, so each instance can be created differently as the code is converted into a circuit.

It is difficult to implement a graphic language with top-down programmability because interactive use may overwhelm the system with cell variations. Noninteractive languages can take any amount of time and use much memory while generating a finished circuit, but interactive systems must respond quickly to parameter changes. If there are many different instances with diverse parameter values, it may be necessary to retain the cell definition in many different forms, which can potentially consume excessive amounts of storage. The alternative is to evaluate the definition of a parameterized cell whenever a reference is made to an instance, and this consumes excessive time. In either case, the time and space advantage of hierarchical representation is compromised. Although the parameterized cells in real circuits typically have only two or three different forms, the question of how they should be implemented still remains. In the Juno system, which is graphic and top-down, each cell instance is recomputed as it is drawn [Nelson]. However, this system is not used for circuit design so there is never very much on the screen. The NS system [Cherry *et al.*] provides top-down programmability by retaining all the different structural forms that instances may take.

Hierarchical programmability demands that the CAD system be efficient and convenient to use. This means that the use of differing hierarchies should be avoided so that the code structure can resemble the circuit structure. It also demands that there be an effective way of passing information across hierarchical levels so that large circuits can be correctly built.

8.5 Summary

Programmability is that extra control over the design process that allows any circuit to be produced. It extends the power of all tools and defines new, special-purpose ones. Given that hardware is often described algorithmically, a CAD system that has the ability to use those algorithms in design can consider all specifications.

Programs for design can exist in many forms, from traditional imperative text to purely graphical and declarative. Graphic programming is particularly attractive in an environment that is inherently spatial. Also, declarative programming is preferred because of its appeal to the nonprogrammers who typically do design. It simplifies issues of hierarchy and merges well with the overall design effort.

Questions

1 What is the danger of using an embedded language for hardware description?

2 Why do designers dislike programming?

3 How would you maintain an association between a conditional expression and its resulting graphics?

4 Why is a collection of linear inequalities a poor way to describe layout?

5 How would you eliminate unnecessary processing when solving constraints with parallel methods?

6 What is the most difficult aspect of implementing parameterized cells?

7 Why is there a need to link textual and graphical descriptions of a circuit?

References

Applicon, *IAGL User's Guide*, Applicon Incorporated, Burlington, Massachusetts, June 1983.

Arnold, John E., "The Knowledge-Based Test Assistant's Wave/Signal Editor: An Interface for the Management of Timing Constraints," Proceedings 2nd Conference on Artificial Intelligence Applications, 130-136, December 1985.

Batali, J. and Hartheimer, A., "The Design Procedure Language Manual," AI Memo 598, Massachusetts Institute of Technology, 1980.

Borning, Alan, "ThingLab—A Constraint-Oriented Simulation Laboratory," PhD dissertation, Stanford University, July 1979.

Borriello, Gaetano, "WAVES: A Digital Waveform Editor for the Design, Documentation, and Specification of Interfaces," unpublished document.

Brown, Harold; Tong, Christofer; and Foyster, Gordon, "Palladio: An Exploratory Environment for Circuit Design," *IEEE Computer*, 16:12, 41-56, December 1983.

Buric, Misha R. and Matheson, Thomas G., "Silicon Compilation Environments," Proceedings Custom Integrated Circuits Conference, 208-212, May 1985.

CAE Corporation, *CAE 2000 Command Language User's Manual*, August 1984.

Calma, *GPL II Programmers Reference Manual*, GE Calma Company, February 1981.

Cherry, James; Shrobe, Howard; Mayle, Neil; Baker, Clark; Minsky, Henry; Reti, Kalman; and Weste, Neil, "NS: An Integrated Symbolic Design System," *VLSI '85*, (Horbst, ed), 325-334, August 1985.

Clarke, Edmund and Feng, Yulin, "Escher—A Geometrical Layout System for Recursively Defined Circuits," Proceedings 23rd Design Automation Conference, 650-653, June 1986.

Computervision, *CADDS II/VLSI Integrated Circuit Programming Language User's Guide*, Computervision Corporation Document 001-00045, Bedford, Massachusetts, April 1986.

Davis, Tom, and Clark, Jim, "SILT: A VLSI Design Language," Stanford University Computer Systems Laboratory Technical Report 226, October 1982.

Gosling, James, *Algebraic Constraints*, PhD dissertation, Carnegie-Mellon University, CMU-CS-83-132, May 1983.

Henderson, Peter, "Functional Geometry," Proceedings ACM Symposium on LISP and Functional Programming, 179-187, August 1982.

Holt, Dan and Sapiro, Steve, "BOLT—A Block Oriented Design Specification Language," Proceedings 18th Design Automation Conference, 276-279, June 1981.

Hsueh, Min-Yu and Pederson, Donald O., "Computer-Aided Layout of LSI Circuit Building-Blocks," Proceedings International Symposium on Circuits and Systems, 474-477, July 1979.

Johnson, Stephen C., "Hierarchical Design Validation Based on Rectangles," Proceedings MIT Conference on Advanced Research in VLSI (Penfield, ed), 97-100, January 1982.

Kingsley, C., *Earl: An Integrated Circuit Design Language*, Masters Thesis, California Institute of Technology, June 1982.

Lipton, Richard J.; North, Stephen C.; Sedgewick, Robert; Valdes, Jacobo; and Vijayan, Gopalakrishnan, "ALI: A Procedural Language to Describe VLSI Layouts," Proceedings 19th Design Automation Conference, 467-473, June 1982.

Mathews, Robert; Newkirk, John; and Eichenberger, Peter, "A Target Language for Silicon Compilers," Proceedings 24th IEEE Computer Society International Conference, 349-353, February 1982.

Mayo, Robert N., "Combining Graphics and Procedures in a VLSI Layout Tool: The Tpack System," University of California at Berkeley Computer Science Division technical report, January 1984.

Mosteller, R. C., "REST—A Leaf Cell Design System," *VLSI '81* (Gray, ed), Academic Press, London, 163-172, August 1981.

Nelson, Greg, "Juno, a constraint-based graphics system," *Computer Graphics*, 19:3, 235-243, July 1985.

North, Stephen C., "Molding Clay: A Manual for the Clay Layout Language," Princeton University Department of Electrical Engineering and Computer Science, VLSI Memo #3, July 1893.

Rosenberg, Jonathan B. and Weste, Neil H. E., "ABCD—A Better Circuit Description," Microelectronics Center of North Carolina Technical Report 4983-01, February 1983.

Saito, Takao; Uehara, Takao; and Kawato, Nobuaki, "A CAD System For Logic Design Based on Frames and Demons," Proceedings 18th Design Automation Conference, 451-456, June 1981.

Sastry, S. and Klein, S., "PLATES: A Metric Free VLSI Layout Language," Proceedings MIT Conference on Advanced Research in VLSI (Penfield, ed), 165-169, January 1982.

Stallman, R.M. and Sussman, G.J., "Forward Reasoning and Dependency Directed Backtracking in a System for Computer-Aided Circuit Analysis," *Artificial Intelligence*, 9:2, 135-196, October 1977.

Steele, G. L. Jr., *The Definition and Implementation of a Computer Programming Language Based on Constraints*, PhD dissertation, Massachusetts Institute of Technology, August 1980.

Sussman, Gerald Jay, "SLICES—At the Boundary between Analysis and Synthesis," AI Memo 433, Massachusetts Institute of Technology, 1977.

Sutherland, Ivan E., *Sketchpad: A Man-Machine Graphical Communication System*, PhD dissertation, Massachusetts Institute of Technology, January 1963.

Trimberger, Stephen, "Combining Graphics and A Layout Language in a Single Interactive System," Proceedings 18th Design Automation Conference, 234-239, June 1981.

Turing, A. M., "Computing Machinery and Intelligence," *Mind*, 59:236, 433-460, October 1950.

Weste, Neil, "Virtual Grid Symbolic Layout," Proceedings 18th Design Automation Conference, 225-233, June 1981.

Wilcox, C. R.; Dageforde, M. L.; and Jirak, G. A., *Mainsail Language Manual*, Version 4.0, Xidak, 1979.

Williams, John D., "STICKS—A graphical compiler for high level LSI design," Proceedings AFIPS Conference 47, 289-295, June 1978.

Zippel, Richard, "An Expert System for VLSI Design," Proceedings IEEE International Symposium on Circuits and Systems, 191-193, May 1983.

GRAPHICS

9.1 Introduction

Graphics, although often overlooked, is a significant part of many programming efforts. For the CAD system builder there is no escaping graphics because even design systems that use purely textual specification must also plot their circuits. Without graphics, computers would be exceedingly boring and difficult to use.

The two important factors in graphics are speed and clarity. Drawing speed can make or break an entire system because it is the first attribute that a user encounters. Systems that display slowly are ignored immediately, regardless of their other salient features. Conversely, there are systems that do very little, but display so quickly that users feel good about them. An interactive CAD system must be able to display large parts of complex circuits without losing the user's attention, and produce hardcopy plots without tying up the system for long periods of time. Extended delays will cause frustration, encourage distraction, and cut productivity.

Display clarity is a broad issue that includes ease of perusal, correctness, and even proper sequence of drawing. For the overall content of an image to be understandable, the individual parts must be distinguishable. Proper color and texture are important, as is informative and uncluttered labeling. The correctness of a display is more important than most people think because the human eye can detect one incorrect point on an otherwise perfect screen [Catmull]. If the displayed artifact is not as good or better than the actual part, the designer will fear for the circuit's quality. Also, the order of drawing should make some sense on interactive displays so that the designer can quickly tell what is going on. This means that redraw should cover current objects first and that all of an object, or at least a distinguishing portion of it, should be drawn before the system moves on to draw another object.

In this everchanging world, there is little chance that the display device used today will still be there tomorrow. New hardware is constantly appearing that can perform complex graphic functions with ease. For this reason, an important attribute of graphics programming is modularity. There should be no special-purpose calls to the display system, and all graphics control should be centrally located in the code. Any graphics functions that are unusual and not portable should be coded both in their efficient, display-specific way and in an alternative way for systems without the feature. The best approach is to use a well-known virtual-graphics protocol such as GKS [X3H3], Core [GPSC], or PHIGS [ANSI]. By doing this, you are better assured that control of new displays will fit into the existing scheme, because these protocols are well planned. You may also find that the protocol has been implemented directly on some displays, which makes porting very simple.

This chapter discusses techniques for two-dimensional graphics display and hardcopy plotting. In addition, some of the basics of graphic input will be mentioned. The chapter is primarily concerned with actual graphics functions; proper use of these functions will not be emphasized here. Chapter 10, Human Engineering, will show how to employ these techniques effectively. Also, this chapter is focused primarily on graphics techniques that are relevant to VLSI design. For a more thorough coverage, including such subjects as three-dimensional transformations, hidden-surface removal, and light-source shading, the reader is encouraged to consult texts devoted to graphics [Foley and Van Dam; Newman and Sproull]. These texts contain more than enough information for VLSI design system construction, and interested programmers will want to be familiar with them because of their general usefulness.

9.2 Display Graphics

When a designer uses an interactive display to edit a circuit, that display is connected to the main computer via a **frame buffer**. A frame buffer is simply a section of memory that the computer fills with an image. Connected to the frame buffer is a **display processor** that converts the image memory into a video signal. There may also be a **graphics accelerator** under the control of the main computer for rapid manipulation of the frame-buffer memory. There are many different types of frame buffers, display processors, graphics accelerators, and displays. Simple frame buffers are **iconic**, containing an equal amount of memory for each **pixel**, or point on the screen. These iconic frame buffers are merely arrays with no other structure.

More advanced frame buffers use complex **display lists**, which encode the image in a form similar to that of a computer instruction set. The amount of memory required to hold such an image varies with the complexity of the image, but it is usually less than would be required by an iconic frame buffer. Although using the display processor for display lists is more complicated and expensive, it allows the main computer to manipulate complex images more rapidly because less frame-buffer memory has to be modified to make a change.

Another distinction is whether the display is **raster** or **calligraphic**. Raster displays, which are the most common, have electron guns that sweep the entire surface of the screen in a fixed order. Calligraphic displays, sometimes called **vector** displays, are increasingly rare; they have electron guns that move randomly across the screen to draw whatever is requested. Needless to say, calligraphic displays never use iconic frame buffers but instead are driven from display lists. The calligraphic-display list processor is simpler than the raster-display processor, however, because it can usually send drawing instructions directly to the electron guns. When memory was expensive, this tradeoff of more complex display hardware for less memory made sense; today the most expensive component is the calligraphic display tube, so these are typically faked with display processors that drive iconic frame buffers connected to less expensive raster displays. A final problem with calligraphic displays is that they have difficulty filling a solid area on the screen, which makes them unpopular for IC design.

Once the type of display is understood, the quality must be determined. The three main factors in image quality are the availability of color, the number of pixels on the screen, and the number of bits for each pixel. A first and obvious choice is between black-and-white (sometimes called **monochrome**) and color displays. For IC design, color is very important because there are many layers that must be distinguished. Schematics designers have less of a need for color but can still use it to reduce the clutter of complex designs.

The next factor in display quality is its resolution in pixels per inch. The **spatial resolution**, the number of points on the screen, is usually lower for raster displays than for calligraphic displays. Raster displays typically have about 500 pixels across and down. Less expensive displays, such as are found on personal computers, have half that number, and high-resolution displays, which are becoming more common, have twice that number. Calligraphic displays can typically position the electron gun at 4000 or more discrete points across the screen.

The number of bits available for each pixel determines the **intensity resolution**, which is just as important as the spatial resolution. Many black-and-white displays have only two intensity levels, on or off, making them **bilevel** displays. They can therefore store a complete pixel in one bit of memory. Simple color displays have eight colors determined by the bilevel combinations of red, green, and blue display guns. In order to have shading, however, more bits are needed to describe these three primary colors. For viewing by humans, no more than 8 to 10 bits of intensity are needed, which becomes 24 to 30 bits in color. Thus a high-resolution color display can consume over 3 megabytes of memory. For VLSI design, however, only 8 or 9 bits per pixel are needed; these can be evenly distributed among the color guns or mapped to a random selection. **Color mapping** translates the single value associated with a pixel into three color values for the display (see Fig. 9.1). In a typical map, 9 bits of pixel data address

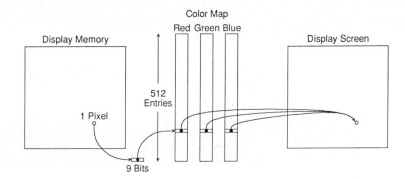

FIGURE 9.1 Color mapping. One 9-bit pixel addresses a full color value in the map.

three tables, each 512 entries long. The individual entries can be 8 or more bits of intensity, allowing a full gamut of colors to be produced, provided no more than 512 different colors are displayed at any time.

Ideally, programming a frame buffer should be fairly simple, with the graphics accelerator doing all the difficult work of setting individual pixels. When such is the case, the controlling program needs to supply only vector endpoints, polygon vertices, and text strings to have these objects drawn. However, in less ideal situations, the CAD programmer may have to code these algorithms. The next section describes typical features of graphics accelerators and the following sections discuss the software issues.

9.2.1 Hardware Functions

There are a few operations that the frame buffer can typically do with or without a graphics accelerator. One feature that is often available is the ability to set the background color. On color-mapped displays, this simply means setting the map value for entry 0 of the table. On nonmapped displays, the entire surface must be filled with the desired background color and it must also be used whenever an object is erased.

One of the most useful features of many raster frame buffers is the **write mask**. This is a value that selects a subset of the bit-planes to use when changing the value of a pixel. For example, if the value 511 is written to a pixel in a 9-bit frame buffer and the write mask has the value 257, then only the top and bottom bits will be set and the others will be left as they were. Given this facility and a color map, one can assign a different bit-plane to each layer of a VLSI design and modify each plane independently of the others by masking them out (see Fig. 9.2). The drawing or erasing of layers takes place exclusively in the dedicated bit-plane without regard for the graphical

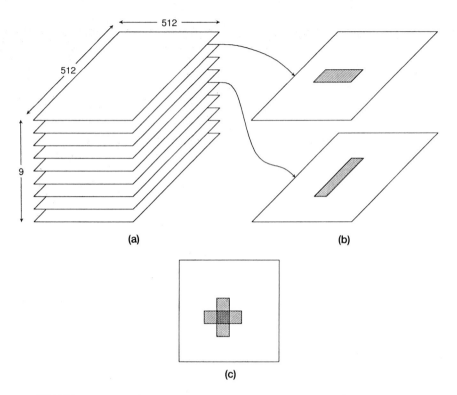

FIGURE 9.2 Individual bit-plane control allows display memory to be organized by layer: (a) Display memory (b) The contents of two sample bit planes (c) The resulting display.

interactions occurring with other layers in other bit-planes. With proper selection of color-map entries, each layer or bit-plane can be a different color and all combinations of layer overlap can be sensibly shaded.

For systems that have graphics accelerators, a number of hardware functions become available. One very common feature that is provided on raster displays is the ability to move a block of data on the screen quickly. This operation, called **block transfer** (or **bitblt** or **BLT**), is useful in many graphic applications. Block transfer can be used to fill areas of the screen and to move objects rapidly. Complex graphic operations can be done with the block transfer because it typically provides a general-purpose arithmetic and logical engine to manipulate the data during transfer. So, for example, an image can be moved slowly across the display without obscuring the data behind it. This is done by XORing the image with the display contents at each change of location, and then XORing it again to erase it before advancing. The block-transfer operation is usually fast enough that it can be used to pan an image, shifting the entire screen in one direction and filling in from an undisplayed buffer elsewhere in memory. Such panning presumes that an overall iconic image, larger than what is displayed, is available somewhere else in memory. Of course, all this data moving is necessary only on raster displays, because display-list frame buffers can pan without the need for block transfer or large image memories.

Another feature possible with the block transfer operation is the rapid display of hierarchical structures. When the contents of a cell is drawn, the location of that cell's image on the screen can be saved so that, if another instance of that cell appears elsewhere, it can be copied. This **display caching** works correctly for those IC design environments in which all instances of a cell have the same contents and size. It suffers whenever a fill pattern is used on the interior of polygons because the phase of the pattern may not line up when shifted to another location and placed against a similar pattern. A solution to this problem is to draw an outline around each cell instance so that patterns terminate cleanly at the cell border [Newell and Fitzpatrick].

Of course, there are many special-purpose graphics accelerators in existence, due particularly to the advances in VLSI systems capabilities. Video terminals have accelerators with built-in typefaces. Modern raster displays have accelerators that draw lines, curves, and polygons. Very advanced accelerators, beyond the scope of this chapter, handle three-dimensional objects with hidden-surface elimination. Control of these devices is no different than the software interface described in the following sections.

9.2.2 Line Drawing

Lines to be drawn are always presented as a pair of endpoints, (x_1, y_1) and (x_2, y_2), which define a line segment on the display. A simple line-drawing algorithm divides the intervals $[x_1\ x_2]$ and $[y_1\ y_2]$ into equal parts and steps through them, placing pixels. This is problematical in many respects, including low speed and line-quality degradation. The problem of speed is that the interval and increment determination requires complicated mathematics such as division (especially painful when done in hardware). The line quality is also poor for such a technique because the choice of interval may cause bunching of pixels or gaps in the line. If the interval is chosen to be 1 pixel on the longest axis, then the line will be visually correct but still hard to compute.

The solution to drawing lines without complex arithmetic is to use scaled-integer values for the determination of pixel increments. The most popular technique also does all initial calculations in terms of these scaled integers [Breshenham 65]. The algorithm makes use of an indicator that initially has the value $D = 2\ \Delta y\ -\ \Delta x$ (where

$\Delta y = y_2 - y_1$ and $\Delta x = x_2 - x_1$), and a current point (x, y), which initially is (x_1, y_1). After plotting this point, x is incremented by 1, and y increments by 1 only if $D > 0$. If y increments, then D has the value $2(\Delta y - \Delta x)$ added to it; otherwise, it has the value $2\Delta y$ added to it. This continues until (x, y) reaches (x_2, y_2).

The Breshenham algorithm is very fast because the terms can be precomputed and all of them require only addition and subtraction. The version described is, of course, for lines with a slope of less than one because x increments smoothly. For other lines, the coordinate terms are simply reversed.

Many CAD systems need textured lines to distinguish different objects. These can easily be drawn by storing a bit pattern that is rotated with the Breshenham counter, suppressing pixels that are not in the pattern. Thick lines can also be drawn by offsetting the startpoints and endpoints so that multiple parallel lines are drawn to fill in a region. However, unless these lines are very close together, there will be gaps left when the pixels of adjoining lines fail to fall on every position. This phenomenon is called a **möire** pattern and can occur wherever discrete approximations are made to geometry (see Fig. 9.3). Thus for very thick lines it is better to use a filled polygon.

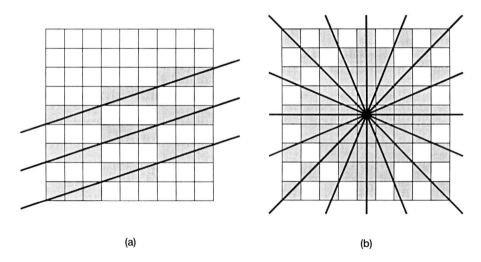

(a)   (b)

FIGURE 9.3 Möire patterns: (a) Möire patterns in three parallel lines (b) Möire patterns in radiating lines.

9.2.3   Polygon Drawing

Polygons come in three styles: **opened**, **closed**, and **filled**. The first two
are simply aggregates of lines but the third type is a separate problem
in graphics. A filled polygon is trivially rendered if it is rectangular
and has edges parallel to the axes. Since such polygons are common
in VLSI design, special tests are worth making. Also, there have been
some graphics accelerators that, when built into frame buffers, allow
these rectangles to be drawn instantly. One system has a processor
at each pixel that sets itself if its coordinate matches the broadcast
polygon boundary [Locanthi]. Another system redesigns standard
memory organization to allow parallel setting of rectangular areas
[Whelan]. Rectangle filling is important because it is a basic operation
that can be used to speed up line drawing, complex polygon filling,
and more.

The problem of filling arbitrary polygons is somewhat more difficult.
As with line drawing, there are many methods that produce varying
quality results. One method is the **parity-scan** technique that sweeps
the bounding area of the polygon, looking for edge pixels. The premise
of this algorithm is that each edge pixel reverses the state of being
inside or outside the polygon. Once inside, the scanner fills all pixels
until it reaches the other edge. There is an obvious problem with this
method: Quantization of edges into pixels may leave an incorrect
number of pixels set on an edge, causing the parity to get reversed.
Attempting to use the original line definitions will only make the
mathematics more complex, and there may still be vertices of one
point that confuse the parity. A variant on this technique, called **edge
flag** [Ackland and Weste], uses a temporary frame buffer and com-
plements the pixels immediately to the right of the polygon edges.
This method works correctly because double complementing, which
occurs at vertices, clears the edge bit and prevents runaway parity
filling. When all edges have been entered into the temporary frame
buffer, parity scan can be used to fill the polygon correctly.

Another polygon-fill method that has problems is **seed fill**. Here,
a pixel on the interior is grown in a recursive manner until it reaches
the edges. As each pixel is examined, it is filled, and any neighboring
pixels that have not been filled are queued for subsequent examination.
This method can use large amounts of memory in stacking pixels to
be filled and it also may miss tight corners (see Fig. 9.4). Additionally,
there is the problem of finding an interior point, which is far more
complex than simply choosing the center. To be sure that a point is
inside a polygon, a line constructed between it and the outside must
intersect the polygon edges an odd number of times. This requires a
nontrivial amount of computation because the process is expensive

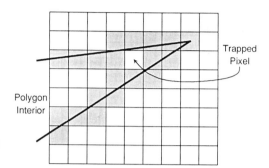

FIGURE 9.4 Polygon seed fill misses pixels in corners.

and still does not suggest a way of finding an interior point, only of checking it. Techniques for finding the point include (1) starting at the center, (2) advancing toward vertices, and (3) scanning every point in the bounding rectangle of the polygon.

The best polygon-fill algorithm is dependent on the available hardware. If special-purpose hardware can be constructed, polygons can be filled almost instantly. For example, there is a hardware frame buffer that does a parallel set of all pixels on one side of a line [Fuchs *et al*.]. This can be used to fill convex polygons by logically combining interior state from each polygon edge.

The hardware for setting pixels on one side of a line must compute the equation

$$Ax + By + C = 0$$

for each pixel. The values for Ax and By are broadcast down the rows and across the columns to a simple add-compare unit at each pixel (see Fig. 9.5). This pixel processor sums the values and compares the result with a constant $(-C)$. Computation of Ax and By occurs rapidly with tree-structured summation units on the sides. All arithmetic is done in bit-serial to keep interconnect down to one wire and to reduce the complexity of each pixel processor.

When there is no fancy hardware for edge filling, an alternative is to analyze the polygon and to reduce it to a simpler object. For example, convex polygons can be decomposed into trapezoids by slicing them at the vertices. The result is a left- and right-edge list that, when scanned from top to bottom, gives starting and ending coordinates to be filled on a horizontal line (see Fig. 9.6). At each horizontal scan line there is only one active edge on each side and

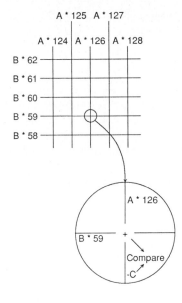

FIGURE 9.5 Parallel computation of line drawing.

these edges can be traversed with Breshenham line drawing. Pixels between the edges can be filled with size $n \times 1$ rectangle operations.

Nonconvex polygons must first be decomposed into simpler ones. This requires the addition of cutting lines at points of inflection so that the polygon breaks into multiple simpler parts. Alternatively, it can be done using the trapezoid method described previously with a

Trapezoid	Left Edge	Right Edge
1	D	A
2	D	B
3	C	B

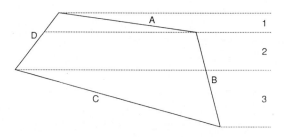

FIGURE 9.6 Horizontal slices decompose a polygon into trapezoids. Left- and right-edge lists delimit each trapezoid.

little extra bookkeeping for multiple fill pieces on each scan line [Newell and Sequin].

9.2.4   Clipping

Clipping is the process of removing unwanted graphics. It is used in many tasks, including complex polygon decomposition, display-window bounding, and construction of nonredundant database geometry. Two clipping algorithms will be discussed here: polygon clipping and screen clipping.

In **polygon clipping**, the basic operation is to slice an arbitrary polygon with a line and to produce two new polygons on either side of that line. Figure 9.7 shows how this operation can be used to determine the intersection of two polygons. Each line of polygon A is used to clip polygon B, throwing out that portion on the outside of the line. When all lines of polygon A have finished, the remaining piece of polygon B is the intersection area.

The polygon-clipping algorithm to be discussed here runs through the points on the polygon and tests them against the equation of the clipping line [Catmull]. As mentioned before, a line equation is of the form:

$$Ax + By + C = 0$$

and defines a set of points (x, y) that reside on the line because they satisfy the equation. All points that are not on the line will cause the right side of the equation to be nonzero, with the sign as an indication of the particular side.

The clipper takes each point in the polygon and uses the equation to place it in one of two output lists that define the two new split polygons. Three rules determine what will happen to points from the original polygon as they are sequentially transformed into the two clipped polygons. These rules use information about the current point and the previous point in the original polygon (see Fig. 9.8):

1 If the current point is on the same side as the previous point, then place it in the output list associated with that side. This happens to points B, D, and E in the figure.

2 If the current point is on the other side of the previous point, then (a) find the intersection of that edge and the clipping line, (b) place the intersection point in both lists, and (c) place the current point on the list associated with its side. This happens to point A in the figure (the last point processed).

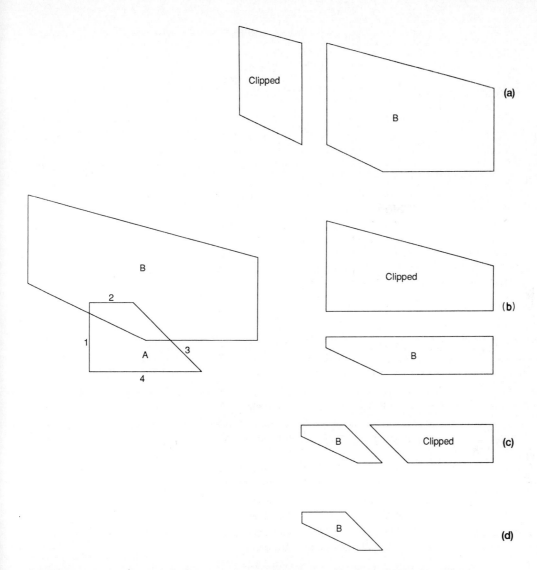

FIGURE 9.7 Polygon intersection done with clipping: (a) Polygon B clipped against line 1 (b) Polygon B clipped against line 2 (c) Polygon B clipped against line 3 (d) Polygon B clipped against line 4.

3 If the current point is on the clipping line, then place it in both lists. This happens to point C in the figure.

This clipping operation runs through the input polygon only once and produces two clipped output polygons in linear time.

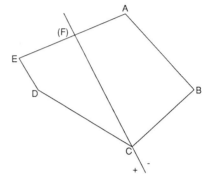

+ Polygon	- Polygon
C	B
D	C
E	F
F	A

FIGURE 9.8 Clipping a polygon against a line. Two split polygon lists are constructed as the original points are sequentially examined.

There are other clipping methods that apply to different needs, the most common being screen clipping. In **screen clipping**, vectors must be clipped to a rectangular window so that they do not run off the edge of the display area. Although this can be achieved with four passes of polygon clipping, there is a much cheaper way to go. The **Sutherland-Cohen clipping** algorithm [Newman and Sproull] starts by dividing space into nine pieces according to the sides of the window boundaries (see Fig. 9.9). A 4-bit code exists in each piece and has a 1 bit set for each of the four window boundaries that have been crossed. The central piece, with the code 0000, is the actual display

FIGURE 9.9 Space codes for the Sutherland-Cohen clipping algorithm. Each bit position is associated with one edge of the central window.

1 0 1 0	0 0 1 0	0 1 1 0
1 0 0 0	←Window 0 0 0 0	0 1 0 0
1 0 0 1	0 0 0 1	0 1 0 1

area to which lines are clipped. Computing the code value for each line endpoint is trivial and the line is clearly visible if both ends have the code 0000. Other lines are clearly invisible if the bits of their two endpoints are nonzero when ANDed together. Lines that are not immediately classified by their endpoint codes require a simple division to push an endpoint to one edge of the window. This may have to be done twice to get the vector completely within the display window.

9.2.5 Text

Display of text is quite simple but can be complicated by many options such as scale, orientation, and readable placement. The basic operation copies a predesigned image of a letter to the display. Each collection of these letters is a **typeface**, or **font**, and there may be many fonts available in different styles, sizes, and orientations. Although these variations of the letter form may be simple parameters to algorithms that can be used to construct the font [Knuth], it is easier and much faster to have fonts precomputed.

When placing text on top of drawings it is important to ensure the readability of the letters. If it is possible to display the text in a different color than the background drawing, legibility is greatly enhanced. An alternative to changing the text color is to use **drop shadows**, which are black borders around the text (actually only on the bottom and right sides) that distinguish the letters in busy backgrounds. A more severe distinction method is to erase a rectangle of background before writing the text. This can also be achieved by copying the entire letter array instead of just those pixels that compose the letter.

Of course, proper placement of text gives best results. For example, when text is attached to an object, it is more readable if the letters begin near the object and proceed away. Text for lines should be placed above or below the line. Excessive text should be truncated or scaled down so that it does not interfere with other text. Text becomes excessive when its size is greater than the object to which it attaches. All text should be optional and, in fact, the display system should allow any parts of the image to be switched on and off for clutter reduction. As anyone who has ever looked at a map can attest, text placement is a difficult task that can consume vast resources in a vain attempt to attain perfection. A CAD system should not attempt to solve this problem by itself, but should instead provide enough placement options that the user can make the display readable.

9.2.6 Circular Curves

There are many curved symbols that need to be drawn on the screen. The simplest is the circle, which can be drawn with the same method as a line: **Breshenham's algorithm** [Breshenham 77]. The technique is to draw only one octant (the upper-right, from twelve o'clock to one-thirty) and mirror it eight times to produce a circle. For the purpose of algorithm description, the circle is centered at (0, 0) so that for a radius R the octant that is drawn is from (0, R) to (R/$\sqrt{2}$, R/$\sqrt{2}$). For each point (x, y) in this octant, the seven other points $(-x, y)$, $(x, -y)$, $(-x, -y)$, (y, x), $(-y, x)$, $(y, -x)$, and $(-y, -x)$ are also drawn.

The Breshenham method proceeds as follows: Initially start at $x = 0$, $y = R$, and $D = 3 - 2R$. At each step, increment x and conditionally decrement y if $D \geq 0$. When y is decremented, add $4(x - y) + 10$ to D and when y is not decremented, add $4x + 6$ to D. Proceed until $x \geq y$ at which point the octant is drawn.

Filled circles work the same way except that for each pair of octants at the same horizontal level, an $n \times 1$ rectangle is drawn between the points. Thus for point (x, y) on the octant, fill from $(-x, y)$ to (x, y), from $(-x, -y)$ to $(x, -y)$, from $(-y, x)$ to (y, x), and from $(-y, -x)$ to $(y, -x)$.

Arcs of circles are difficult to draw because they can start or end anywhere. For the Breshenham method to be used, the number of full and partial octants must be determined and carefully tracked to know where to draw. Also, the Breshenham code must be able to end in mid-octant or to run backwards as needed by the arc. Arcs can be better drawn if they are first converted to vectors. There are many published sets of algorithms for vector approximation of curved shapes [Rogers and Adams].

9.2.7 Spline Curves

Noncircular curves can be drawn in any number of ways, but the most common technique is the **spline**. The spline is a smooth curve that follows a set of **control points**. Unlike other curve methods, the spline can handle any number of points and produce a curve that is continuous in both the first and second derivatives. The disadvantage of the spline is that the curve does not necessarily pass through the control points, only near them. Other curve methods such as **Hermite** and **Bezier** [Bezier] have the opposite properties: They do pass through the control points but do not extend as cleanly to large numbers of points.

The **B-spline** (not to be confused with the β-spline [Barsky and Beatty]) is defined by two functions: $X(t)$ and $Y(t)$. As the value of t goes from 0 to 1, these functions describe the curve. It is necessary to break the control points into groups of four where each group defines a different equation for $X(t)$ and $Y(t)$. After the curve for points 1,2,3,4 is drawn, the points are advanced by one and the curve for points 2,3,4,5 is drawn. This continues until points $n - 3$, $n - 2$, $n - 1$, n are drawn at which time the entire spline has been rendered.

Formally, $X(t) = T M G_x$ and $Y(t) = T M G_y$ where T, M, G_x, and G_y are matrices. The matrix T is a simple shorthand for quadratic expansion:

$$T = [t^3 \ t^2 \ t \ 1]$$

The matrices G_x and G_y are also shorthand to define the four control points of interest:

$$G_x = \begin{bmatrix} X_i \\ X_{i+1} \\ X_{i+2} \\ X_{i+3} \end{bmatrix} \qquad G_y = \begin{bmatrix} Y_i \\ Y_{i+1} \\ Y_{i+2} \\ Y_{i+3} \end{bmatrix}$$

These matrices are the only ones that change as the spline advances through its points. Finally, the matrix M is called the **blending matrix** because it combines the current control points into a quadratic equation:

$$M = \frac{1}{6} \begin{bmatrix} -1 & 3 & -3 & 1 \\ 3 & -6 & 3 & 0 \\ -3 & 0 & 3 & 0 \\ 1 & 4 & 1 & 0 \end{bmatrix}$$

Putting these matrices together produces the following functions:

$$X(t) = \frac{X_i(-t^3 + 3t^2 - 3t + 1) + X_{i+1}(3t^3 - 6t^2 + 4) + X_{i+2}(-3t^3 + 3t^2 + 3t + 1) + X_{i+3}t^3}{6}$$

$$Y(t) = \frac{Y_i(-t^3 + 3t^2 - 3t + 1) + Y_{i+1}(3t^3 - 6t^2 + 4) + Y_{i+2}(-3t^3 + 3t^2 + 3t + 1) + Y_{i+3}t^3}{6}$$

Frame buffer programming can be a bottomless pit of effort because the user is never satisfied and always wants more. However, there is a standard tradeoff of speed and functionality that must be balanced, and the CAD system with too many cute display options may begin to be too slow for comfort. Users become accustomed to anything in time, so the display should do its best and leave it there.

9.2.8   Transformation

Transformations are used to modify coordinates prior to rendering them, so that the figures change position, size, or orientation. Although some displays have built-in transformation facilities, programmers may have to do these operations explicitly. Position transformations are trivially implemented with addition, and scale transformations simply require multiplication. The only difficult transformation is rotation.

To rotate a point (x, y) by Θ degrees about the origin, compute the point:

$$x' = x \cos \Theta$$
$$y' = y \sin \Theta$$

Since sine and cosine always return values between -1 and $+1$, rotation can be done using fixed-point integers and table lookup. For example, the sine and cosine values can be stored in 16-bit integers with 14 of the bits to the right of the decimal point. The two bits to the left allow values of $+1$ and -1 (one significant bit and one sign bit). In this scheme, the value 1.0 is represented as 4000_{16}, the value 0.5 is 2000_{16}, and the value -1.0 is $C000_{16}$. Multiplication of these values simply requires a shift of 14 bits when done. Also, sine and cosine values can be stored in a one-quadrant table that is small (only 90 entries for 1-degree accuracy) and easily indexed. This is a common trick used to avoid floating-point operations.

9.3   Hardcopy Graphics

Every CAD system needs to provide hardcopy versions of circuits. Whether on paper or film, the permanent image is useful for offline consideration and for documentation. Although hardcopy graphics techniques are essentially the same as those found in display programming, there are other considerations. As a first step, it is necessary to understand the nature of these plotting devices.

Like displays, plotters fall into two broad categories: raster and calligraphic. Raster devices such as laser and ink-jet printers cover every pixel on the page and can produce arbitrarily complex images. Calligraphic devices such as pen plotters and some film recorders have a single writing head that must be moved to draw.

Unlike displays, the intensity resolution of plotters is generally low. Most hardcopy devices are strictly bilevel, able only to plot a black dot or to leave the spot white. Color printers have additional sources of ink but can still place the inks with only one intensity. Pen plotters can sometimes fake intensity by drawing over an area multiple times, but there is a limit to what the paper can endure, so these plotters still have only a few intensity levels. Essentially, the variations of layer appearance must be faked with patterned areas.

9.3.1 Patterned Fill

To give the appearance of varying shades of gray, a pattern of black and white dots can be printed. When precise intensity levels are needed, a process called **dithering** is used; it computes the proper number and placement of these dots [Jarvis, Judice, and Ninke]. In the plotting of VLSI circuits, however, it is sufficient to have a set of precomputed patterns that are associated with the different layers.

Selection of patterns should be done in such a way that the overlap of multiple layers is distinguishable. This overlap is generally an ORing of the bits to produce the union of black dots in all patterns. Therefore the individual patterns should be kept sparse so that overlaps do not become too dense. Since there are IC processes with over a dozen layers, and since as many as half of them may overlap at a point, it is best to be generous in the use of white space in a pattern.

Another feature of patterns is their **gestalt**, or subimage, which consists of recognizable symbols in the pattern that help with identification. In Fig. 9.10, the top two patterns are composed of subelements that remain distinct when the pattern is tiled over an area. Users can see these elements on the plot and identify the layers. The bottom patterns have no gestalt since they are designed to abut smoothly, generating a uniform field with a particular density.

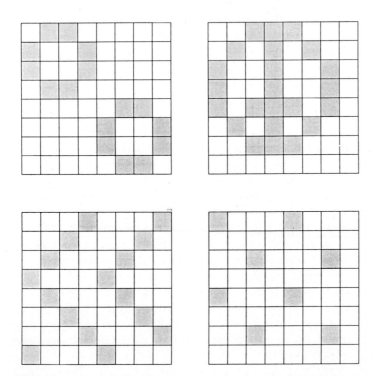

FIGURE 9.10 Stipple patterns.

Pattern sizes do not need to be very large; they typically range from 8 to 32 bits on a side. The advantage to using a power of two for the size is that, for any coordinate on the page, the pattern element to use can be found by masking the low bits of the x or y value and indexing the pattern array. This has the side effect that the pattern is always positioned independent of the polygon boundary, so that patterns in abutting polygons will merge correctly.

9.3.2   Display Consistency

Another issue in hardcopy graphics is the correct duplication of the display when making a plot. Inconsistency between hardcopy and display is a problem that is caused by differences in graphics hardware. The three main culprits are pixel size, text drawing, and color. The ratio of pixel height to pixel width may not be the same on any two devices and compensation should be made to keep the scale right. Also, text that is placed on the screen so that it fits properly may have a different size on the hardcopy, which lessens readability. Finally, color accuracy is very difficult in light of the different methods used to show color. These problems must be solved to produce correct displays.

The CAD programmer should be aware that many displays do not have square pixels. The **aspect ratio** of the screen can often be adjusted to squeeze or to stretch either axis and make the image square. If this adjustment is not possible, the software must scale the coordinates so that the layout looks right. Hardcopy devices are more difficult to adjust and may have rectangular pixels as a permanent feature. If this is the case, then these devices must also be scaled in software. The important thing is to keep the image in step with the manufactured artifact, which always has uniform x-to-y ratios.

Less significant but still disturbing to the designer is the difference in appearance between text on a display and text on a plot. Typically, the hardcopy is destined to be published, presented, or otherwise preserved and must look its best. The designer may spend much time adjusting the text on the display for maximum readability only to discover that the hardcopy version looks very different.

The problem with text is that most displays have a limited number of font sizes but a continuum of graphics scales. This means that as the circuit zooms in or out, the text will stay the same size. Only PostScript systems allow a true continuum of raster text scales, and they are typically found only on hardcopy devices [Adobe]. On calligraphic displays, text is represented with vectors and can scale continuously, but these displays are rarely used today. An alternative is to describe all text calligraphically and convert it to pixels before plotting or displaying. Although this will scale properly, it will look bad because vector characters must be hand cleaned when converted to raster.

The simplest solution is to choose a range of text scales on each device that correspond in size. This will keep the text right to a small margin of error, and it will look good on each device. Remember also that some fonts are variable width and so their x axis must be adjusted to get the correct scale. A consistent hardcopy can save designers much time at that critical point: when the design is done and the details begin.

9.3.3 Color Inversion

A significant difference between color displays and color plots is the nature of color combinations. A display tube emits light and therefore, when multiple colors combine, they are added. The three-color display primaries—red, green, and blue—are called the **additive color space** because their absence is black and their sum is white.

On paper, color is reversed because ink does not emit but rather absorbs light. This means that more ink darkens rather than brightens, so two colors inked over each other will subtract. Paper plots therefore use **subtractive color space**, in which the primaries are cyan, magenta, and yellow, the absence of which is white and the presence of which is black. The color wheel of Fig. 9.11 shows how both sets of primaries relate: Green and blue on a display will yield cyan; yellow and magenta on paper will yield red. Notice that these two spaces are inverses in a common space so, to convert colors, these simple formulas are used:

$$Red = 1 - Cyan$$
$$Green = 1 - Magenta$$
$$Blue = 1 - Yellow$$

assuming that all values range from zero to one.

Although these formula appear to be simple solutions to the problem of hardcopy and display color consistency, much more is necessary. Displays can typically use a full range of intensities in each primary color, whereas hardcopy printing is all or nothing. When a dot of ink hits the paper, it fills that pixel with the full intensity of its color. Thus to achieve a range of intensity, it is necessary to paint patterns that have the proper proportion of on and off color. Precomputed patterns do not work as well as proper dithering techniques in this case.

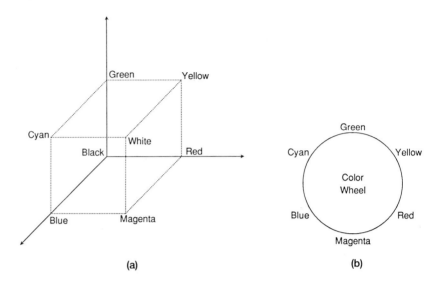

FIGURE 9.11   Color space: (a) A three-dimensional volume (b) A wheel.

Another hitch in these color-conversion equations is the need to use a pure black ink that is available on some plotters. The sad truth is that when cyan, magenta, and yellow ink overlap, the result is not a black but a dark brown. Since black is desired in this situation, many paper plotters have a fourth ink that is really black. Thus the color conversion must have special code to detect the simultaneous use of all three primaries and replace them with black ink. It is only this **four-color-printing** process that can reproduce an acceptable range of images.

9.3.4   Pen Motion

When a calligraphic hardcopy device is being driven, the motion of the writing head may be of concern. The obvious example is the pen plotter that takes time to move the pen from one point to another. The programmer must ensure minimal pen motion both within and between the graphic objects.

Reducing pen motion within an object simply requires specifying lines in sequence. Thus a square that is 100 on a side and centered at (50,50) should be specified as the lines (0,0) to (0,100); (0,100) to (100,100); (100,100) to (100,0); and (100,0) to (0,0). Although this may seem obvious, it is easy to make the simple mistake of reversing the "from" and "to" coordinates when specifying the lines, which will double the time needed to draw the square.

Reducing pen motion between objects is a much more difficult problem that cannot be solved perfectly. After finishing one object, it is best to draw another that is close. Although this random walk through the image will reduce pen motion in the early stages of plotting, it may cause long jumps near the end when there are only a few scattered objects that remain unplotted. Coherence of drawing is also important because the designer may be watching the plot and will be confused if some objects are skipped for nonobvious reasons. Thus a linear order, imposed as a weighting factor on proximity, produces an image that is drawn more or less from top to bottom.

Pen motion should also be considered when calligraphic techniques are used to fill an area. The optimal motion is a back-and-forth raster scan of the area that leaves parallel lines for a fill pattern. For variation, the lines can be spaced differently, run at different angles, or even make a cross-hatched area.

Another consideration in plotting time is pen changing. This usually takes more time than any drawing operation and occasionally requires motion so that the writing head can get to where the pens are stored. Thus it is best to plot all of one color before moving to the next pen.

9.4 Input Devices

The final aspect of graphics covered in this chapter is the input of information to the system. Most graphic-input devices require no programming, so their interfaces are quite simple. Two-dimensional devices such as tablets, joysticks, and mice provide two-dimensional values; One-dimensional devices such as knobs and thumbwheels provide single values. The only possible software operations are scaling and relative-motion tracking. This section will therefore concentrate on the physical workings of various graphic-input devices to provide a feel for their use.

9.4.1 Valuators

One-dimensional input devices, such as knobs and thumbwheels, are called **valuators**. Typically, **knobs** are aggregated in large sets, each one having a specific function associated with it. **Thumbwheels** are large-diameter knobs turned on their side and sunk into a panel so that the exposed edge can be rolled like a sliding surface. Typically there are two thumbwheels placed at right angles so that x- and y-axis specification can be done.

Valuators can have fixed ranges of rotation or they can roll infinitely. If there are stops at the ends of the rotation range, then the programmer will be given absolute coordinate values from the device. If the valuator turns without stops, however, each turn will provide a relative-motion value that must be accumulated to determine the overall value. Relative-motion devices are more flexible because they can give a greater range of values, can be scaled for fine tuning, and can even be treated nonlinearly. A popular use of nonlinear input handling is **timed acceleration**, in which rapid changes scale more than slow motion does. This enables a great distance to be covered with a quick hand motion.

9.4.2 Tablets

Early graphic-input devices used a special pen and a special surface that would track the pen position. The first tablets were called **spark pens** because the pen emitted clicking sounds by generating sparks. The tablet surface had rows of microphones along its edges and it triangulated the pen position from the speed of sound delay between the pen and the microphones. These devices are rarely used today because of the advent of the **Rand tablet**. In the Rand tablet, a mesh of wire runs under the surface and the pen acts as an antenna for signals that are sent through the wires. Rand tablets are less expensive and quieter than spark pens.

Tablets typically have one or more buttons on the pointing device. When an actual pen is used there is a switch in the tip that can be activated by pressing down on the tablet surface. There may also be a **puck** that has a set of buttons on it (as many as 16). These devices can be programmed to do separate actions on the downstroke of a button, while it is down, when it is released, and while it is up. The coordinate values are, of course, absolute within the range of the tablet surface.

Tablet positions are displayed on the screen by drawing a **cursor** at the current position. This cursor is a symbol such as an X, an arrow, or **crosshairs**, which consist of two perpendicular lines drawn from one edge of the screen to another. Although some displays handle cursor tracking in hardware, many require saving, setting, and restoring of the image at each advance of the cursor. For the cursor to be visible over any part of the screen, it must be a unique color. This color can simply be the inverse of the value that is normally there so that each pixel of the cursor will stand out. Beware, however, of color maps that use the same color in complementary positions; a more suitable cursor-color algorithm may be necessary. Proper cursor tracking of the tablet will free the user from the need to look at the tablet, because the hand-eye coordination keeps the user's attention on the screen.

9.4.3   The Screen Surface

An obvious solution to the problem of having a special tablet surface is to use the display screen as a surface. The **light pen** does just that by having a photocell in the tip of the pen that is synchronized with the screen refresh. When the pen is touched to a point on the screen, the display determines its position because the pen signals when it detects the sweep of the electron gun. On calligraphic displays that cannot guarantee a gun hit at every point, a cursor is drawn under the pen so that motion in a particular direction signals the cursor to move that way, keeping some graphics under the pen at all times.

But why use a pen? The **touch screen** allows a finger to point on the display. Acoustic methods can be used to triangulate the finger position. Of course the finger is the least sensitive pointing device, so design cannot get very far with such input. In general, however, it is uncomfortable to use screen pointing for long periods, so few light pens or touch screens are found in CAD systems.

9.4.4   Mice, Tracker Balls, and Joysticks

The **mouse** is the most common two-dimensional-input device today because it is inexpensive, compact, and easy to use. There is no special surface; only a puck with a large ball bearing in it that sticks out of the bottom. The puck is rolled over any surface and sensors detect motion on the bearing. There are also buttons on top of the puck, giving it the same power as a tablet but without the cumbersome surface.

In actuality, the mouse must be rolled over a special surface so that it can get proper traction. Also, these surfaces get dirty and the dirt collects inside the bearing. To solve this, the **optical mouse** has no moving parts but instead tracks patterns of light and dark on a textured surface [Lyon]. The surfaces used with mice can be smaller than tablet surfaces because the devices use relative motion, so when the user hits the edge he or she need merely pick up the mouse and drop it back in the center for more range.

The **tracker ball** is simply an upside-down mouse (actually, tracker balls came first so the converse is more correct). Placing the ball bearing on top and having the user run a hand over it allows the same control to be achieved. Although no surface is needed, the buttons are harder to hold onto when the hand is moving, and there is no optical version to prevent the problem of dirt accumulation. Tracker balls are rare today.

The **joystick** is a relative input device consisting of a handle that can be moved in two axes. When the user pulls or pushes the handle, y control is achieved; when he or she moves it left or right, x is controlled. Some joysticks spring back to a center position and others must be returned manually. Since the center is a fixed concept, the joystick is typically used for velocity control rather than position information. It may also have buttons on the handle.

9.4.5 Unusual Input Devices

A novel concept in computer input is the notion of a **feedback input device** [Noll; Atkinson *et al.*]. When motors are included along with the sensors, the computer can not only sense position but also change it. Obviously, this works only with devices that can be physically moved by hand, such as joysticks and mechanically tracked tablets. The motors can be programmed to push back with an equal force so that the user "feels" a solid barrier. Such an effect could be used to stop design-rule violations, for example, and in other situations in which the CAD system can guide or restrict the design process.

An ever-popular notion in computer design is the possibility of using speech input. There are even CAD systems today that use a headset microphone to take commands such as "place," "rotate," and so on. Although this looks sexy to the nonuser, a few hours spent "speaking" a circuit will convince anyone that there must be better ways. Speech input is best used for only infrequent conversations, and in those situations in which manual input is not possible. For example, during the processing of integrated-circuit wafers, an engineer may have both hands in a glove box and still need to direct a computer. This is the place for speech input.

9.5 Summary

Graphics is a necessary part of any CAD effort. Although most of the display techniques covered in this chapter are automatically handled by ever more sophisticated devices, it is important to be able to fill in when a needed function is missing. It is also useful for the designer to understand the basics of graphic input and output hardware, so that he or she can make proper packaging decisions when a complete CAD system is being specified. The next chapter builds on this one by showing how advanced graphics techniques can make the user interface comfortable and productive.

Questions

1 Why are calligraphic displays a poor choice for VLSI design?

2 How can block-fill operations speed the drawing of lines?

3 How can clipping be used to find the union of two polygons?

4 How can the block-transfer function be used to write patterned areas?

5 What other color representations exist beside the additive (red, green, blue) and the subtractive (cyan, magenta, yellow)?

6 Why is the mouse the most popular graphic-input device?

7 How can a spline curve be extended so that it passes through its first and last control points?

References

Ackland, Bryan and Weste, Neil, "Realtime Animation Playback on a Frame Store Display System," *Computer Graphics*, 14:3, 182-188, August 1980.

Adobe Systems Incorporated, *PostScript Language Tutorial and Cookbook*, Addison-Wesley, Reading, Massachusetts, 1985.

ANSI, *Programmer's Hierarchical Interactive Graphics Standard (PHIGS)*, American National Standards Institute X3H3/84-40, February 1984.

Atkinson, William D.; Bond, Karen E.; Tribble, Guy L.; and Wilson, Kent R., "Computing with Feeling," *Computers and Graphics*, 2:2, 97-103, 1977.

Barsky, Brian A. and Beatty, John C., "Local Control of Bias and Tension in Beta-splines," *Computer Graphics*, 17:3, 193-218, July 1983.

Bezier, P, *Numerical Control—Mathematics and Applications*, (A. R. Forest, trans), Wiley, London, 1972.

Breshenham, J. E., "Algorithm for Computer Control of Digital Plotter," *IBM Systems Journal*, 4:1, 25-30, 1965.

Breshenham, J. E., "A Linear Algorithm for Incremental Digital Display of Circular Arcs," *CACM*, 20:2, 100-106, February 1977.

Catmull, Edwin, "A Hidden Surface Algorithm with Anti-Aliasing," *Computer Graphics*, 12:3, 6-11, August 1978.

Foley, J. D. and Van Dam, A., *Fundamentals of Interactive Computer Graphics*, Addison-Wesley, Reading, Massachusetts, 1982.

Fuchs, Henry; Poulton, John; Paeth, Alan; and Bell, Alan, "Developing Pixel-Planes, A Smart Memory-Based Raster Graphics System," Proceedings MIT Conference on Advanced Research in VLSI (Penfield, ed), 137-146, January 1982.

GPSC, "Status Report of the Graphic Standards Planning Committee," *Computer Graphics*, 13:3, August 1979.

Jarvis, J. F.; Judice, C. N.; and Ninke, W. H., "A Survey of Techniques for the Image Display of Continuous Tone Pictures on Bilevel Displays," *Computer Graphics and Image Processing*, 5:1, 13-40, March 1976.

Knuth, Donald E., *TEX and METAFONT—New Directions in Typesetting*, Digital Press, Bedford, Massachusetts, 1979.

Locanthi, Bart, "Object Oriented Raster Displays," Proceedings 1st Caltech Conference on VLSI (Seitz, ed), 215-225, January 1979.

Lyon, Richard F., "The Optical Mouse, and an Architectural Methodology for Smart Digital Sensors," Proceedings C-MU Conference on VLSI Systems and Computations (Kung, Sproull, and Steele, eds), 1-19, October 1981.

Newell, Martin E. and Fitzpatrick, Daniel T., "Exploiting Structure in Integrated Circuit Design Analysis," Proceedings MIT Conference on Advanced Research in VLSI (Penfield, ed), 84-92, January 1982.

Newell, Martin E. and Sequin, Carlo H., "The Inside Story on Self-Intersecting Polygons," *Lambda*, 1:2, 20-24, 2nd Quarter 1980.

Newman, William M. and Sproull, Robert F., *Principles of Interactive Computer Graphics*, 2nd Edition, McGraw-Hill, New York, 1979.

Noll, A. Michael, "Man-Machine Tactile Communication," *Society for Information Display Journal*, 1:2, 5-11, July/August 1972.

Rogers, David F. and Adams, J. Alan, *Mathematical Elements for Computer Graphics*, McGraw-Hill, New York, 1976.

Whelan, Daniel S., "A Rectangular Area Filling Display System Architecture," *Computer Graphics*, 16:3, 147-153, July 1982.

X3H3/83-25r3 Technical Committee, "Graphical Kernel System," *Computer Graphics* special issue, February 1984.

HUMAN ENGINEERING

10.1   Introduction

The success of a system is often more dependent on its ease of use than on its power. Good systems not only address the user's needs, but also provide rapid and considerate interaction. This chapter discusses **human engineering**, the tailoring of systems so they can be used to maximal advantage by all humans. Included in this effort is the development of a good **user interface**, the part of the system that communicates with the human. In addition, there must be an underlying integrity to the system so that new and unusual needs can be properly met.

This chapter discusses four areas that compose human engineering: task and user modeling, information display, command language, and feedback. **Task modeling** is the aspect of human engineering that runs deepest in the code since it demands a view of the task that is consistent with the user's actions and perception. This also includes a **user model** to predict accurately those actions and perceptions. The **information display** must be designed properly to prevent confusion, and is especially significant in graphics editors. Good **command languages** are necessary since they are the control path from the user and can create the largest bottlenecks. Finally, informative **feedback** keeps users happy and prevents confusion on the part of newcomers.

Human engineering must be built into systems from the start and cannot be tacked on top like some kind of macro package. Although this chapter is the last in a string of CAD system discussions, its subject matter should be considered first. The theme, however, will come as no surprise because many of the preceding chapters have echoed this cry for better thought-out systems that can be used well.

10.2   Task and User Modeling

The prerequisite for proper human engineering is a deep understanding of the user's particular goals and operating methodologies. Also useful is a general understanding of the way humans interact with machines. When programmers are unaware of these issues, they make assumptions that are intuitive to them in the hope that all will work right. The result is often inconsistent both at the user-interface level and internally in the code. Attempts to rectify the user interface are fraught with difficulties because the code becomes complicated, hard to maintain, and full of bugs.

When the task has been modeled correctly, the other aspects of human engineering proceed easily. Command-language syntax becomes obvious because it maps directly from the task to the program internals. The same is true of the information display and the nature of the feedback. Also, a truly consistent task model agrees with the users' notions of what is difficult and what is simple, reducing the frustration of using time-consuming operations.

Thus the first step in building a VLSI design system is developing an understanding of electronics and design so that a sensible model can be made. Chapter 1 discussed the circuit-design process and Chapter 3 showed some representations for VLSI CAD: topological, geometrical, and hierarchical data structures. Every synthesis and analysis tool needs a combination of these data structures to work best.

10.2.1 Timings for Humans

The next step in building any user interface is developing an understanding of the way humans interact with machines. This interaction has only recently been analyzed and some quantitative results are now available [Card, Moran, and Newell]. Essentially, the human performs three different operations when interacting with a computer: perception, cognition, and motion. A **perception** is the operation of observing the screen to obtain some information. Experiments have shown that a single perception action takes from 50 to 200 milliseconds, averaging 100 milliseconds or 0.1 seconds. This means, for example, that any display action such as motion that occurs at least 10 times per second will appear continuous to an observer. The **cognition** operation is the step that humans make when deciding how to react to a perception. After a cognition step, which averages 70 milliseconds, the human may do another perception or, more likely, a motion. **Motion** actions include striking a key or moving the pointing device. Basic motion actions take only 70 milliseconds, but this presumes that the user's hand is already on the device. If the user's hands are incorrectly placed and must switch devices, the delay is 400 milliseconds. Thus, while the mouse or joystick may be faster to use, the setup time is so high that for short operations preceded and followed by keyboard activity, the keyboard is still the preferred device.

Card, Moran, and Newell also describe larger operations that relate more directly to the use of a design system. Typing, which combines perception, cognition, and motion, averages 0.2 seconds per key. Moving the cursor to a spot on the display takes 1.1 seconds. Mental preparation for a new command or new part of a command takes 1.35 seconds. These figures are certainly simplifications, but they allow design functions to be evaluated for their relative efficiencies.

These timings can be used to evaluate the relative merits of different human-interaction methods. For example, the "replace" command substitutes a new object in place of the currently selected object. This command can be issued in the traditional way by typing the string `rep new-object` or it can be issued with menu picks. The menu method starts by selecting the "replace" entry and then selecting the new object from an object menu (which may actually have to be two menu hits to select a class of objects and then an object in that class). The typing method requires four steps: type the command keyword (`rep`), terminate the keyword (a space), type the new object name (average of 10 characters), and terminate the command (a carriage-return). Each of these operations requires a mental operation to initiate it followed by a number of keystroke operations (see Fig. 10.1). The menu-pick method requires three button pushes to select the command, the new object class, and the new object. Each pick involves a mental operation, a cursor movement, and a button push. The figure shows that this is slightly faster than the keyboard method, but many factors

Typing: `rep new-object`

Action	Operation	Time
Choose command	Mental	1.35
Type "rep"	3(key)	3×0.2
Terminate keyword	Mental	1.35
Type space	Key	0.2
Choose replacement object	Mental	1.35
Type object name	10(key)	10×0.2
Terminate command	Mental	1.35
Type carriage-return	Key	0.2
Total		8.6 seconds

Pointing: "replace," "new object class," and "new object"

Action	Operation	Time
Choose command	Mental	1.35
Point to "replace"	Cursor	1.1
Push button	Button	0.2
Choose new object class	Mental	1.35
Point to "class"	Cursor	1.1
Push button	Button	0.2
Choose new object	Mental	1.35
Point to "object"	Cursor	1.1
Push button	Button	0.2
Total		7.95 seconds

FIGURE 10.1 Comparative timings of keyboard and mouse commands.

could alter this, such as the need for an additional menu hit, delays of machine response time, or the need to move hands between the keyboard and the mouse.

10.2.2   Terminology

Another aspect of proper task modeling is the correct use of terminology. Programmers should avoid making up new words to describe a situation and should instead find out what the right jargon is. This will make the code more readable and also will keep internal error messages from being totally obtuse. The user will then be able to help in those unusual circumstances in which otherwise hidden parts of the system suddenly surface.

In addition to using correct terminology, the programmer must consider the best way to keep the terms clear in the user's mind. The problem of terminology confusion is called **interference** and it occurs in two forms: acoustic and semantic [Card, Moran, and Newell]. **Acoustic interference** is caused by having two words sound alike. This problem occurs more often in short-term memory, when new words are presented for immediate use. For example, a library with the cell names "hit," "fit," and "kit" may make sense to the user when their functions are first introduced, but the acoustic interference between them will damage the user's ability to recall the names a short while later.

Semantic interference is caused by having two words or concepts share their meaning. This problem appears in long-term memory, where word meanings are learned. For example, teaching a second computer language causes semantic interference that damages the ability to use the first computer language.

There is a very simple test of the quality of a task model. Look at the sequence of commands for a complex operation and decide whether they are comfortable and intuitive or are merely a recipe of mindless commands. Avoid having to tell the user to "issue these commands without worrying about what they mean" because users do worry about the meaning of what they are typing. Recipes are bad and encourage ignorance on the part of users and maintainers alike. When a correct task model is provided, the need for complex and confusing sequences is greatly reduced.

10.3   Information Display

The first thing that a user notices about a VLSI design system is its display. If that user is going to continue to look at the display, the viewing position must be comfortable. Card, Moran, and Newell report that the most comfortable vision is 15 degrees down from the plane of the head, and that the head is most comfortable when tilted 20 degrees down from the horizon. Therefore the display should be 35 degrees down from the eyes.

The availability of color displays is generally considered to be an advantage. However, there are certain graphics that gain no advantage by being colored. For example, using different colors in different areas of the display does not improve user response time. The only positive value of color in such situations is that it makes users happier because they find their work area to be more cheerful [Varner]. Beware, also, of color displays that cannot focus as well as black and white displays: Make sure the color screen looks clear and crisp when the system is drawing noncolored objects.

The organization of the display contents is also very important to a VLSI designer. There are typically four areas on the display: editing windows with circuitry, menus of command options, permanent status information, and temporary messages. To avoid confusion and provide ease of use, these parts of the display should be uncluttered, distinguishable, and sensibly placed (see Fig. 10.2). A well-designed **floor-plan** effectively presents this information to the user.

When the system uses only one screen, that surface must contain all the displays mentioned. Systems that have multiple screens can divide the display areas sensibly, providing more room for each. In two-screen systems, there is typically a graphics display that has editing windows and menus, and a video terminal that can handle messages (see Fig. 10.3).

In some systems, the user is given the option of rearranging the size and location of windows on the screen. In the absence of such a facility, however, the floor-plan must be determined in advance by

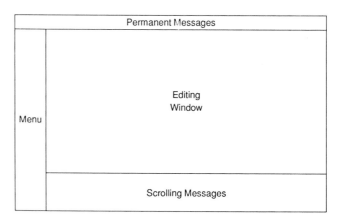

FIGURE 10.2 Typical information display layout.

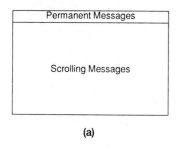

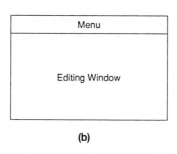

FIGURE 10.3   Typical two-screen information display layout: (a) Text screen (b) Graphics screen.

the programmer. Consider the problem of excessive eye movement when designing the layout; for example, place menus of commands near the message areas that indicate command response. Also reduce hand motion on the pointing device by keeping menus and editing windows close. When two screen systems are used, the relative place-ment of the monitors should be known in advance so that a sensible layout can be made.

Each display area should be distinct so that the user knows where to look. Therefore, do not let the different parts of the display overlap; for example, do not print messages over the editing window. Make sure that each section of the screen takes an appropriate area, and keep the number of different sections low, combining closely related functions into one.

10.3.1   Editing Windows

The editing window that displays circuitry should be the largest part of the screen. Although many CAD systems allow only one window, the ideal circuit display is divisible into multiple windows for parallel viewing of different aspects of the design. This allows the user to see different levels of the hierarchy, different views of the same circuit, different pieces of the same cell, or different display scales. Windows should be able to show any kind of design information including layout, schematics, waveforms, and text.

Although complete flexibility of editing windows is an ideal goal, it can be expensive to implement, and tempting to ignore. Rather than stop at a one-window system, however, there is a middle ground: a few special-purpose windows can give much advantage. For example, the Icarus system [Fairbairn and Rowson] uses one editing window and a smaller world-view window. Although changes cannot be made in the world-view window, it shows the entire circuit currently being edited and allows the user to select easily subareas for display in the main window. This same facility appears in text windows that have **scroll-bar** subwindows on the side. The bar shows the location of the current window within the file and allows direct access to other portions of the text.

One decision that must be made in window layout is whether multiple windows should overlap or abut. Overlapping windows are a popular notion because they simulate a desk surface with arbitrarily placed pieces of paper. Unfortunately, the use of overlapped windows means that there will always be some that are partially occluded and this can be annoying. Also, the code complexity and time required to draw in an occluded window is high because simple clipping methods such as the Sutherland-Cohen algorithm [Newman and Sproull] cannot be used.

The alternative is to divide the screen into nonoverlapping windows that abut and fill the display [Gosling]. The user will not have to waste time rearranging the order of overlap, the graphics will be easier to program, the system will run faster, and the user will still be able to see an arbitrary number of different things at once.

10.3.2   Menus

Menus are lists of options that appear on the screen. By moving the cursor of the pointing device over a menu entry and pressing a button, the user invokes that option. Early menus were simply lists of words in blocks; the words were abbreviated commands and the blocks delimited the area that would be recognized during menu selection. This basic formula has been updated in many respects.

The most important improvement to menus is the use of **icons** (sometimes called **glyphs**, or graphical symbols), instead of text in the menu boxes. The problem with words is that they are ambiguous and evoke semantic interference in the minds of each designer. It often happens that the precise meaning of textual menu entries is forgotten between sessions and must be relearned. Icons, on the other hand, are unique in their appearance and are easily distinguishable in the menu. Experiments have shown that viewers recognize previously seen pictorial information far better than they do previously seen words [Haber] and so the learning of icons is faster.

One concern in human engineering is the dual role of the cursor as a pointer in the editing window and as a selector of commands in the menu. How can the user point to a place in the circuit and then point to a corresponding menu entry without losing the circuit position? The obvious answer is to have the command be on a button of the pointing device so that the position in the circuit is retained. Another solution is to have two cursors, one that points in the editing window and a second that appears in the menu to choose a command. When a button is pushed over the desired piece of circuitry, the first cursor freezes and a second cursor appears in the menu area to roam through it while the button is held down. When the designer releases the button, the menu action under the second cursor is performed, that cursor disappears, and the original cursor in the editing window is reactivated. This reduces hand motion from the editing window to the menu area and clearly shows the command and its intended subject.

A more popular solution to the problem of effective menu selection is to have **pop-up menus** that appear on top of the editing window exactly where the cursor is located at the time (see Fig. 10.4). These menus occlude the circuit only while a command is being selected, and then they are removed. They do not take up permanent screen area, which allows more room for the circuit display. Perhaps the best feature of pop-up menus is that they can implement defaults by drawing the preferred entry directly under the cursor. The user does not need to move the cursor at all to select this option; he or she merely pushes buttons on the pointing device.

Pop-up menus can contain arbitrarily complex command options by being hierarchical. This is accomplished by placing option classes in the initial menu so that selection of an entry causes another pop-up menu to appear with suboptions. Each lower-level pop-up menu is drawn offset from the previous one, enabling the user to see the progression of options that have been selected. The notion of hierarchical menus can also be found in non-pop-up environments such as **marching menus** [Applicon] that appear from left to right across the bottom of the screen below the editing window (see Fig. 10.5).

Hierarchy is not the end-all in command languages and the programmer should be careful that the complexity of command selection does not get out of hand. There are systems that require five or more menu selections before a complete command is invoked, and that is

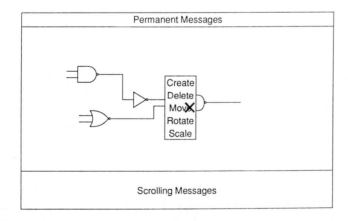

FIGURE 10.4 Pop-up menus appear at the location of the cursor (X).

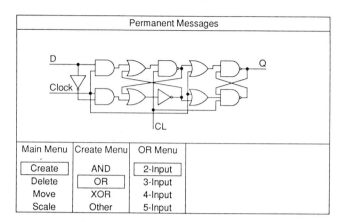

FIGURE 10.5 Marching menus show hierarchy and build from left to right.

too much. With the proper use of defaults, remembered state, and task modeling, much of this complexity can be reduced.

Another implementation of menus, possible only when the pointing device is a tablet, draws the options on the tablet surface (rather than the screen) in a location that has no correspondence on the screen (see Fig. 10.6). This is desirable because it does not use screen space for fixed menus and also because the printed labels on the tablet can contain more detail than can be displayed. From a human-engineering standpoint, however, this is a bad idea because it causes excessive eye and even head movement when the user repeatedly turns from the screen to the tablet. Tablet use does not normally require that the user watch the hand on the tablet because the movement of the cursor on the screen is adequate feedback. The use of tablet menus, too

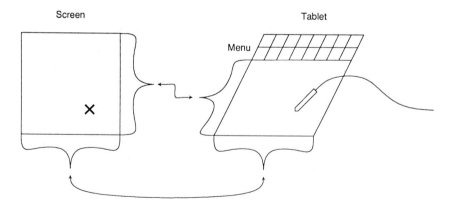

FIGURE 10.6 Tablet menus are in locations on the tablet surface that have no correspondence on the screen.

many buttons on the puck, or any other scheme that causes a shift in attention, will slow the design process.

10.3.3   Message Display

In all interactive systems, even those that are highly graphics oriented, there is a textual conversation taking place. Although the use of graphics and menus can reduce the need for a keyboard, there will always be some things that are typed. In addition, systems need to print status, error, and information messages somewhere on the screen. Two kinds of text must be displayed: fixed position information and scrolled messages.

Fixed position information is important enough to demand that the user always be able to see it. For example, current status such as the circuit being edited and the machine's run state are frequently requested by the user. This information belongs in a separate area of the screen for fixed messages so that each piece of information can be found immediately. The programmer must allow enough space for the largest message that may appear, and still have provisions for messages that are too large. For example, when a text field overflows it can be chopped off at the right but when a numeric field overflows it cannot be abbreviated and there should be a definite indication that the datum is wrong.

Less important information can be placed in another part of the screen that scrolls. **Scrolling** is the action that takes place on most video terminals in which new lines are added to the bottom of the screen as the display is shifted up. Because the top line is lost from view when a new one is added, good systems will pause when the scrolling area fills with new messages. The user can then acknowledge this pause to allow the system to continue displaying messages. It is necessary to give some indication that the pause has happened, such as a beep or the word "more" displayed at the bottom. Both the pause indication and the method used to acknowledge the pause should be similar to equivalent functions found elsewhere in the operating system. Another useful feature of scrolling control is an option to set how much more will be scrolled before the next pause. Typically, the user is willing to allow all the current messages to scroll away, but may sometimes want only one more line before the system pauses again.

Scrolling may be difficult to achieve on some displays because it requires moving large numbers of data. In these circumstances, the information can simply rotate through the area, displaying each new message over the oldest one. This **rolling** is sometimes confusing to the user and so it requires the display of a line cursor to indicate the most recent message. It is also helpful to erase the line below the newest message so that a clear break can be seen between it and the oldest one below it.

For text input, the scrolling area should be used like a terminal. This demands a character cursor so the user can see progress along the line. Some prompt characters should be printed so that the user knows when input is expected and so that input can be distinguished from output.

10.4   Command Language

The **command language** is the channel for user input and it must be capable of expressing all possible transactions, whether they are issued from the keyboard or from some graphical pointing device. Even when the number of different devices grows large, it is necessary for the command language to have a uniform design so that it is easy to learn and extends cleanly. Important functions should be redundantly available on different devices for ease of invocation, and there should be options for rearranging the location of different commands. Other desirable features of a command language are minimal use of modes, convenient abort, online help, and proper error recovery. The next two sections discuss device-specific aspects of command languages, followed by a general discussion of command-language features.

10.4.1   Textual Command Languages

The keyboard is and will remain the most powerful input device because it can transmit the most complex and specific commands. Since the user of a CAD system typically types a certain number of operating-system commands in order to use the machine (login, run programs, and so on), it is to the advantage of the CAD system to mimic that command language. This does not mean duplicating commands, but rather using the same style of command structure. For example, on a UNIX operating system [Ritchie and Thompson], switches are prefixed with a ``-'', and on a DEC operating system [Digital] the switches begin with a ``/''.

Since CAD is very much an editing task, another possible model that can be used for the command interface is to mimic the current text editor. For example, if the CAD system runs on a machine that provides the EMACS editor [Stallman], it might allow similar concepts such as dynamic key binding, single-key commands, and the use of special keys as a prefix for longer commands. General editing concepts—

for example, the semantics of action verbs in the command language and the syntax of operands—can be shared.

The preceding discussion raises the issue in keyboard command languages of whether or not to use the keys individually as complex function buttons. Many systems allow a single keystroke to do a full action such as creation and deletion. Systems without this capability require a full line of text to be typed including a carriage-return. Since full commands are more powerful they must always be available, even in single-keystroke environments. This option can be provided by having a special single key that prefixes the typing of longer commands, as in Caesar [Ousterhout] and Electric.

When typing long commands, it is helpful to have command completion. With this feature, originally found in the TENEX operating system [Bobrow *et al.*], the command interpreter fills out any partially typed keyword if enough letters have already been entered to identify that keyword uniquely. For example, if the system expects the commands ''compile,'' ''compose,'' or ''create'' and the user has typed cr, then the system knows which command has been selected and can append the letters eate when a carriage-return or other delimiter is typed. Command-completion systems also provide special keys for immediate completion and for help. The special completion key, often an *escape*, fills out the keyword without terminating the line or taking any other action. The help key, often a ''?'', lists all options at the current point. If, in the preceding example, the user were to type co and an *escape*, the system would append mp but nothing more because there is ambiguity after that. The system could also give some indication that the keyword is not complete (a beep or flash). If the user then types a ''?'', the system will list the options—''compile'' or ''compose''—and then return the user to the input after ''comp.''

This example shows the need to select the keywords in a command language carefully. If the user is going to have the advantage of abbreviating commands for reduced typing, then a comfortable command set will be unique in the first few letters. Having two commands that both begin with same four letters requires that the user type five letters in order to abbreviate minimally, and that is not much of a saving over typing the full command. In the example, the programmer would be well advised to change ''compose'' to ''assemble'' or any other word that does not conflict with ''compile.''

10.4.2 Graphical Command Languages

There is a feeling among computer users that the keyboard is an inappropriate input device for what is essentially graphical activity, and that its use should be minimized. Instead, a mouse or other pointing device would seem better suited to the job. Actually, the keyboard is potentially very powerful and can provide much more bandwidth than can a mouse, but this requires that the user be experienced. To make novices more comfortable, a mouse is convenient and experiments indicate that it is faster to use [Card, Moran, and Newell]. Of course, the keyboard cannot be totally replaced because there will always be names and numbers to type, but command input can be shifted away, making the keyboard a secondary device. In some systems, the keyboard is used to enhance the pointing device by having single keystrokes act as pointer buttons, thus giving many more options to a pointing action. However, this requires a shift of attention from the screen to the keyboard, a delay that human engineering seeks to avoid. In a proper graphical command language, the input device must be programmed for maximum convenience or else the user will demand to use the keyboard again.

Menus are an obvious choice for implementing a graphically based command language. Although the time taken to select a menu entry was reported earlier to be 1.1 seconds, this time is actually dependent on the menu-entry size and its distance from the current cursor position [Card, Moran, and Newell]. To understand this dependency better, it is necessary to examine menu picking in more detail.

Returning to the model of human activity, it can be seen that the action of moving the cursor toward a menu entry requires one perception action (to see where the hand is going), one cognition action (to determine what to do), and one motion action (to do it). These three operations require 240 milliseconds, according to the time values given earlier. However, this does not put the cursor in the menu, but rather moves it some fixed percentage toward its goal. The actual time to get to the menu is therefore an exponential function based on distance and accuracy. This function, called **Fitt's Law**, states that the time to a menu entry will be:

$$I \times \log_2\left(\frac{2D}{S}\right)$$

where D is the size of the menu, S is the distance to the menu, and I is a constant: 100 milliseconds. Thus menu entries should be close and large for best response, and there is a linear tradeoff between size and distance.

There are, of course, other ways to indicate a command graphically, such as cursive-script recognition. The Ledeen character recognizer [Curry], a simple and effective method of distinguishing handwritten letters, can be used to identify a large set of commands that are the result of stroke combinations. The Bravo3 design systems use this as their primary command interface [Applicon].

Although graphical command languages have the advantage of easily specifying screen positions, they are unable to give precise coordinates or other exact numeric values. One solution is to provide a number menu that mimics familiar devices such as telephones or adding machines. This menu works like its physical analog and allows the user to "punch up" a desired value. Further, it supplies the same functionality as a real device including sign entry, decimal point, backup, and abort.

An alternative to the tedium of selecting individual digits when defining an exact value is to use a graphical **number wheel** [Thornton]. The screen displays a wheel and the current value on that wheel, so that when the user points to it and moves the cursor, the wheel is "rolled" and the value changes. The wheel can keep spinning after a "pull," and additional pulls can speed it up. Slower pulls will advance the value more slowly and a single grab will stop the spinning, thus completely mimicking a physical wheel and giving the user a familiar feel to number selection.

For some tasks, the inaccuracy of the cursor can be made more precise in special-purpose ways. If a drawing task has known regularity, it may be helped by enforcing that regularity. For example, when objects are drawn, their placement can be snapped to a grid that is used to align everything on the screen. The size of the grid should be controllable, as should the option of displaying it so the user can see the spacing. As another example, when drawing demands that all lines be at right angles or at 45-degree increments, the cursor position can be adjusted from where the cursor is to the closest point that meets this requirement.

When the cursor approaches an object, it is often meant to touch that object. Thus, when the user is pointing on the screen, the cursor can snap to the closest other object if a connection is desired. In situations in which a connection may or may not be as necessary, the cursor can snap to another object only when near that object. This proximity area, called a **gravity field**, is larger around vertices than around edges because the vertices are the preferred cursor locations (see Fig. 10.7).

10.4.3   General Command-Language Features

There are certain attributes that all command languages should possess, regardless of whether they are issued from a keyboard, a pointing device, or even speech input. The interface should be friendly, mode-free, and able to abort, undo, or recover from errors.

The most important command language feature is friendliness.

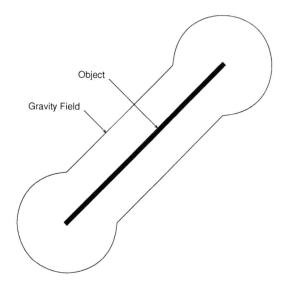

Object

Gravity Field

FIGURE 10.7 The "dumb-bell" shape of a gravity field. When the cursor enters this area, it is drawn to the object. Vertices have a greater gravity field than lines do.

Error messages must be positive in nature, suggesting alternatives to the user's situation. The negative message "cannot do . . . " should instead be "must do . . . first."

In addition, help needs to be available at every opportunity for user input. Not only should online help be available for all commands, but each prompt should be prepared to explain more fully what is wanted. The online help information is best when it is structured in the same way as the actual commands being explained. This means that tree-structured command languages should have tree-structured help systems. Also, the method of obtaining help should be consistent and easy to guess—for example, the "?" key. Besides basic help, there can be help guidance to suggest the proper questions to ask for given subjects and there can even be a tutorial available to present basics. The ultimate system has an intelligent agent that watches and critiques user actions. With such a facility, a total neophyte can use the system without any confusion or human assistance.

One aspect of command languages that programmers attempt to avoid but rarely do is modes. **Modes** are collections of commands that are tailored for specific tasks. Although they are convenient for that particular task, they are confusing to learn and to use because they tend to overload manual operations, redefining them at different times. For example, the user might learn that the "X" key means *erase* when in "placement mode" and it means *undefined* when in "simulation mode." This inevitably leads to issuing the right command in the wrong mode, which does arbitrarily bad things.

If modes are to be used (and they are nearly impossible to avoid), then they should be kept to a minimum. The current mode should be clearly displayed in the permanent message area, and those functions that have meanings that change in each mode should have their current meaning displayed. Most important, the user must be able to recover from mistakes, whether they be due to mode confusion or the result of just plain stupidity. This means allowing the abort of any partially completed action and the undo of any finished mistake.

Abort options do not belong only in likely trouble spots; they must exist in all prompts to the user. They should be an obvious option and should revert to a sensible state for the continuation of work. Undo control is harder to implement because it may require the tracking and resetting of large amounts of state. In addition to undoing the last command, it should be possible to undo many previous actions and to undo an undo, which amounts to a redo. Beware of complex undo control that allows the user to make inconsistent change requests.

10.4.4   Crash Recovery

The ultimate error recovery is, of course, safety from crashes. It is no small thing to insist that a large computer system be immune to crashes. People spend years debugging programs and writing diagnostics, yet systems still fail. Even after the perfect, bulletproof program has been written, there will always be external sources of error such as operating-system faults and hardware failure. In short, no program is safe from crashes and the only way to make a program robust is to have it recover properly.

When a change is made to a circuit, that change is probably not permanent because most systems change only the memory representation and not the disk version. It is only when the circuit is saved on disk that the changes are **committed**, or made permanent. Highly robust database systems have explicit commit steps after every change so that a secure disk record is made of the action. In most CAD systems this is too expensive. Also, users are not willing to commit until they finish an editing session.

In systems that update only memory, a crash will cost all of the changes made since the last save. One way to provide recovery in this situation is to **checkpoint** the circuit automatically by saving it on disk every few commands. The user will then be able to use the most recently checkpointed version, which will be only a few changes behind. Unfortunately, VLSI design files may be so large that saving them frequently is more annoying than is redoing lost work.

An alternative to saving the design is to save changes to the design. This information is much more compact and can be written to disk frequently and quickly. Since a design system transforms **user input** to **database changes**, either end of that processing pipe can be saved. If the user input is saved, then each keystroke, pointer position, and button push is recorded in a session log file. If the database changes are saved, then each new object, modification of an object, or object deletion is recorded.

Recovery using change files requires that the files be "played back" to re-create the last state of the database. User input playback will more precisely show the steps that led to the crash and are thus more helpful in debugging. In addition, user input files are smaller than database change files because users typically give very little information in relation to what is produced. After all, it is the job of the CAD system to translate brief user requests into circuits that are increasingly complex. Thus user-input logging is the most concise and effective way to recover from crashes. It also makes a nice demonstration of a design system to see an entire circuit built at the high-speed of session playback.

The one thing to ensure in user-input files is that there is enough information recorded to reproduce the session correctly. If, for example, a "select" command is issued when the cursor is over two identical objects, the database state will determine which object is selected first. If that database state cannot be guaranteed during playback, a different object may be selected, causing incorrect recovery. To overcome this problem, additional command-specific information may have to be recorded in the logging file.

10.5 Feedback

It is the task of human engineering to keep the user happy during sessions with the system. This means that whenever the user does something, there needs to be a definite and friendly response. Messages should always be positive in nature and must acknowledge success or failure, and progress on long tasks should be encouraging. Requests for input need identifiable prompts so that the user is aware at all times of the activity of the system.

To provide the best feedback, consider where the user's eyes are directed and work in that area. When a command has been invoked from the keyboard, a message should be printed in the scrolling area to indicate success, failure, or progress. When a command has been invoked from the pointing device, the user's attention is on the display and feedback belongs there. When an unexpected condition occurs, the user's attention must be grabbed from wherever it is—for example, by beeping or by flashing the screen. Keep the number of beeps low, however, or else users will ignore them and eventually may disconnect the speaker.

When input is from the keyboard, feedback is textual and its nature depends on the result of the input. For a successful result, there should be a short message indicating any important state changes that have occurred. Failure messages should suggest alternatives to the erroneous input so that the desired task can be completed. Long operations should acknowledge that they have started and that they have completed. Very long tasks should print some indication of progress along the way. These progress messages may indicate what part of the task is being done and what percent of the effort is complete, or they may merely print a single character (such as an "X") so that the progression of these letters can be watched. Users quickly learn the nature of all feedback and interpret it correctly.

Input from the cursor should elicit feedback near the cursor. In some cases, the normal progression of display changes provides adequate feedback. For example, with pop-up menus, the menu disappears after selection is made, and with commands that change the display, completion of the change acknowledges the command. Thus implicit feedback is available because the system does what is requested in a graphical manner. Failure messages must be displayed near the cursor or else they must be treated as exceptional conditions that are indicated with a beep.

One of the best ways to indicate progress and completion graphically is to change the cursor shape. The NEWSWHOLE system [Tilbrook] changes the cursor from an X to the image of a Buddha to request patience on the part of the user. Hourglasses convey the same meaning and indicate that no new commands may be entered. Another method is to erase the cursor and then repaint it slowly as progress is made.

The cursor can also be used to indicate what kind of action is expected from the user. When the user is pointing to single objects, the cursor is an X; when he or she is drawing a sketch, the cursor is the shape of a pen; and when keyboard input is expected, the cursor changes to a typewriter icon to indicate that a different device should be used. When requesting specific input, it is far better to give a sensible prompt than simply to lock up all other devices and ignore their input.

Another feedback option that can be used to make menu selection responsive is to highlight the menu entries as the cursor passes over them. Displays with block-transfer capabilities can quickly invert the menu entry when the cursor enters. Another option is to draw the entries without enclosing boxes and then show a box around the current entry as the cursor passes over it.

Good feedback can keep users happy even in the worst conditions because it lets them know what to expect—and when to go get coffee. Poor feedback, however, will make system use difficult even when the machine response is adequate, because the users will never be sure of their next step. Never let the user wonder, "Now what?"

10.6   Summary

Human engineering comprises both the innermost and the outermost levels of a good computer system. At the deepest level is the task and user model that permeates the system design to correspond with the user's abilities and expectations. At the top level is a sheen of friendliness and cleverness that comforts the user by interacting intelligently. Between these are the many other system functions that constitute a VLSI CAD system.

Questions

1 What is wrong with a system containing the commands "transform" and "transpose"?

2 Why should the display be organized to position commonly used areas close together?

3 What are the relative merits of marching menus and pop-up menus?

4 What problems will be encountered when single keystrokes are used as abbreviations for full commands?

5 Which will better reduce menu selection time: increasing the size of the menu or decreasing the distance of the menu entry from the cursor?

6 What database organization guarantees that keystroke playback of a previous session will work correctly?

7 Why do advanced users prefer less feedback?

References

Applicon, *Bravo3 User's Guide*, Applicon Incorporated, Ann Arbor, Michigan, 1986.

Bobrow, Daniel G.; Burchfiel, Jerry D.; Murphy, Daniel L.; and Tomlinson, Raymond S., "TENEX: A Paged Time Sharing system for the PDP-10," *CACM*, 15:3 135-143, March 1972.

Card, Stuart K.; Moran, Thomas P.; and Newell, Allen, *The Psychology of Human-Computer Interaction*, Lawrence Erlbaum, Hillsdale, New Jersey, 1983.

Curry, James E., "A Tablet Input Facility for an Interactive Graphics System," Proceedings IJCAI '69, 33-40, May 1969.

Digital, *PDP10 Timesharing Handbook*, Digital Press, Maynard, Massachusetts, 1970.

Fairbairn, D. G. and Rowson, J. A., "ICARUS: An Interactive Integrated Circuit Layout Program," Proceedings 15th Design Automation Conference, 188-192, June 1978.

Gosling, James, personal communications.

Haber, Ralph Norman, "How We Remember What We See," *Scientific American*, 222:5, 104-112, May 1970.

Newman, William M. and Sproull, Robert F., *Principles of Interactive Computer Graphics*, 2nd Edition, McGraw-Hill, New York, 1979.

Ousterhout, J. K., "Caesar: An Interactive Editor for VLSI Layouts," *VLSI Design*, II:4, 34-38, 1981.

Ritchie, D. M. and Thompson, K., "The UNIX Time-Sharing System," *Bell Systems Technical Journal*, 57:6, 1905-1929, 1978.

Stallman, Richard M., "The Extensible, Customizable Self-Documenting Display Editor," Proceedings ACM SIGPLAN SIGOA Symposium on Text Manipulation, Portland Oregon, 147-156, June 1981.

Thornton, Robert W., "The Number Wheel: A Tablet Based Valuator for Interactive Three-Dimensional Positioning," *Computer Graphics*, 13:2, 102-107, August 1979.

Tilbrook, David M., *A Newspaper Pagination System*, Masters Thesis, University of Toronto Department of Computer Science, 1976.

Varner, Denise, "Color Avionics," unpublished manuscript.

ELECTRIC

11.1 Introduction

After many chapters of CAD in the abstract it is time to discuss a specific and real design system called "Electric" (known commercially as "Bravo3 VLSI"). This system is instructive because its source code is available to universities so that students may study its workings. Electric is also worthy because it embodies many of the notions described in this book. However, the most compelling reason for discussing this system is simply that it was programmed by the author of this book [Rubin].

Electric was built as an alternative to traditional VLSI design systems that did not combine graphics, connectivity, and accurate geometry. Many textual systems existed for VLSI design [Lipton *et al.*; Johnson; Kingsley] and there were also graphics systems with no notion of connectivity [Ousterhout; Batali and Hartheimer]. Some systems attempted to merge graphics and connectivity, but they always abstracted the graphics somewhat [Weste; Williams]. The goal of Electric was to combine these into one highly flexible system.

To implement connectivity, the Electric design system has an extendible database, built on a network structure. Nodes of the network correspond to components in the circuit, and arcs of the network are connecting wires. In addition to this network information, there are geometric data associated with every component and wire so that correct layout can be represented. Since the database is extendible, additional structures can be stored to describe behavior, power consumption, design rules, and so on.

The most interesting aspect of Electric is its programmability: Users can specify layout relationships that will be continuously enforced during design. This is achieved with a constraint system that hierarchically propagates layout relationships through the design. The constraint system strengthens the ability to do hierarchical design because different levels of the circuit always remain connected regardless of changes to the layout. Although these constraints are all spatial in nature, a separate programming system built on Prolog [Clocksin and Mellish] allows more flexible programmability by providing an interpretive language for textual design. It also allows some parts of the design system to be coded in Prolog.

Electric accepts the fact that there is an unlimited number of environments for doing design, and the system integrates them in a uniform manner. A design environment is a single module describing primitive components and wires that can be composed into circuits. Included in the module is all the environmental information such as graphic attributes and simulation behavior. To create a module, one simply expresses the environment in network form so that it can be manipulated as a collection of nodes and arcs.

In addition to providing for any design environment, Electric also allows an unlimited number of synthesis and analysis tools. The database oversees their execution in the fashion of an operating system so that each tool can operate in turn and keep up with the activities of the others. This flexibility allows tools to operate incrementally or in a batch style, and also keeps all activity consistent. Change control is managed by the database so that any incorrect tool activity can be undone. In this context of tools as processes, even the user interface is a tool, separate from the database.

This chapter provides detail on the four main aspects of Electric described here: the representation, programmability, design environments, and tools. Specific aspects of the user interface can be found in Appendix F, which is a tutorial for designers. Electric is written in C [Kernighan and Ritchie] and its source code is available to universities through the author of this book. The code comes with three internals manuals that explain the data structures, database subroutines, and the construction of new design environments. A more extensive and robust version of Electric, called Bravo3 VLSI, is sold commercially by Applicon, a Schlumberger company.

11.2 Representation

In Electric, circuits are composed of **nodes** and **arcs** that represent components and connecting wires. For example, a transistor component connected to a contact component will be represented by two nodes and an arc (see Fig. 11.1). In addition, nodes have **ports** on them that are the sites of arc connections. There are two ports in this figure, one on each node at the connection site of the arc. Nodes, arcs, and ports are dynamically allocated objects in Electric, with pointers being used to link them together into networks. Each object has a basic set of attributes that hold vital information such as physical position and network neighbors. This set is extendible and can admit any attribute with any value. For example, a new attribute called "delay-time" could be created on the arc that holds a floating-point number. Also, a new attribute called "gate-contact" could be created on the transistor to hold a direct pointer to the contact node.

To represent the geometry of a layout, Electric stores actual coordinates on each node and arc, expressed in centimicrons. Organization of this information is done with an R-tree (see Chapter 3) that locates every node and arc in a balanced geometric space [Guttman]. Thus

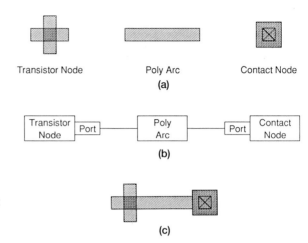

Transistor Node   Poly Arc   Contact Node

(a)

Transistor Node | Port — Poly Arc — Port | Contact Node

(b)

FIGURE 11.1 Electric's node and arc representation: (a) Objects (b) Representation (c) Actual layout.

(c)

there are two ways to access objects in the database: by topology and by geometry.

To make the representation more flexible, Electric uses the distinction between instance and prototype to aggregate common information about classes of objects. For example, the transistor node in Fig. 11.1 is a **node-instance** object, one of which exists for every different node placed in a circuit. There is also a transistor **node prototype**, which exists only once to describe all its instances. Node prototypes have port prototypes on them, which are templates for possible port instances in a circuit. These port prototypes define the physical locations on the node that can be used for connection, and also the allowable arcs that may be connected. To complete the instance-prototype representation, there are arc prototypes that describe defaults for every different kind of arc. Figure 11.2 illustrates how instances and prototypes relate. Notice that port prototypes may have multiple instances on the same node (the contact node instance in this example).

A **technology** in Electric is an object that aggregates node, port, and arc prototypes to describe a particular design environment. For example, the nMOS technology object has three arc prototypes for metal, polysilicon, and diffusion wires. It has node prototypes for enhancement and depletion transistors in addition to a series of node prototypes for interwire connections (butting, buried, and cut contacts) and intrawire connections, or pins. Instances of these prototypes can be assembled into an nMOS circuit (much like in Figs. 2.18 and 2.19). Since technologies are objects, they can also hold additional attributes. Initially, the added information includes design-rule tables and shape descriptions, but more could be handled; for example, one could add an attribute called "toxicity-level," which contains a paragraph of text

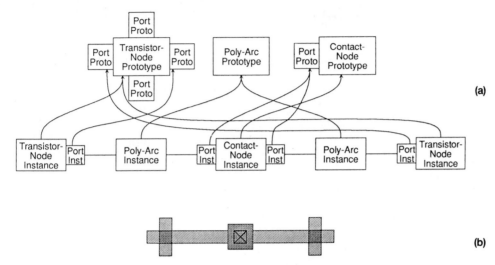

FIGURE 11.2 Instance and prototype representation: (a) Representation
(b) Layout.

describing the environmental ramifications of this semiconductor fabrication process.

The representation of hierarchy is done through a twist of the
instance-prototype scheme. Rather than have a separate class of objects
to represent cells of circuitry, the node-instance object can point to
either primitive or complex prototypes. Whereas **primitive** node prototypes are defined in the environments as described, **complex** node
prototypes are actually cells that contains other node, arc, and port
instances (see Fig. 11.3). Port prototypes on complex node prototypes
(that is, ports on cells) are actually references to ports on nodes inside
of the cell (see the heavy line in the figure). This notion of "exporting"
ports combines with the complex node-prototype scheme to provide
a uniform representation of components in a hierarchical circuit. Although the prototype objects vary slightly in their information content,
the node- and port-instance objects are the same regardless of whether
they describe primitive or complex prototypes.

Everything in Electric is an object with extendible attribute-value
pairs. This section has already mentioned node, arc, and port instances;
node, arc, and port prototypes; and technologies. There are also **library**
objects that aggregate complex node prototypes, and **tool** objects that
operate on the circuitry (see Fig. 11.4). Tool objects contain many
attributes that are actually code routines for performing the essential
synthesis and analysis functions.

As an example of how attribute flexibility works, the operation
of a typical functional simulator will be described. Functional simulators

(a)

(b)

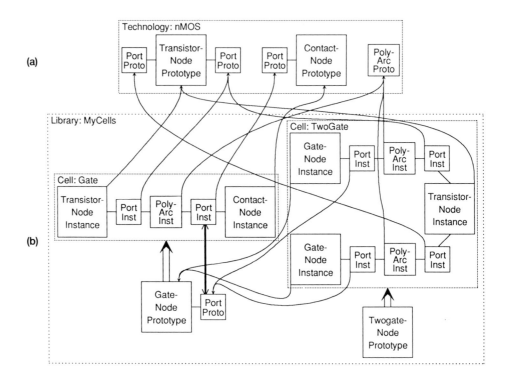

(c)

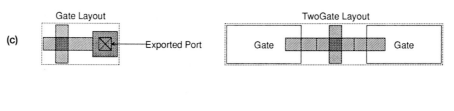

(d)

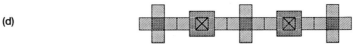

FIGURE 11.3 Hierarchical representation: (a) Primitive prototypes in a technology (b) Complex node prototypes, or cells (c) Cell layouts (d) Exploded layout.

can either simulate the contents of a cell or use a behavioral description, which more efficiently describes the contents. All primitive node prototypes start with such a behavioral description so that the simulator will not attempt to examine their contents. Whenever a cell is to be behaviorally described, the designer enters that information in the same place on the complex node prototype. The simulator merely checks for the existence of the behavior attribute on the prototype

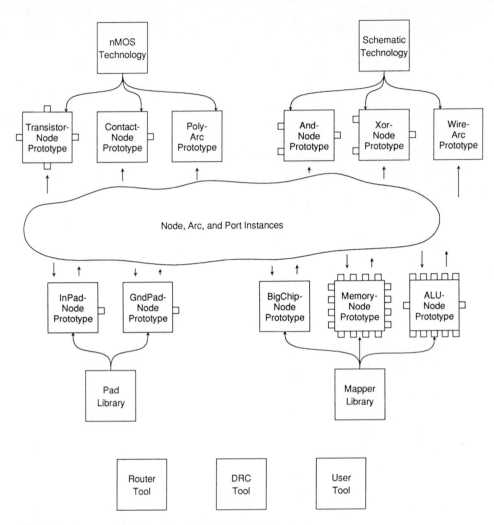

FIGURE 11.4 All objects of the Electric database.

when deciding how to handle a node. Thus, the Electric representation is powerful enough to contain any piece of information and flexible enough to organize that information sensibly.

11.3 Programmability

The most interesting aspect of Electric is its powerful facility for programming designs. Besides providing a built-in Prolog interpreter [Clocksin and Mellish], Electric has a few constraint systems available in the database. One constraint system is based on **linear inequalities** of wire lengths, as Chapter 8 described in detail. The primary constraint system, unique to Electric, works both within the cell and incrementally, propagating changes upward in the hierarchy. This **hierarchical-layout** constraint system enables designers to work top-down in a graphical editor.

Although the hierarchical-layout constraint system was initially an integral part of the database, the need for multiple constraint solvers fostered a more modular scheme. Each constraint solver provides interface routines for the possible database changes: creation, deletion, and modification of nodes, ports, arcs, and cells. Additional interface routines must exist for initialization and special control directives, and to inform the constraint system that a batch of changes has been described and can be solved. With such a scheme, any constraint system can be incorporated in the database.

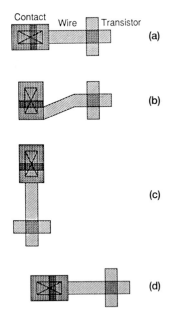

FIGURE 11.5 The effect of rigid wires: (a) Original (b) Contact component rotated, unconstrained wire (c) Contact component rotated, rigid wire (d) Contact component moved right, rigid wire.

11.3.1   Hierarchical-Layout Constraints

Hierarchical-layout constraints are always placed on wires in Electric. They set properties so that a change to a component on either end of the wire affects the component on the other end. In the absence of constraints, wires rotate and stretch as necessary to connect their two components whenever either moves. Thus the essential premise of Electric is that wires always remain properly connected through all changes to the database.

The first hierarchical-layout constraint is **rigidity**, which, when applied to a wire, causes the length to remain constant and fixes the orientation with respect to both components. If either component connected to a rigid wire moves, then both the wire and the other component move similarly. If a component rotates, then the wire rotates and the component on the other end spins about the end of the wire. It can be seen that a rigid wire essentially creates a single object out of the two components that it connects (see Fig. 11.5). To allow for unusual constraint situations, rigidity may be set or unset temporarily, which means that the property overrides the current constraint for the next change only.

When a wire is not rigid, there are two other constraints that may apply: **Manhattan** and **slidable**. The Manhattan constraint forces a wire to remain at its current angle: either horizontal or vertical. If one component on a horizontal Manhattan wire moves up, then the other will also. If that component moves to the left, however, the wire will simply stretch without affecting the other component. The rotation or mirroring of a component does not affect a Manhattan wire unless the wire is attached off-center, in which case a slight translation occurs (see Fig. 11.6).

The other constraint that can be applied to nonrigid wires is the slidable factor, which affects components with nonpoint connection sites. When a connection has a positive area, the wire connecting to it can slide within that area. This means that small changes to the component may not affect the wire at all because the connection area still properly contains the end of the wire. The slidable constraint allows the wire to make these independent movements, whereas without this constraint both component and wire must move together. In Fig. 11.7, the 6 × 4 contact component has a port area that begins 1 unit in from its edge. The wire is thus connected in the middle of a 2-wide port and can slide by 1 in either direction. The figure also illustrates an AND gate the port of which is the entire left side (a 0 × 5 port). The wires remain connected at the same place because they are not slidable.

Most design environments allow slidability so that the wires can adjust to detailed changes of connection configuration. An example of the need to turn off this constraint is in schematic gates such as AND and OR, in which the inputs are all attached to one large connection on the side. If slidability is allowed, then small motions of the gate component will bunch the inputs together.

The most powerful of the hierarchical-layout constraints is one that relates hierarchical levels of a layout. All exported connection sites in a cell definition are bound to their actual connections on instances of the cell, higher up in the hierarchy. This means that, if a component in a cell has an exported connection and the component moves, then that connection moves and any wires attached on instances of the cell also move (see Fig. 11.8). This is a natural extension of the Electric philosophy that everything must remain connected after a change.

The order of constraint execution is critical to ensuring a proper solution and correct layout. Electric solves all constraints inside a cell before moving up the hierarchy to other cells. When the hierarchy is more complex than a simple tree organization, care must be taken to ensure proper traversal so that lower-level cells are solved before the higher-level ones. For example, if cell C in Fig. 11.9 is changed, then cell B must be reevaluated before cell A, even though both contain instances of cell C.

Within a cell, the constraints are solved in two passes: first the rigid wires and then the nonrigid wires. This gives rigidity the priority it needs to work correctly. A time-stamp mechanism prevents constraint loops from running amok, by detecting reapplication of constraints and forcing a quiescing action. If a reapplied constraint is consistent with the layout, then no change is necessary and the loop terminates.

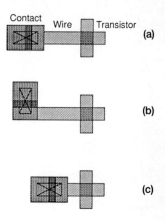

FIGURE 11.6 The effect of Manhattan wires: (a) Original (b) Contact component rotated, Manhattan wire (c) Contact component moved right, Manhattan wire.

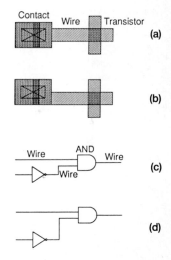

FIGURE 11.7 The effect of slidable wires: (a) Original (contact has connection area that is inset 1 from the top and bottom) (b) Contact component moved up by 1, slidable wire. (c) Original (d) AND moved up, Manhattan nonslidable wires.

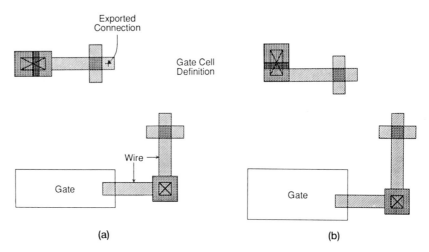

FIGURE 11.8 Propagation of hierarchical-layout constraints: (a) Original (b) Contact component in gate cell rotated, all three wires Manhattan.

Otherwise, there is an overconstrained situation and Electric jogs the wire, replacing it with three that properly connect. This scheme for propagation of constraints proceeds quickly and effectively in the domain of VLSI design.

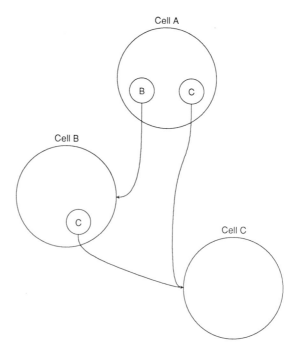

FIGURE 11.9 Nontree hierarchy must be traversed carefully.

11.3.2   Prolog Programming

When more powerful programmability is needed, the user can write a piece of Prolog to generate any layout. This language system does not function incrementally, reacting to changes as they occur, but rather acts like a standard imperative language that manipulates the database when executed. Prolog code can create circuitry by invoking predicates to define new cells and place objects in cells. It is also possible for Prolog code to invoke the constraint system by setting wire properties and changing component locations. These capabilities are identical to those available in C programs linked with Electric, except that the Prolog code runs interpretively and thus provides a better development environment. Work is under way to merge the two programming facilities, providing arbitrary Prolog-coded constraints that react to incremental changes in the circuit.

11.4   Environments

Electric contains a number of different design environments. As described earlier, these environments exist as **technology** objects in the database. The objects contain both procedural and tabular information that describes their contents. For example, the primitive components are represented by a linked list of node-prototype objects. However, the graphical description of these primitives is done procedurally with one routine that returns a count of polygons, and another routine that is repeatedly called to describe each polygon. This allows primitives to be arbitrarily parameterized for maximum flexibility.

Adding environments simply requires that the necessary tables and routines be coded into a module, and a technology object be created with pointers to this module. The routines are for initialization, special control, node description, arc description, and port description. The node-description routines actually take two forms, depending on whether the description is for display or for circuit analysis. Tabular information includes design rules, simulation characteristics, default sizes, connectivity information, behavior, reformatting data, layer patterns, and colors. Much of the information handled by routines is also coded tabularly.

Most of the environments in Electric are for layout of integrated circuits, there being many different processes and many variations within a process. For example, there are four different CMOS environments: an idealized version with one layer of metal [Griswold], a double-metal version for MOSIS design rules [Mukherjee], a double-

polysilicon version for Canadian foundry design rules, and a special
MOSIS version with round geometry. In addition to those for CMOS,
there are environments for nMOS and Bipolar layout.

In non-IC design, there are environments for schematic capture
and printed-circuit-board layout, and some high-level architectural environments such as one for digital filters [Kroeker]. It is even possible
to do nonelectronic design with two environments: artwork for arbitrary
graphics, and GEM for concurrent event specification [Lansky and
Owicki]. The most interesting environment, however, is called "generic"
because it contains many useful primitives for flexibility of design.

11.4.1   Generic

The generic environment allows the intermixing of all other environments
by providing universal components and wires that have unrestricted
connectivity. With the **universal** wire, a gate from the schematics
environment can be connected to a transistor of the nMOS environment.
The use of this wire is necessary because, although the two components
can reside in the same cell, neither the basic wire of schematics nor
the polysilicon or diffusion wires of nMOS can legally make the connection. The same reasoning applies to the use of the universal pin
in connecting wires from different environments. The mixing of environments is useful for a number of purposes including process experimentation (mixing multiple IC-layout environments), IC-design
shorthand (mixing layout and schematics), and system-level simulation
(mixing IC and PC layout).

Another feature of the generic environment is the **invisible** wire,
which can connect any components but does not make an electrical
connection. This is useful for placing constraints on two objects regardless of their network relationship. An invisible wire can be constrained just like any other wire, but it carries no signals and does
not appear in the fabrication output.

Also in the generic environment are special primitives that are
useful in design. The Cell-Center component, for example, is a fiducial
mark that, when placed in a cell, defines the origin for cursor-based
manipulation of cell instances. There are also primitives in the generic
environment that define various menu icons for the user interface.

11.4.2   MOS

The MOS environments contain the high-level layout primitives that
have been described in Chapter 2 (see Figs. 2.18 and 2.21). A transistor
component is used to represent the intersection of polysilicon and
diffusion so that the network can be easily maintained. There are two

transistor primitives in each MOS environment: enhancement and depletion in nMOS; *n*-type and *p*-type in CMOS. There are also a host of contacts for connecting wires on the different layers, and a set of pins for connecting wires on the same layer.

An important escape hatch for IC design is the availability of pure layer components that allow arbitrary geometry to be produced when the higher-level primitives are not flexible enough. Of course, in an attempt to avoid such use, much attention has gone into planning the primitives to make them usable and powerful. For example, the contact-cut primitives automatically add extra cut polygons as their size grows large so that multiple components do not have to be used. Also, the transistor primitives allow an arbitrary path to be specified for serpentine description. It is therefore only the most unusual layout, typically analog, that needs the pure-layer components.

Another high-level aspect of MOS environments is the automatic placement of implants. In nMOS, this simply means that the depletion transistor includes the depletion-implant layer as part of its description. In CMOS however, all diffusion activity takes place in substrate or well implants, so all the transistors, diffusion wires, diffusion contacts, and diffusion pins must include implant information. Thus there are two sets of these components, one for each implant, and the correct version must always be used. Given this feature, the designer is able to work without consideration for implant placement because there is always enough of it automatically placed to encapsulate the circuitry properly.

11.4.3 Bipolar

The bipolar environments are significantly different from MOS environments, even though they are all for IC layout. Rather than having a few geometries for the simple field-effect transistor, it is necessary to provide total flexibility in the layout of junction transistors by allowing arbitrarily shaped bases, emitters, and collectors. Thus there are three primitives that combine to form a transistor. These primitives are complex since they describe the many bipolar layers necessary for base, emitter, and collector construction. There are also contacts and vias that make connections between conducting layers.

11.4.4 Schematics

The schematic-capture environment contains the necessary components for digital and analog diagrams of circuits. On the digital side there are the logic gates AND, OR, XOR, Flip-Flop, and Buffer, all of which can be customized to handle any number of inputs and negate any signal. Negation is accomplished by placing an inverting bubble on the connecting wire rather than on the component. To the designer, this provides sufficient flexibility for the specification of negation and significantly reduces the number of primitives. In the database, there is simply a negating bit set on the wire, which makes the code for extracting logic somewhat more complex although not unmanageable. The Flip-Flop can be parameterized in its triggering (master-slave, positive, or negative) and in its function (RS type, JK type, D type, and so on). The Buffer primitive, when negated, becomes an inverter and can also have an enable line on it.

Analog components are provided in the schematics environment and can connect freely with the digital primitives. Included in this set is the Transistor (*n*-type, *d*-type, *p*-type, *pnp*, *npn*, JFET, or MESFET), Resistor, Capacitor, Inductor, and Diode. All components can be parameterized with ratios and other relevant values.

There are some miscellaneous primitives useful in all schematic design. The Black-Box can contain any function and connect arbitrarily. The Power, Ground, Source, and Meter primitives are similarly general and can be used in simulation to indicate input and output graphically. Finally, an Off-Page connector is provided for network continuity between cells of the circuit. For truly unusual symbols, users can sketch their own bodies using the artwork environment described next.

The schematic-capture environment also provides buses and bus taps for aggregating signals. Although the logic gates can accept buses, it is up to the various analysis and synthesis tools to handle this correctly and ensure that the signal counts are sensible. Individual network naming is available from the network-maintainer tool described later.

11.4.5   Nonelectrical

As an exercise in expanding the power of Electric, some nonelectrical environments were created. The most interesting is the artwork environment, which contains graphic primitives for arbitrary image creation. Not only is this useful for the generation of sketches, but also it can augment VLSI design by adding artwork to layout or by defining new graphic symbols. This environment is so powerful that it was used to generate all the figures in this book. The availability of a general-purpose drawing facility also allows the simulators to represent waveform plots in the database.

The bulk of the graphics in the artwork environment is done with primitive components that appear as polygons, text, circles, arrowheads, and so on. There are also wires that can draw lines and curves but, in general, the notion of connectivity is less important to a sketching system.

When the artwork environment is used, many of the user interface settings must be changed because the command set for VLSI design is not the same as that is needed for sketching. This argument is also true for the distinction between layout and schematics: No two design environments can conveniently share exactly the same set of commands for their use. Thus, for each environment that is not standard VLSI layout, there is a command file that retailors the user interface for the best use. This file changes defaults, rearranges menus, and creates new commands appropriate to the different style of design.

11.5 Tools

It is not enough for a CAD system to provide a full spectrum of design environments; there must also be a set of tools for working in these environments to manipulate the circuitry. These synthesis and analysis tools must be available in an easy-to-use way, with common interfaces for both humans and machines. To achieve this integration, Electric has provided a standard structure of subroutine calls that each tool provides and invokes.

11.5.1 Integration

The tool interface in Electric resembles a miniature operating system with round-robin control. Electric continuously cycles through each tool, calling its **turn** routine, which gives it a chance to do some processing. Some tools may be switched off, in which case they do not function incrementally and do not receive a turn. During a tool's turn, it may examine the database and change it; the former being done by subroutine calls or direct data structure access and the latter being done exclusively by subroutine calls. Electric accumulates changes made during a tool's turn and **broadcasts** them to all tools when the turn is over. A broadcast consists of a set of subroutine calls to report new, deleted, and modified nodes, arcs, ports, and cells. Because of the constraint systems in the database, the actual changes may be more extensive than the requested changes, so the complete set is broadcast to every tool, including the one that issued the original changes. During the broadcast, a tool may take local action but it may not issue additional changes; these must be delayed until its turn.

The turn-broadcast cycle of tool invocation allows powerful control of design activity in a highly modular environment. An indication of the modularity is the fact that the entire user interface acts as a tool and works within this framework rather than being part of the database. Thus it is no different from any other tool and can interact uniformly with all of them.

Not only does the database manage the broadcast of changes, but it also handles the changes' undoing. All changes made by a tool during its turn are collected in a **batch** that is retained so that it can be reversed if undo is requested. The number of retained batches is usually more than the number of tools, so that a complete action and all reactions can be undone. The undo mechanism simply reverses the sense of every change and broadcasts it to each tool. Since this is a broadcast, it is a change and is aggregated in a new batch. This means that an undo can be undone repeatedly and in complex combinations with previous changes. It also means that interlocking mechanisms must be used to prevent unusual undo sequences from generating nonsensical changes.

11.5.2   Example

An example of tool interaction will illustrate the flexibility of this scheme. Assume that there are four tools: the user interface, the router, the design-rule checker, and the simulator (see Fig. 11.10). All but the last are on, and it is currently the user interface's turn (action 1 in the figure). During its turn, the user interface waits for a single command and executes it; in this case a component is created. The user interface merely sends this change to the database (action 2) and then returns control to Electric—its turn is over. This new component is then broadcast to the three "on" tools (action 3) so they can display it (the user interface), and queue it for further analysis (router and design-rule checker).

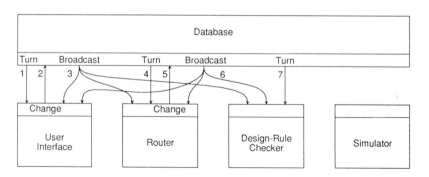

FIGURE 11.10 Sample Electric control sequence.

When the router gets a turn (action 4) it examines its queue of newly created or manipulated objects and checks to see whether it needs to make any connections. Routing action is not always incremental, but can be in Electric and is in this example. Any changes made (action 5) are again broadcast (action 6) and handled appropriately by each tool. For example, the user interface shows the new wires. The router knows not to examine its own changes when they are broadcast.

The final tool to run in this example is the design-rule checker (action 7), which looks for layout errors in the objects that have changed. All changes made by previous tools are checked here and any error messages are printed. Note that a correcting design-rule checker would be perfectly feasible to implement in this system but the current tool makes no changes. The simulator tool, being off, is skipped and control returns to the user interface.

The simulator tool can be made to run simply by turning it on. Once on, it will receive a turn immediately (since the command to start simulation completes the user interface's turn, and the absence of change causes control to roll to the simulator quickly). During its turn, the simulator will produce a netlist and/or run, depending on the particular simulation being done. The final activity of the simulator is to turn itself off since it has performed a one-shot activity. Thus even batch tools can function within the Electric control scheme.

Although this example describes the overall nature of some tools in Electric, there is much more to say about them. The rest of this section discusses the salient features of the individual tools of the Electric VLSI design system.

11.5.3 User Interface

The user interface is the most intelligent of Electric's tools; it is able to do both synthesis and analysis. During its turn, it waits for a single command from the keyboard or the pointing device and performs that activity. Commands can be issued as single keystrokes, button pushes on the pointing device, menu selections (fixed and pop-up), or full commands typed in with their parameters. The first three forms are shorthand for the last form, and dynamic binding can attach any full command to a single key, button, or menu. Combined with a parameterized macro facility, variables, and conditionals, the user interface can be tailored to resemble any system that is familiar to or comfortable for the user.

There are a number of different kinds of commands that can be issued to the user interface. Change commands such as "move," "create," and "undo" result in the change being sent to the database for constraint analysis and subsequent display when broadcast back.

Nonchange commands, such as the selection of a current object or a request to display a grid, affect the user interface and the screen without altering the database. The user interface relinquishes control and ends its turn even when nonchange commands are issued.

Another class of commands is one that affects other tools. The user interface can turn on or off any tool (except itself), thus controlling the entire system. When an incremental tool such as the design-rule checker is turned off, that tool will no longer examine each change. However, the database tracks which cells have been changed while a tool is off, and the tool will be able to catch up when turned back on, thus acting in a batch mode.

Many commands that appear to be standard functions of a design system are actually implemented as invocations to other tools. For example, reading and writing libraries is done by directing the I/O tool, and network highlighting is done through a request to the network-maintenance tool.

In addition to being very flexible and powerful, the user interface is coded in a highly modular style that permits ease of porting to any graphic display. All graphics input and output is done through a device-independent subroutine package that is modeled after the SIGGRAPH Core [GPSC]. The package is easy to modify for new devices and currently supports many displays.

11.5.4   Input and Output

The reading and writing of the database in various formats is the job of the input/output tool. Currently this system can handle two Electric-specific formats, one binary and one text. The binary file format is more compact but less portable to machines with different bit orga-nization. In addition, the I/O system can read and write CIF [Hon and Sequin], although CIF input appears as a collection of pure-layer components that are totally unconnected. A final format that the I/O system handles is a plot output that can be read by a separate program to produce hardcopy on a number of devices. For maximum generality, I/O of EDIF [EDIF committee] is expected soon.

11.5.5   Design-Rule Checking

The Electric design-rule checker runs incrementally, checking each change as it is made to the circuit. It operates by comparing the boundaries of all polygons on new and changed components. Although this polygonal comparison is less efficient than are raster methods when examining entire circuits, it is more efficient when checking small numbers of changes. To make the tool even smarter, each polygon is tied to a particular network path so that appropriate rules can be used when components are actually connected. The use of connectivity makes the design-rule checker more accurate and less prone to finding spurious errors. The only drawback of the current design-rule checker is that it is not hierarchical because incremental checking of all subcells would take too much time.

Another facility of the design-rule tool is a global short-circuit detector, which finds geometric connections that are not specified in the network. This is done by flattening the hierarchy and associating each polygon with a network node number. The list is then sorted to find polygon overlaps on different nets. In addition to finding short-circuits, this also helps to verify Electric's network information by finding components that should be, and even appear to be, connected, but simply are not.

11.5.6   Simulation

Electric currently has seven simulator interfaces that generate netlists for various simulators. ESIM, RSIM, and RNL [Baker and Terman] simulate MOS at a switch level as does MOSSIM [Bryant]. CADAT [HHB] is a functional simulator and MARS [Singh] is an experimental hierarchical simulator that can handle arbitrary behavioral specification.

Perhaps the most interesting simulation environment in Electric is SPICE [Nagel], which can graphically specify all parameters. Connecting meter and source components to a circuit, and parameterizing them appropriately, allows a complete input and output to be determined (see Fig. 11.11). When the simulation runs, the output values are used to generate a waveform plot with artwork primitives. Since this plot is a cell like any other, it can be saved as documentation, zoomed into for closer examination, and even altered with the normal editing commands to fudge the data!

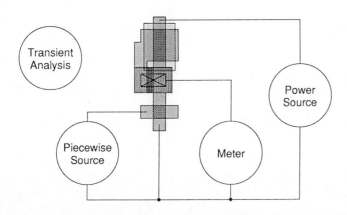

FIGURE 11.11 Graphical specification of SPICE simulation. Source components indicate inputs and meter components produce output plots. The transient-analysis component indicates the nature of the simulation.

11.5.7   Routing

The routers in Electric can handle many functions. A maze router is available to make point-to-point connections, a channel router connects multiple sets of points on opposite sides of a rectangular area, and a global router is under development to decompose irregular routing needs into the simpler channel problem.

When array-based layout is being done, implicit connections exist where two cells abut. These connections must be explicit for Electric to maintain its topological view of the circuit. To help do this, two stitching functions exist for the explicit connection of implicitly joined circuitry. The auto-stitcher looks for connection sites that adjoin or overlap geometrically and connects them explicitly. The mimic-stitcher watches user activity as wires are placed, and creates other wires in similar situations throughout the layout. These two stitchers work incrementally, adding wires as they are needed.

11.5.8   Compaction

Electric has a one-dimensional compacter that can work in a batch style or can function incrementally. When turned on, it adjusts the layout to its minimal design-rule separation. If the compacter remains on, it will close up the circuit when objects are deleted and will spread open the layout to make room for new components.

A more advanced compacter based on simulated annealing is being developed in Electric. It will work in a batch mode and will also interact with the user interface to maintain compact layout.

11.5.9   PLA Generation

There are two Programmable-Logic-Array generators in Electric: one specifically tailored for nMOS layout and another that is a general-purpose tiling engine. Both work from personality tables that describe an array of elements. The nMOS PLA generator produces all the surrounding circuitry, including drivers and power connections, so that a complete functioning module is available. The general-purpose PLA generator produces only those arrays of logic without any interconnect. However, it works from a user-specified library of array elements and orientation examples, so this library can include any peripheral circuitry.

11.5.10   Gate-Matrix Generation

The gate-matrix-generation tool produces CMOS layout from schematics. The input is a cell with schematic gates that are analyzed to produce Boolean equations. These equations are then transformed into a series of NAND expressions so that they can be uniformly implemented. The final step produces a regular array of CMOS layout, equivalent to the schematic. This tool is written in Prolog and yet is integrated uniformly in the Electric environment [Liu].

11.5.11   Verification

The verifier is able to aggregate symbolic descriptions of individual components to produce an overall circuit function that is then compared with a user specification of that circuit's behavior [Barrow]. The model of circuitry that is used can handle schematic gates as well as MOS transistors [Gordon]. As is the gate-matrix tool, the verifier is written in Prolog but runs well in the Electric system.

11.5.12   More Tools

There is a network-maintainer tool that merely propagates node information throughout the circuit whenever a change is made. It keeps lists of net names and can highlight any connected path.

Many other tools could fit easily in Electric: Analysis tools such as power estimation, timing verification, and test-vector generation would all be useful. More synthesis tools such as floor-planners, silicon compilers, and better routers will someday be added. Not only is the list of different tools unbounded, but the number of tools in a given class can grow to accommodate experimental versions of any CAD algorithm. Electric's integration scheme is designed to be able to incorporate modularly any VLSI CAD tool.

11.6   Designing a Chip

To illustrate the use of Electric, this section will describe the construction of a static-memory chip [Lyon and Schediwy]. This chip makes use of a novel four-transistor bit of static memory. The fundamental memory cell is logically designed using the schematics environment (see Fig. 11.12). An equivalent CMOS layout is then produced without concern for compact spacing (see Fig. 11.13). For proper layout efficiency, alternate bits of memory are different, so two bits define the leaf cell of the design. These bits can be compacted by the one-dimensional compacter (see Fig. 11.14) and then compacted further by rearranging components and recompacting (see Fig. 11.15).

To create a 128 × 32-bit array of memory, six levels of hierarchy are employed to build a 4 × 2 array, a 4 × 4 array, a 16 × 8 array, a 64 × 8 array, a 64 × 32 array, and a 128 × 32 array (see Fig. 11.16).

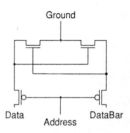

FIGURE 11.12 Schematic for four-transistor static-memory cell.

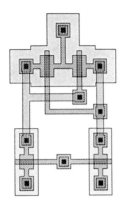

FIGURE 11.13 CMOS layout for four-transistor static-memory cell.

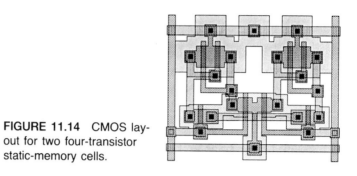

FIGURE 11.14 CMOS layout for two four-transistor static-memory cells.

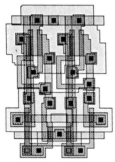

FIGURE 11.15 Compacted CMOS layout for two four-transistor static-memory cells.

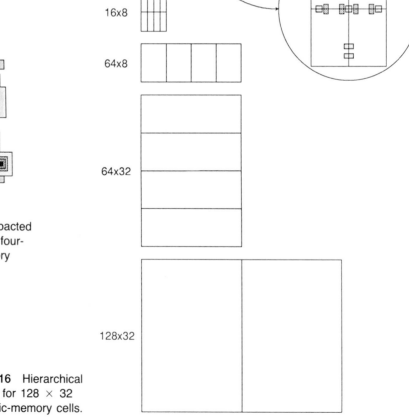

FIGURE 11.16 Hierarchical organization for 128 × 32 array of static-memory cells.

Note that each level of hierarchy actually connects its subcells with little stitches so that the overall connectivity is maintained. These stitches are automatically created by the router. Also, each level of hierarchy must export all unstitched ports to the next level. This is done automatically by the array-based layout commands.

Once the basic memory array is created, driving circuitry must be placed on the edges. A word (32-bit) driver for a single word is designed to pitch match the memory (see Fig. 11.17) and the driver is arrayed using hierarchy. The block of 128 drivers attaches to the bottom of the memory array. Similar drivers and decoders are built on the sides. The overall floor-plan, including pads, is shown in Fig. 11.18. This layout contains 32,650 transistors, described with 110 cells.

As an indication of the complexity of Electric, this circuit consumes 1.4 megabytes of disk space. Although large, it does contain more information than typical design databases. It can be read in 50 seconds and written in 30 seconds (on a SUN/3 workstation). Netlist generation takes only 15 seconds.

11.7 Summary

Electric is a workbench for the exploration of CAD algorithms. Because of its representational facilities, it can keep all kinds of information—much is already available. All the essential CAD system functions exist and are easily used. Thus node extraction and rasterization do not need to be implemented since the data are there. Input and output facilities are also available, as are powerful graphic display and editing.

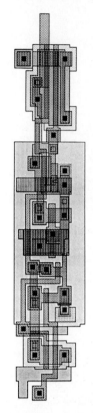

FIGURE 11.17 Word line driver for static-memory array.

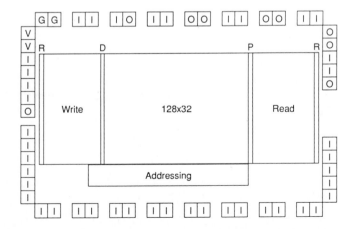

FIGURE 11.18 Floor-plan of static-memory chip.

The CAD tool programmer needs to write only the inner loop of new code, and it can be easily plugged into the Electric framework. The system's powerful integration allows the addition of tools, environments, constraint systems, graphic interfaces, and much more.

Questions

1 Why is Electric's user interface not more tightly coupled with the database?

2 What is the difference between a complex node instance and a primitive node instance?

3 Where would you place information about the power dissipation of wires?

4 How can overconstrained situations arise in the hierarchical-layout constraint system?

5 During change broadcasts to tools in Electric, why can further changes not be made?

6 How can routing needs be specified graphically?

References

Baker, Clark M. and Terman, Chris, "Tools for Verifying Integrated Circuit Designs," *Lambda*, 1:3, 22-30, 4th Quarter 1980.

Barrow, Harry G., "VERIFY: A Program for Proving Correctness of Digital Hardware Designs," *Artificial Intelligence*, 24:1-3, 437-491, December 1984.

Batali, J. and Hartheimer, A., "The Design Procedure Language Manual," AI Memo 598, Massachusetts Institute of Technology, 1980.

Bryant, Randal Everitt, *A Switch-Level Simulation Model for Integrated Logic Circuits*, PhD dissertation, Massachusetts Institute of Technology Laboratory for Computer Science, report MIT/LCS/TR-259, March 1981.

Clocksin, W. F. and Mellish, C. S., *Programming In Prolog*, 2nd Edition, Springer-Verlag, Berlin, 1984.

Electronic Design Interface Format Steering Committee, *EDIF—Electronic Design Interchange Format Version 1 0 0*, Texas Instruments, Dallas, Texas, 1985.

Gordon, M., "A Very Simple Model of Sequential Behaviour of nMOS," *VLSI '81* (Gray, ed), Academic Press, London, 85-94, August 1981.

GPSC, "Status Report of the Graphic Standards Planning Committee," *Computer Graphics*, 13:3, August 1979.

Griswold, Thomas W., "Portable Design Rules for Bulk CMOS," *VLSI Design*, III:5, 62-67, September/October 1982.

Guttman, Antonin, "R-Trees: A Dynamic Index Structure for Spatial Searching," *ACM SIGMOD*, 14:2, 47-57, June 1984.

HHB, *CADAT User's Manual*, Revision 5.0, HHB-Softron, Mahwah, New Jersey, June 1985.

Hon, Robert W. and Sequin, Carlo H., "A Guide to LSI Implementation," 2nd Edition, Xerox Palo Alto Research Center technical memo SSL-79-7, January 1980.

Johnson, Stephen C., "Hierarchical Design Validation Based on Rectangles," Proceedings MIT Conference on Advanced Research in VLSI (Penfield, ed), 97-100, January 1982.

Kernighan, Brian W. and Ritchie, Dennis M., *The C Programming Language*, Prentice-Hall, Englewood Cliffs, New Jersey, 1978.

Kingsley, C., *Earl: An Integrated Circuit Design Language*, Masters Thesis, California Institute of Technology, June 1982.

Kroeker, Wallace I., *Integrated Environmental Support for Silicon Compilation of Digital Filters*, Masters Thesis, University of Calgary Computer Science Department, March 1986.

Lansky, A. L. and Owicki, S. S., "GEM: A Tool for Concurrency Specification and Verification," Proceedings 2nd Annual ACM Symposium on Principles of Distributed Computing, 198-212, August 1983.

Lipton, Richard J.; North, Stephen C.; Sedgewick, Robert; Valdes, Jacobo; and Vijayan, Gopalakrishnan, "ALI: A Procedural Language to Describe VLSI Layouts," Proceedings 19th Design Automation Conference, 467-473, June 1982.

Liu, Erwin S. K., "A Silicon Logic Module Compiler," Project Report, University of Calgary Department of Computer Science, April 1984.

Lyon, Richard F. and Schediwy, Richard R., "CMOS Static Memory with a New 4-Transistor Memory Cell," Proceedings Stanford Conference on Advanced Research in VLSI, March 1987.

Mukherjee, Amar, *Introduction to nMOS and CMOS VLSI Systems Design*, Prentice-Hall, Englewood Cliffs, New Jersey, 1986.

Nagel, L. W., "Spice2: A Computer Program to Simulate Semiconductor Circuits," University of California at Berkeley, ERL-M520, May 1975.

Ousterhout, J. K., "Caesar: An Interactive Editor for VLSI Layouts," *VLSI Design*, II:4, 34-38, 1981.

Rubin, Steven M., "An Integrated Aid for Top-Down Electrical Design," *VLSI '83* (Anceau and Aas, eds), North Holland, Amsterdam, 63-72, August 1983.

Singh N. "MARS: A Multiple Abstraction Rule-Based Simulator," Stanford University Heuristic Programming Project HPP-83-43, December 1983.

Weste, Neil, "Virtual Grid Symbolic Layout," Proceedings 18th Design Automation Conference, 225-233, June 1981.

Williams, John D., "STICKS—A graphical compiler for high level LSI design," Proceedings AFIPS Conference 47, 289-295, June 1978.

APPENDIX:
GERBER FORMAT

This appendix describes the format of The Gerber Scientific Instrument Company's photoplotters. Although Gerber is only one of the manufacturers of photoplotters, many other companies have used **Gerber format** for their photoplotters, making it somewhat of an industry standard. Within the Gerber product line, there are many different models and they all read slight variations of this format. Some machines read a binary version of the format, some a text version, and some can accept either. For the sake of simplicity, only the common attributes of the text format will be described. More detail can be obtained from the Gerber Scientific Instrument Company, 83 Gerber Road West, South Windsor, Connecticut 06074.

A.1 Overall Format

Figure A.1 gives the BNF syntax of Gerber format. Each line of text contains a series of commands and their numeric parameters, terminated with an asterisk. A command consists of a single letter and the parameter is a number that immediately follows the letter. All numbers are specified with implied decimal points, such that a sequence of $m + n$ digits forms a number that has m digits to the left of the decimal point and n digits to the right. It is possible to shorten a number to fewer than $m + n$ digits by specifying that leading or trailing zeros are omitted. If, for example, $m = 3$, $n = 2$, and leading zeros are omitted, then the digits 12345 form the number 123.45 and the digits 400 represent the number 4.

The most important command letter is the G command, which is called the **preparatory function code**. The number that follows the G sets a mode for the data that follow (see Fig. A.2). For example, G57 is the comment command that allows documenting text to follow, up to the asterisk. There are six different drawing functions that these codes can introduce: lines, circular arcs, parabolas, cubic curves, text, and miscellaneous. In addition, the style of plotting coordinates can be declared with these preparatory function codes (absolute or relative).

Before the drawing operations are described, some background must be provided. The photoplotter writes in a random-access style by moving a **pen** to arbitrary locations on the film. This pen is actually a photographic writing head but the analogy to real pens is correct. When the pen is moved, it can be either up or down. Moving the pen while it is up simply causes the pen to be repositioned on the film. Moving the pen while it is down causes the pen to write. All the drawing commands presume that the pen is down.

file	::=graphics | graphics file
graphics	::=line | arc | parabola | cubic | text | misc
line	::=firstline [nextlines]
firstline	::=[seq] linetype coordinate
nextlines	::=nextline | nextline nextlines
nextline	::=[seq] coordinate
linetype	::=G12 | G11 | G01 | G10 | G60
arc	::=[seq] circletype X x Y y I i J j [draft] *
circletype	::=G02 | G03
parabola	::=firstparabola [nextparabolas]
firstparabola	::=[seq] G06 coordinatepair
nextparabolas	::=nextparabola | nextparabola nextparabolas
nextparabola	::=[seq] coordinatepair
coordinatepair	::=coordinate [seq] coordinate
cubic	::=[seq] G07 coordinatelist
coordinatelist	::=coordinate | coordinate [seq] coordinatelist
text	::=[seq] G56 [rotation] [scale] font message *
rotation	::=W [sign] digit digit digit digit digit digit
scale	::=M50 | M51 | M52 | M53 | M54
font	::=D10
message	::=messageletter | messageletter message
messageletter	::=*any letter but* *
misc	::=[seq] G54 [miscellaneous] [draft] | [seq] drawtype *
miscellaneous	::=M00 | M30
drawtype	::=G90 | G91
coordinate	::=X x Y y [draft] *
draft	::=D01 | D02 | D04 | D05 | D06 | D07
seq	::=*n digits*
x	::=[sign] digits
y	::=[sign] digits
i	::=digits
j	::=digits
sign	::=+ | -
digits	::=digit | digit digits
digit	::=0 | 1 | 2 | 3 | 4 | 5 | 6 | 7 | 8 | 9

FIGURE A.1 Gerber format.

G12	Line drawing, 0.01x scale
G11	Line drawing, 0.1x scale
G01	Line drawing, 1x scale
G10	Line drawing, 10x scale
G60	Line drawing, 100x scale
G02	Circular-arc drawing, clockwise
G03	Circular-arc drawing, counterclockwise
G06	Parabola drawing
G07	Cubic-curve drawing
G56	Text drawing
G54	Miscellaneous drawing
G57	Comment
G90	Absolute coordinates
G91	Relative coordinates

FIGURE A.2 Gerber format preparatory function codes.

A.2   Lines

To draw lines, the preparatory function codes G01, G10, G11, G12, or G60 must appear followed by coordinate information. For example, to draw a 3 × 3 square starting at the lower-left corner, the following codes can be used (presuming that *m.n* is 2.1):

```
G91*                G57 Set Relative Coordinates *
G01X030Y000*        G57 Draw bottom *
   X000Y030*        G57 Draw right *
   X-030Y000*       G57 Draw top *
   X000Y-030*       G57 Draw left *
```

Notice that the preparatory function code does not have to be repeated to remain in effect. The X and Y commands specify the actual drawing coordinates. They can accept signed values but the number of digits must fit the *m.n* format.

The line-drawing commands can be used to position the pen without drawing. This is done by placing the pen-up drafting code D02 on the drawing command line. Figure A.3 lists the possible drafting codes.

D01	Drawing on
D02	Drawing off
D04	Dashed lines Type 1
D05	Dashed lines Off
D06	Dashed lines Type 2
D07	Dashed lines Type 3
D10	Standard font
D11–D99	Alternate fonts

FIGURE A.3   Gerber format drafting codes.

For example, to move the pen to location (12,-3) use the following command (assuming absolute coordinate mode and that *m.n* is 2.2):

```
G01X1200Y-0300D02*
```

A.3   Circular Arcs

Drawing circular arcs requires that a center, a starting point, and an ending point be specified. The current pen position is used as the arc starting point; the X and Y parameters specify the arc endpoint; and the I and J parameters specify the arc center. The direction of drawing the arc is taken from the particular arc command used: G02 or G03. One restriction on the drawing of arcs is that they must start and end in the same quadrant. This is because the I and J parameters, which specify the arc center, are unsigned absolute distances relative to the starting point. Therefore the arc must remain in the same quadrant to prevent ambiguity in the center location. Drawing a complete circle requires that four of these commands be issued.

As an example of circular-arc drawing, suppose that the arc in Fig. A.4 is to be drawn. This arc crosses a quadrant boundary and so must be drawn in two pieces. Further assume that the *m.n* format is 2.2 and that the coordinate mode is absolute. These commands will draw the arc:

```
G01X-0400Y0300D02*        G57 Move to (-4, 3) *
G02X0000Y0500I0400J0300D01*  G57 Draw the left side to (0, 5) *
   X0300Y0400J0500*        G57 Draw the right side to (3, 4) *
```

Notice that the second line does not have an I command because the value is zero, which is the default.

The restriction of drawing only one quadrant at a time could be lifted if the I and J commands had signs. Then the circle center could

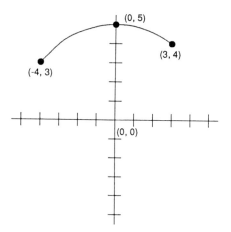

FIGURE A.4   Circular arc must be drawn in two pieces when it crosses a quadrant boundary.

be unambiguously specified and the quadrant would not be needed as a clue to the photoplotter. Some photoplotters do allow signs in the arc-center specification and these plotters can draw full circles at once. However, for the sake of compatibility, use of this option is not advised. A CAD system can break a circle into quadrants with little effort and the resulting description will not be significantly larger.

A.4   Curves

Parabolic curves can be specified by providing a start point, an endpoint, and a vertex through which the curve must pass. The start point is the current point; the vertex is specified by the first set of X and Y coordinates on that line; and the endpoint is specified by a second set of X and Y coordinates on the next line. For example, to draw the parabola in Fig. A.5, assuming relative coordinates, a current pen

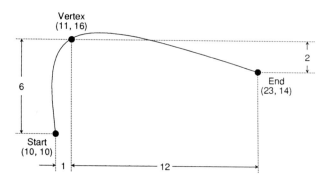

FIGURE A.5   Parabolic arc.

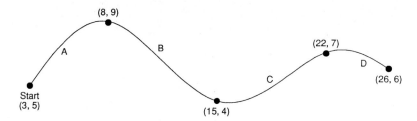

FIGURE A.6 Cubic curve.

position of (10,10), and *m.n* is 2.0, use:

```
G06X01Y06*     G57 The vertex *
  X12Y-02*     G57 The endpoint *
```

Note that it takes two lines in the file to draw the parabola and that the second line has no G code. The preparatory function code for parabolas sets a mode so that every pair of lines after the first parabola will be interpreted as another parabola until a different preparatory function code is introduced.

Cubic, or third-order, curves can be generated by giving a sequence of four or more points. Parabolic curve fitting is used on the ends and cubic fitting is used in the middle segments. For example, to draw the five-point curve in Fig. A.6, the following commands can be used (assuming absolute coordinates, *m.n* is 3.0 and current pen position is (3,5)):

```
G07X008Y009f*     G57 Draw segment A *
   X015Y004*      G57 Draw segment B *
   X022Y007*      G57 Draw segment C *
   X026Y006*      G57 Draw segment D *
```

The two end segments (A and D) will be drawn using the three endpoints in a parabolic curve, but the middle two segments (B and C) will be drawn using the four surrounding points in a cubic curve.

A.5 Text

Text can be drawn using the G56 preparatory function code. Following this code are two optional modifier codes, a nonoptional font code, and the text string to draw. All text is drawn so that the lower-left corner of the string is at the current pen position.

The first optional modifier of text is the W code, which specifies a rotation for the string. Following the W is an optional sign and a

M00	Program stop
M30	Rewind tape
M50	Scale text by 10
M51	Scale text by 25
M52	Scale text by 50
M53	Scale text by 75
M54	Scale text by 100

FIGURE A.7   Miscellaneous Gerber commands.

number in 3.3 format (that is, six digits with the implied decimal in the middle). This is a rotation in degrees counterclockwise from normal, rotated about the current pen position. Thus text rotated 90 degrees will read from bottom to top rather than from left to right.

The second optional modifier is the scaling factor, which is specified as an M code (see Fig. A.7) followed by a scaling number. The unscaled size of text depends on the plotting unit and is 0.015 inches square if plotting in inches or 1.5 millimeters square if plotting in millimeters.

The text font code specifies a character set to use when drawing text. This code is not optional and should usually be the drafting code D10 because font 10 is the only standard font. The text to be plotted follows the D10 immediately and proceeds up to the asterisk at the end of the line.

Some examples of text usage are in order. Figure A.8 shows the results of the following command (assuming that the current pen position is (5,7) and that the drawing units are inches):

```
G56M52D10HELLO THERE*
```

Figure A.9 shows the result of the following command (assuming that the current pen position is (6,8) and that the drawing units are millimeters):

```
G56W-090000D10AMY*
```

FIGURE A.8   Text plotting.

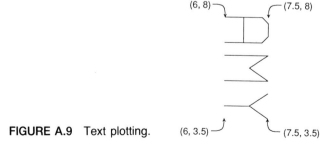

FIGURE A.9   Text plotting.

A.6   Miscellaneous

The final preparatory function code is G54, which performs miscellaneous control. Following that code is a D (drafting) code from Fig. A.3. These codes are used to raise or lower the pen and to select a pen drawing style. Notice that this code also selects the text font. Most usage of the drafting code is incorporated into other preparatory function codes. However, if the drafting code is to be issued independently of drawing, that can be done with the G54 code. For example, to draw the symbol in Fig. A.10, use these commands (assuming that the current position is (5,5), the drawing mode is relative, and the $m.n$ is 2.0):

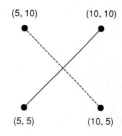

FIGURE A.10   Drawing example.

```
G01X05Y05*       G57 Lower-left to upper-right *
   X00Y-05D02*   G57 Move down without drawing *
   X-05Y05D04*   G57 Lower-right to upper-left (dashed) *
G54D05*          G57 Turn off dashed lines *
```

In addition to the preparatory function codes, a sequencing code can be used to order the lines in a file. The N command is used for sequencing followed by any digits (up to three) that form a sequence number. This number is not used by the plotter and is for organizational purposes only. Presumably, it dates back to the days of punched cards, when a dropped deck had to be reordered. Most data transfer is done by magnetic tape these days and many plotting houses have modems so that the data can be shipped by telephone.

There are many other options in Gerber format. However, most of them are restricted to certain plotter models and others pertain to obscure tasks. In general, task-specific parameters are set up before a plot is done. These parameters describe overall scaling, character fonts, dashed line styles, digit format ($m.n$), absolute or relative coordinates, and metric or English units. Some parameters can be changed during the plot but most are constant and apply uniformly to the task at hand.

APPENDIX: CALTECH INTERMEDIATE FORMAT

Caltech Intermediate Format (CIF) is a recent form for the description of integrated circuits. Created by the university community, CIF has provided a common database structure for the integration of many research tools. CIF provides a limited set of graphics primitives that are useful for describing the two-dimensional shapes on the different layers of a chip. The format allows hierarchical description, which makes the representation concise. In addition, it is a terse but human-readable text format. CIF is therefore a concise and powerful descriptive form for VLSI geometry.

Each statement in CIF consists of a keyword or letter followed by parameters and terminated with a semicolon. Spaces must separate the parameters but there are no restrictions on the number of statements per line or of the particular columns of any field. Comments can be inserted anywhere by enclosing them in parenthesis.

There are only a few CIF statements and they fall into one of two categories: geometry or control. The geometry statements are: LAYER to switch mask layers, BOX to draw a rectangle, WIRE to draw a path, ROUNDFLASH to draw a circle, POLYGON to draw an arbitrary figure, and CALL to draw a subroutine of other geometry statements. The control statements are DS to start the definition of a subroutine, DF to finish the definition of a subroutine, DD to delete the definition of subroutines, 0 through 9 to include additional user-specified information, and END to terminate a CIF file. All of these keywords are usually abbreviated to one or two letters that are unique.

B.1   Geometry

The LAYER statement (or the letter L) sets the mask layer to be used for all subsequent geometry until the next such statement. Following the LAYER keyword comes a single layer-name parameter. For example, the command:

```
L NC;
```

sets the layer to be the nMOS contact cut (see Fig. B.1 for some typical MOS layer names).

The BOX statement (or the letter B) is the most commonly used way of specifying geometry. It describes a rectangle by giving its length, width, center position, and an optional rotation. The format is as follows:

B *length width xpos ypos* [*rotation*] ;

Without the *rotation* field, the four numbers specify a box the center of which is at (*xpos*, *ypos*) and is *length* across in *x* and *width* tall in *y*. All numbers in CIF are integers that refer to centimicrons of distance, unless subroutine scaling is specified (described later). The optional *rotation* field contains two numbers that define a vector endpoint starting at the origin. The default value of this field is (1, 0), which is a right-pointing vector. Thus the *rotation* clause 10 5 defines a 30-degree counterclockwise rotation from the normal. Similarly, 10 –10 will rotate clockwise by 45 degrees. Note that the magnitude of this rotation vector has no meaning.

The WIRE statement (or the letter W) is used to construct a path that runs between a set of points. The path can have a nonzero width and has rounded corners. After the WIRE keyword comes the width value and then an arbitrary number of coordinate pairs that describe the endpoints. Figure B.2 shows a sample wire. Note that the endpoint and corner rounding are implicitly handled.

The ROUNDFLASH statement (or the letter R) draws a filled circle, given the diameter and the center coordinate. For example, the statement:

R 20 30 40;

will draw a circle that has a radius of 10 (diameter of 20), centered at (30, 40).

The POLYGON statement (or the letter P) takes a series of coordinate pairs and draws a filled polygon from them. Since filled polygons must be closed, the first and last coordinate points are implicitly connected

NM	nMOS metal
NP	nMOS polysilicon
ND	nMOS diffusion
NC	nMOS contact
NI	nMOS implant
NB	nMOS buried
NG	nMOS overglass
CMF	CMOS metal 1
CMS	CMOS metal 2
CPG	CMOS polysilicon
CAA	CMOS active
CSG	CMOS select
CWG	CMOS well
CC	CMOS contact
CVA	CMOS via
COG	CMOS overglass

FIGURE B.1 CIF layer names for MOS processes.

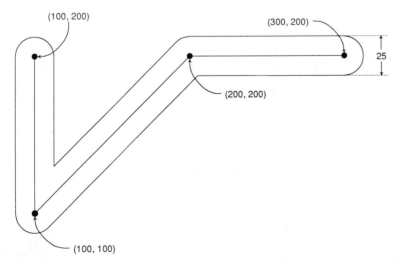

FIGURE B.2 A sample CIF "wire" statement. The statement is: W25 100 200 100 100 200 200 300 200;

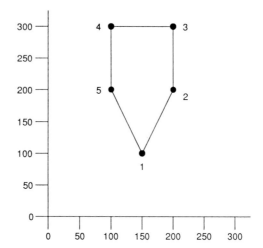

FIGURE B.3   A sample CIF "polygon" statement. The statement is: P 150 100 200 200 200 300 100 300 100 200;

and need not be the same. Polygons can be arbitrarily complex, including concavity and self-intersection. Figure B.3 illustrates a polygon statement.

B.2   Hierarchy

The CALL statement (or the letter C) invokes a collection of other statements that have been packaged with DS and DF. All subroutines are given numbers when they are defined and these numbers are used in the CALL to identify them. If, for example, a LAYER statement and a BOX statement are packaged into subroutine 4, then the statement:

C 4;

will cause the box to be drawn on that layer.

In addition to simply invoking the subroutine, a CALL statement can include transformations to affect the geometry inside the subroutine. Three transformations can be applied to a subroutine in CIF: translation, rotation, and mirroring. Translation is specified as the letter T followed by an x, y offset. These offsets will be added to all coordinates in the subroutine, to translate its graphics across the mask. Rotation is specified as the letter R followed by an x, y vector endpoint that, much like the *rotation* clause in the BOX statement, defines a line to the origin. The unrotated line has the endpoint (1, 0), which points to the right. Mirroring is available in two forms: MX to mirror about the x axis and MY to mirror about the y axis. The geometry is flipped about the axis by negating the appropriate coordinate.

Any number of transformations can be applied to an object and their listed order is the sequence that will be used to apply them. Figure B.4 shows some examples, illustrating the importance of ordering the transformations (notice that Figs. B.4c and B.4d produce different results by rearranging the transformations).

Defining subroutines for use in a CALL statement is quite simple. The statements to be packaged are enclosed between DS (definition start) and DF (definition finish) statements. Arguments to the DS statement are the subroutine number and a subroutine scaling factor. There are

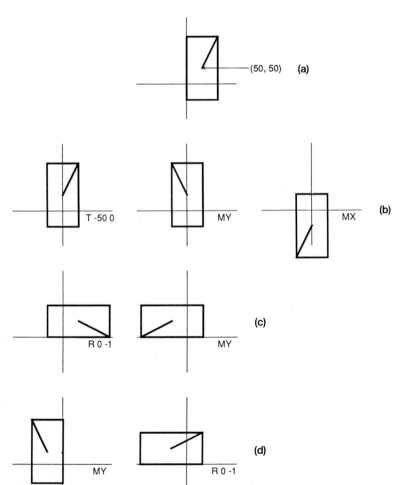

FIGURE B.4 The transformations of a CIF "call": (a) Subroutine 10: BOX 100 200 50 50; WIRE 10 50 50 100 150; (b) Invocation: C 10 T −50 0 MY MX; (c) Invocation: C 10 R 0 −1 MY; (d) Invocation: C 10 MY R 0 −1;

no arguments to the DF statement. The scaling factor for a subroutine consists of a numerator followed by a denominator that will be applied to all values inside the subroutine. This scaling allows large numbers to be expressed with fewer digits and allows ease of rescaling a design. The scale factor cannot be changed for each invocation of the subroutine since it is applied to the definition. As an example, the subroutine of Fig. B.4 can be described formally as follows:

```
DS 10 2 20;
  B10 20 5 5;
  W1 5 5 10 15;
DF;
```

Note that the scale factor is 2/20, which allows the trailing zero to be dropped from all values inside the subroutine. CIF subroutines may be nested in declaration and invocation. Forward references are allowed provided that a subroutine is defined before it is used. Thus the sequence:

```
DS 10;
  . . .
  C 11;
DF;
DS 11;
  . . .
DF;
C 10;
```

is legal, but the sequence:

```
C 11;
DS 11;
  . . .
DF;
```

is not. This is because the actual invocation of subroutine 11 does not occur until after its definition in the first example.

B.3   Control

CIF subroutines can be overwritten by deleting them and then redefining them. The DD statement (delete definition) takes a single parameter and deletes every subroutine that has a number greater than or equal to this value. The statement is useful when merging multiple CIF files because designs can be defined, invoked, and deleted without causing naming conflicts. However, it is not recommended for general use by CAD systems.

Extensions to CIF can be done with the numeric statements 0 through 9. Although not officially part of CIF, certain conventions have evolved for the use of these extensions (see Fig. B.5).

The final statement in a CIF file is the END statement (or the letter E). It takes no parameters and typically does not include a semicolon.

0 x y layer N name ;	Set named node on specified layer and position
0V x1 y1 x2 y2 ... xn yn ;	Draw vectors
2A "msg" T x y ;	Place message above specified location
2B "msg" T x y ;	Place message below specified location
2C "msg" T x y ;	Place message centered at specified location
2L "msg" T x y ;	Place message left of specified location
2R "msg" T x y ;	Place message right of specified location
9 cellname ;	Declare cell name
94 label x y ;	Place label in specified location

FIGURE B.5 Typical user extensions to CIF.

APPENDIX: GDS II FORMAT

In the design of integrated circuits, the most popular format for interchange is the Calma **GDS II** stream format (GDS II is a trademark of Calma Company, a wholly owned subsidiary of General Electric Company, U.S.A.). For many years, this format was the only one of its kind and many other vendors accepted it in their systems. Although Calma has updated the format as their CAD systems have developed, they have maintained backward compatibility so that no GDS II files become obsolete. This is important because GDS II is a binary format that makes assumptions about integer and floating-point representations.

A GDS II circuit description is a collection of cells that may contain geometry or other cell references. These cells, called **structures** in GDS II parlance, have alphanumeric names up to 32 characters long. A library of these structures is contained in a file that consists of a library header, a sequence of structures, and a library tail. Each structure in the sequence consists of a structure header, a sequence of **elements**, and a structure tail. There are seven kinds of elements: **boundary** defines a filled polygon, **path** defines a wire, **structure reference** invokes a subcell, **array reference** invokes an array of subcells, **text** is for documentation, **node** defines an electrical path, and **box** places rectangular geometry.

C.1   Record Format

In order to understand the precise format of the above GDS II components, it is first necessary to describe the general record format. Each GDS II record has a 4-byte header that specifies the record size and function. The first 2 bytes form a 16-bit integer that contains the record length in bytes. This length includes the 4-byte header and must always be an even number. The end of a record can contain a single null byte if the record contents is an odd number of bytes long. The third byte of the header contains the type of the record and the fourth byte contains the type of the data. Since the data type is constant for each record type, this 2-byte field defines the possible records as shown in Figs. C.1 and C.2.

Magnetic tapes containing GDS II files will have 2048 byte blocks that contain these records. The block size is standardized but has no bearing on record length or position. There is also a capability for circuits that require multiple reels of tape.

File Header Records:	Bytes 3 and 4	Parameter Type
HEADER	0002	2-byte integer
BGNLIB	0102	12 2-byte integers
LIBNAME	0206	ASCII string
REFLIBS	1F06	2 45-character ASCII strings
FONTS	2006	4 44-character ASCII strings
ATTRTABLE	2306	44-character ASCII string
GENERATIONS	2202	2-byte integer
FORMAT	3602	2-byte integer
MASK	3706	ASCII string
ENDMASKS	3800	No data
UNITS	0305	2 8-byte floats

File Tail Records:	Bytes 3 and 4	Parameter Type
ENDLIB	0400	No data

Structure Header Records:	Bytes 3 and 4	Parameter Type
BGNSTR	0502	12 2-byte integers
STRNAME	0606	Up to 32-characters ASCII string

Structure Tail Records:	Bytes 3 and 4	Parameter Type
ENDSTR	0700	No data

FIGURE C.1 GDS II header record types.

C.2   Library Head and Tail

A GDS II file header always begins with a HEADER record the parameter of which contains the GDS II version number used to write the file. For example, the bytes 0, 6, 0, 2, 0, 1 at the start of the file constitute the header record for a version-1 file. Following the HEADER comes a BGNLIB record that contains the date of the last modification and the date of the last access to the file. Dates take six 2-byte integers to store the year, month, day, hour, minute, and second. The third record of a file is the LIBNAME, which identifies the name of this library file. For example, the bytes 0, 8, 2, 6, "C", "H", "I", "P" define a library named "CHIP." Following the LIBNAME record there may be any of the optional header records: REFLIBS to name up to two reference libraries, FONTS to name up to four character fonts, ATTRTABLE to name an attribute file, GENERATIONS to indicate the number of old file copies to keep, and FORMAT to

Element Header Records:	Bytes 3 and 4	Parameter Type
BOUNDARY	0800	No data
PATH	0900	No data
SREF	0A00	No data
AREF	0B00	No data
TEXT	0C00	No data
NODE	1500	No data
BOX	2D00	No data

Element Contents Records:	Bytes 3 and 4	Parameter Type
ELFLAGS	2601	2-byte integer
PLEX	2F03	4-byte integer
LAYER	0D02	2-byte integers
DATATYPE	0E02	2-byte integer
XY	1003	Up to 200 4-byte integer pairs
PATHTYPE	2102	2-byte integer
WIDTH	0F03	4-byte integer
SNAME	1206	Up to 32-character ASCII string
STRANS	1A01	2-byte integer
MAG	1B05	8-byte float
ANGLE	1C05	8-byte float
COLROW	1302	2 2-byte integers
TEXTTYPE	1602	2-byte integer
PRESENTATION	1701	2-byte integer
ASCII STRING	1906	Up to 512-character string
NODETYPE	2A02	2-byte integer
BOXTYPE	2E02	2-byte integer

FIGURE C.2   GDS II element record types.

indicate the nature of this file. The strings in the REFLIBS, FONTS, and ATTRTABLE records must be the specified length, padded with zero bytes.

The parameter to FORMAT has the value 0 for an archived file and the value 1 for a filtered file. Filtered files contain only a subset of the mask layers and that subset is described with one or more MASK records followed by an ENDMASK record. The string parameter in a MASK record names layers and sequences of layers; for example, "1 3 5-7."

The final record of a file header is the UNITS record and it is not optional. The parameters to this record contain the number of user units per database unit (typically less than 1 to allow granularity of user specification) and the number of meters per database unit (typically much less than 1 for IC specifications).

Eight-byte floating-point numbers have a sign at the top of the first byte, a 7-bit exponent in the rest of that byte, and 7 more bytes that compose a mantissa (all to the right of an implied decimal point). The exponent is a factor of 16 in excess-64 notation (that is, the mantissa is multiplied by 16 raised to the true value of the exponent, where the true value is its integer representation minus 64).

Following the file header records come the structure records. After the last structure has been defined, the file terminates with a simple ENDLIB record. Note that there is no provision for the specification of a root structure to define a circuit; this must be tracked by the designer.

C.3 Structure Head and Tail

Each structure has two header records and one tail record that sandwich an arbitrary list of elements. The first structure header is the BGNSTR record, which contains the creation date and the last modification date. Following that is the STRNAME record, which names the structure using any alphabetic or numeric characters, the dollar sign, or the underscore. The structure is then open and any of the seven elements can be listed.

The last record of a structure is the ENDSTR. Following it must be another BGNSTR or the end of the library, ENDLIB.

C.4 Boundary Element

The boundary element defines a filled polygon. It begins with a BOUNDARY record, has an optional ELFLAGS and PLEX record, and then has required LAYER, DATATYPE, and XY records.

The ELFLAGS record, which appears optionally in every element, has two flags in its parameter to indicate template data (if bit 16 is set) or external data (if bit 15 is set). This record should be ignored on input and excluded from output. Note that the GDS II integer has bit 1 in the leftmost or most significant position so these two flags are in the least significant bits.

The PLEX record is also optional to every element and defines element structuring by aggregating those that have common plex numbers. Although a 4-byte integer is available for plex numbering, the high byte (first byte) is a flag that indicates the head of the plex if its least significant bit (bit 8) is set.

The LAYER record is required to define which layer (numbered 0 to 63) is to be used for this boundary. The meaning of these layers is not defined rigorously and must be determined for each design environment and library.

The DATATYPE record contains unimportant information and its argument should be zero.

The XY record contains anywhere from four to 200 coordinate pairs that define the outline of the polygon. The number of points in this record is defined by the record length. Note that boundaries must be closed explicitly, so the first and last coordinate values must be the same.

C.5   Path Element

A path is an open figure with a nonzero width that is typically used to place wires. This element is initiated with a PATH record followed by the optional ELFLAGS and PLEX records. The LAYER record must follow to identify the desired path material. Also, a DATATYPE record must appear and an XY record to define the coordinates of the path. From two to 200 points may be given in a path.

Prior to the XY record of a path specification there may be two optional records called PATHTYPE and WIDTH. The PATHTYPE record describes the nature of the path segment ends, according to its parameter value. If the value is 0, the segments will have square ends that terminate at the path vertices. The value 1 indicates rounded ends and the value 2 indicates square ends that overlap their vertices by one-half of their width. The width of the path is defined by the optional WIDTH record. If the width value is negative, then it will be independent of any structure scaling (from MAG records, see next section).

C.6   Structure Reference Element

Hierarchy is achieved by allowing structure references (instances) to appear in other structures. The SREF record indicates a structure reference and is followed by the optional ELFLAGS and PLEX records. The SNAME record then names the desired structure and an XY record contains a single coordinate to place this instance. It is legal to make reference to structures that have not yet been defined with STRNAME.

Prior to the XY record there may be optional transformation records. The STRANS record must appear first if structure transformations are desired. Its parameter has bit flags that indicate mirroring in *x* before rotation (if bit 1 is set), the use of absolute magnification (if bit 14 is set), and the use of absolute rotation (if bit 15 is set). The magnification and rotation amounts may then be specified in the optional MAG and ANGLE records. The rotation angle is in counterclockwise degrees.

C.7 Array of Structures Element

For convenience, an array of structure instances can be specified with the AREF record. Following the optional ELFLAGS and PLEX records comes the SNAME to identify the structure being arrayed. Next, the optional transformation records STRANS, MAG, and ANGLE give the orientation of the instances. A COLROW record must follow to specify the number of columns and the number of rows in the array. The final record is an XY with three points: the coordinate of the corner instance, the coordinate of the last instance in the columnar direction, and the coordinate of the last instance in the row direction. From this information, the amount of instance overlap or separation can be determined. Note that flipping arrays (in which alternating rows or columns are mirrored to abut along the same side) can be implemented with multiple arrays that are interlaced and spaced apart to describe alternating rows or columns.

C.8 Text Element

Messages can be included in a circuit with the TEXT record. The optional ELFLAGS and PLEX follow with the mandatory LAYER record after that. A TEXTTYPE record with a zero argument must then appear. An optional PRESENTATION record specifies the font in bits 11 and 12, the vertical presentation in bits 13 and 14 (0 for top, 1 for middle, 2 for bottom), and the horizontal presentation in bits 15 and 16 (0 for left, 1 for center, 2 for right). Optional PATHTYPE, WIDTH, STRANS, MAG, and ANGLE records may appear to affect the text. The last two records are required: an XY with a single coordinate to locate the text and a STRING record to specify the actual text.

C.9   Node Element

Electrical nets may be specified with the NODE record. The optional ELFLAGS and PLEX records follow and the required LAYER record is next. A NODETYPE record must appear with a zero argument, followed by an XY record with one to 50 points that identify coordinates on the electrical net. The information in this element is not graphical and does not affect the manufactured circuit. Rather, it is for other CAD systems that use topological information.

C.10   Box Element

The last element of a GDS II file is the box. Following the BOX record are the optional ELFLAGS and PLEX records, a mandatory LAYER record, a BOXTYPE record with a zero argument, and an XY record. The XY must contain five points that describe a closed, four-sided box. Like the boundary, this is a filled figure. Note that this element is no more compact than is a BOUNDARY element and is not allowed by MOSIS. Therefore it is not recommended.

APPENDIX: ELECTRONIC DESIGN INTERCHANGE FORMAT

The **Electronic Design Interchange Format (EDIF)** is a recent effort at capturing all aspects of VLSI design in a single representation. This is useful not only as a communications medium to manufacturing equipment, but also as an interchange format between CAD systems, since none of the high-level information is lost. EDIF is designed to be both easy to read by humans and easy to parse by machines.

EDIF files resemble the LISP programming language because of the use of prefix notation enclosed in parentheses. For example, the CIF polygon:

```
P 100 100 150 200 200 100;
```

looks like this in EDIF:

```
(polygon (point 100 100) (point 150 200) (point 200 100))
```

All EDIF statements consist of an open parenthesis, a keyword, some parameters, and a close parenthesis. The parameters can be other statements, which is what gives EDIF structure. Actually, an EDIF file contains only one statement:

```
(edif parameters)
```

where *parameters* are the described circuit.

Not only does EDIF resemble LISP, but in its highest level it contains all of LISP and is an extension of this highly expressive language. However, in the interest of making parsing simple, there are three levels of EDIF, and lower levels are less powerful. Level 1, the intermediate level, allows variables to be used and cell definitions to be parameterized. EDIF level 0 has no programmability and requires constants in all statements. A LISP preprocessor can translate from EDIF levels 1 or 2 down to level 0, and any given level of EDIF can be read by a parser of a higher level. Since level 0 is all that is necessary for most interchange and all manufacturing specification, only that level will be discussed here. Also, some of the EDIF constructs that deal with simulation, routing, behavior, and other unusual specifications will not be covered in detail.

D.1   EDIF Structure

An EDIF file contains a set of **libraries**, each containing a set of **cells**. Each cell can be described with one or more **views** that show the cell in the form of a schematic, layout, behavioral specification, document, and more. Each view has both an **interface** and a **contents** so that it is cleanly defined and can be linked to other views with a **view map**. Libraries may also contain **technology** information so that defaults can be given for behavior, graphics, and other attributes. The overall structure of an EDIF file looks like this:

```
(edif name
    (status information)
    (design where-to-find-them)
    (external reference-libraries)
    (library name
        (technology defaults)
        (cell name
            (viewmap map)
            (view type name
                (interface external)
                (contents internal)
            )
        )
    )
)
```

The status statement is used to track design progress and contains author names, modification dates, and program versions. Additional status statements may appear in each library, cell, and view. The design statement indicates where a completed design may be found by pointing to the top cell of a hierarchical description. The external statement names libraries that will be used but are not listed in this EDIF file. The library, cell, and view blocks can be repeated as necessary. There is also a comment statement that can be placed at the end of most blocks to add readability to the file. Here is an example of an EDIF file that further illustrates the outer level:

```
(edif my-design
    (status
        (edifversion 1 0 0)
        (ediflevel 0)
        (written
            (timestamp 1985 4 1 11 16 6)
            (accounting author "Steven Rubin")
            (accounting location "Palo Alto")
            (accounting program "Electric")
            (comment "timestamp contains year, month, day, hour,")
            (comment "                        minute, and second")
        )
    )
```

```
(design hot-dog-chip
    (qualify hot-dog-library top-cell)
    (comment "look for top-cell in hot-dog-library")
)
(external pad-library pla-library)
(library hot-dog-library
    (technology 3-micron-nMOS
        . . .
    )
    (cell top-cell
        (viewmap ... )
        (view masklayout real-geometry
            (interface ... )
            (contents ... )
        )
        (view schematic more-abstract
            (interface ... )
            (contents ... )
        )
    )
)
)
```

The written part of a status block may be repeated to show all authors and update events. Also note that the qualify statement, which names a cell in a particular library, is generally useful and can appear anywhere that an isolated name may be ambiguous or undefined.

In the following sections, more information is given to describe the contents, interface, viewmap, and technology blocks.

D.2   Contents

The contents of a cell may be represented in a number of different ways depending on the type of data. Each representation is a different view, and multiple views can be used to define a circuit fully. EDIF accepts seven different view types: **netlist** for pure topology as is required by simulators, **schematic** for connected logic symbols, **symbolic** for more abstract connection designs, **mask layout** for the geometry of chip and board fabrication, **behavior** for functional description, **document** for general textual description, and **stranger** for any information that cannot fit into the other six view types.

The statements allowed in the contents section vary with the view type (see Fig. D.1). The netlist, schematic, and symbolic views are essentially the same because they describe circuit topology. The al-

	Netlist	Schematic	Symbolic	View Mask Layout	Behavior	Document	Stranger
define	X	X	X	X	X		X
unused	X	X	X	X			X
global	X	X	X	X			X
rename	X	X	X	X	X		X
instance	X	X	X	X		X	X
joined	X	X	X	X			X
mustjoin	X	X	X	X			X
criticalsignal	X	X	X				X
required	X	X	X				X
measured	X	X	X				X
logicmodel					X		X
figuregroup				X			X
annotate		X	X				X
wire		X	X				X
section						X	X

FIGURE D.1 Contents statements allowed in EDIF.

lowable statements in these views are the declarations define, unused, global, rename, and instance; the routing specifications joined, mustjoin, and criticalsignal; and the timing specifications required and measured. Schematic and symbolic views also allow the annotate and wire statements. The mask-layout view allows all of the declarations, some of the routing constructs, and the figuregroup statement for actual graphics. The behavior view allows only a few declarations and the logicmodel statement. The document view allows only the instance and section constructs. Finally, the stranger view allows everything but supports nothing. It should be avoided whenever possible.

D.2.1 Declarations

Declarations establish the objects in a cell including signals, parts, and names. Internal signals are defined with the statement:

(define *direction type names*)

where *direction* is one of input, output, inout, local, or unspecified. Only local and unspecified signals have meaning in the contents section of a view; the others are used when this statement appears in an interface section. The *type* of the declaration can be port, signal, or figuregroup, where port is for the interface section, signal is for the contents section, and figuregroup is for the technology section of a library. The *names* being declared can be given as a single name or as a list of names aggregated with the multiple clause. In addition to all these declaration options, it is possible to define arrays by having any name be the construct:

(arraydefinition *name dimensions*)

These arrays can be indexed by using the construct:

(member *name indexes*)

Here are some examples of the define statement:

```
(define input port Clk)
(define unspecified signal (multiple a b c))
(define local signal (arraydefinition i-memory 10 32))
```

After the declaration of signals, a number of other declarations can be made. The unused statement has the form:

(unused *name*)

and indicates that the defined name is not used in this cell view and should be allowed to pass any analysis that might find it to be an error. The global declaration has the form:

(global *names*)

and defines signals to be used inside the cell view and one level lower, in subcomponents that are placed inside the view. Where these subcomponents have ports that match globally declared names, they will be implicitly equated.

Yet another declaration is rename, which can associate any EDIF name with an arbitrary string. This allows illegal EDIF naming conventions to be used, such as in this example:

(rename bwest "B-west{ss}")

D.2.2 Instances

Hierarchy is achieved by declaring an instance of a cell in the contents of another cell. The format of the instance statement is as follows:

(instance *cell view name transform*)

The name of the other cell is in *cell* and the particular view to use is in *view*. This allows the hierarchy to mix view types where it makes sense. A local name is given to this instance in *name* and an optional transformation is specified in *transform*, which looks like this:

(transform *scale rotate translate*)

where *scale* can be:

(scalex *numerator denominator*)

or:

(scaley *numerator denominator*)

rotate can be one of R0, R90, R180, R270, MX, MY, MYR90, or MXR90 and the *translate* clause has the form:

(translate *x y*)

So, for example, the expression:

(transform (scalex 3 10) MX (translate 5 15))

will scale the instance to three-tenths of its size, mirror it about the *x* axis (negate the *y* coordinate) and then translate it by 5 in *x* and 15 in *y*. Although any of the three transformation elements can be omitted, when present they must be given in the order shown, and are applied in that sequence. Unfortunately, there is no provision for non-Manhattan orientation.

Arrays of instances can be described by including a step function in the translate part of the transform clause. This will indicate a series of translated locations for the instances. The format of this iteration is:

(step *origin count increment*)

So, for example, the clause:

(step 7 3 5)

will place three instances translated by 7, 12, and 17 in whichever coordinate this appears. The rotation and scale factors will apply to every array element. Also, it is possible to use different instances in each array location by mentioning multiple cell names in the instance clause. For example,

```
(instance (multiple carry add)
   more-abstract add-chain
   (transform (translate 0 (step 0 16 10))))
)
```

will create an array of 16 instances stacked vertically that alternate between the "more-abstract" view of the "carry" cell and the "more-abstract" view of the "add" cell. This entire instance will be called "add-chain."

D.2.3 Routing and Simulation

To indicate connectivity, the joined construct identifies signals or ports that are connected. The mustjoin construct indicates that signals do not yet connect but should when routing takes place, and the criticalsignal construct establishes priorities for the routing. To illustrate further EDIF's expressive power in routing specification, there is a weakjoined construct that defines a set of joins, only one of which must be connected, and a permutable statement that declares sets of connection points to be interchangeable. These last two statements are found only in the interface section; however, none of the routing constructs will be described in detail.

The final set of constructs in the netlist, schematic, and symbolic views are those concerned with timing. As an example of the level of specification available, the statement:

```
(required (delay (transition H L (minomax 10 20 30)) here there))
```

states that the required delay of a high-to-low transition between points "here" and "there" is between 10 and 30, with a nominal value of 20. The measured statement can be used in the same way to document actual timing. The logicmodel statement, found only in the behavior view, allows a detailed set of logic states and conditions that can control simulation and verification. The EDIF specification should be consulted for full details of these timing, behavior, and routing constructs.

D.2.4 Geometry

In the **mask-layout** view, geometry can be specified with the figuregroup construct, which looks like this:

```
(figuregroup groupname
      pathtype width color fillpattern borderpattern
      signals figures)
```

where the *groupname* refers to a figuregroupdefault clause in the technology section of this library (described later). This set of defaults is available so that the graphic characteristics *pathtype*, *width*, *color*, *fillpattern*, and *borderpattern* need not be explicitly mentioned in each `figuregroup` statement. These five graphic characteristics are therefore optional and have the following format:

```
(pathtype endtype cornertype)
(width distance)
(color red green blue)
(fillpattern width height pattern)
(borderpattern length pattern)
```

The `pathtype` describes how wire ends and corners will be drawn (either `extend`, `truncate`, or `round`). The `width` clause takes a single integer to be used as the width of the wire. The `color` clause takes three integers in the range of 0 to 100, which give intensity of red, green, and blue. The `fillpattern` clause gives a raster pattern that will be tessellated inside of the figure. Two integers specify the size of the pattern and a string of zeros and ones define the pattern. Finally, the `borderpattern` describes an edge texture by specifying a single integer for a pattern length followed by a pattern string that is repeated around the border of the figure. Here are examples of these `figuregroup` attributes:

```
(pathtype round round)
(width 200)
(color 0 0 100)
(fillpattern 4 4 "1010010110100101")
(borderpattern 2 "10")
```

Inside of a `figuregroup` statement, the actual geometry can be specified directly with the *figures* constructs or can be aggregated by signal with the *signals* construct, which has the form:

```
(signalgroup name figures)
```

The *figures* construct in a `figuregroup` can be either `polygon`, `shape`, `arc`, `rectangle`, `circle`, `path`, `dot`, or `annotate`. The polygon is of the form:

```
(polygon points)
```

where each point has the form:

```
(point x y)
```

A `shape` is the same as a `polygon` except that it can contain `point` or `arc` information, freely mixed. The `arc` has the form:

```
(arc start middle end)
```

where these three points are the start point, any point along the arc, and the endpoint. The rectangle takes two points that are diagonally opposite each other and define a rectangle. A circle takes two points that are on opposite sides and therefore separated by the diameter. The path takes a set of points and uses the width and pathtype information to describe the geometry further. The dot construct takes a single point and draws it in a dimensionless way (it should not be used in actual fabrication specifications).

D.2.5   Miscellaneous Statements

In schematic and symbolic views, the annotate clause may be used to add text to a drawing. The form:

(annotate *text corner1 corner2 justify*)

will place *text* in the box defined by the two diagonally opposite corners, and justify the text according to one of nine options: upperleft, uppercenter, upperright, centerleft, centercenter, centerright, lowerleft, lowercenter, or lowerright. For example:

```
(annotate "probe here" (point 50 50) (point 200 100)
            uppercenter)
```

will place the string "probe here" in the upper part of the box $50 \leq x \leq 200$ and $50 \leq y \leq 100$, centered.

Another construct allowed only in schematic and symbolic views is wire. This connects two ports with a wire that can be described graphically. For example:

```
(wire clock.in gated.timer
    (figuregroup metal
        (path (point 10 15) (point 20 15) (point 20 25))
    )
)
```

connects the two points on the metal layer.

The last contents statement to be mentioned is the section construct, which is found only in document views and can hierarchically describe chapters, sections, subsections, and so on. For example:

```
(section "Chapter 1"
    (section "Introduction"
        "This is a book about VLSI CAD tools."
        "I hope you like it."
    )
)
```

D.3 Interface

In addition to there being seven different ways of specifying the contents of a cell, there are the same seven views that apply to the interface of a cell. The interface section is the specification of how a cell interacts with its environment when used in a supercell.

Unlike the contents views, the seven interface views are all essentially the same (see Fig. D.2). The netlist, schematic, symbolic, mask layout, behavior, and stranger views can all contain the same declarations: `define`, `rename`, `unused`, `portimplementation`, and `body`. They also allow the routing statements `joined`, `mustjoin`, `weakjoined`, and `permutable`, in addition to the simulation statements `timing` and `simulate`. The symbolic, mask-layout, and stranger views add the `arrayrelatedinfo` construct, which enables gate-array specification to be handled. The document view offers no constructs as this text rightly belongs in the contents section.

D.3.1 Ports and Bodies

The first interface statement to be discussed is `portimplementation`, which describes the ports and their associated components, graphics,

	Netlist	Schematic	Symbolic	View Mask Layout	Behavior	Document	Stranger
define	X	X	X	X	X		X
rename	X	X	X	X	X		X
unused	X	X	X	X	X		X
portimplementation	X	X	X	X	X		X
body	X	X	X	X	X		X
joined	X	X	X	X	X		X
mustjoin	X	X	X	X	X		X
weakjoined	X	X	X	X	X		X
permutable	X	X	X	X	X		X
timing	X	X	X	X	X		X
simulate	X	X	X	X	X		X
arrayrelatedinfo			X	X			X

FIGURE D.2 Interface statements allowed in EDIF.

timing, and other properties. Although ports can be declared with the `define` statement, `portimplementation` allows more information to be included in the declaration. The format is:

(`portimplementation` *portname figuregroups instances properties*)

where the *portname* is the name of the port as it will be used in supercells. The *figuregroups* describe any graphics attached to the port, the *instances* specify any subcells that describe the port, and the *properties* may indicate power-consumption ratings. Ports that are further described by `instances` of other cells do not need `figure-groups` to define them, so much of the `portimplementation` statement is optional.

The `body` statement is used to describe the external or interfaced aspect of a cell. In mask-layout views, this can describe a protection frame for design-rule checking and compaction. In other views it is simply used to give an external appearance to instances of the cell. The format is:

(`body` *figuregroups instances*)

where *instances* are subcells that can be used to describe the body.

D.3.2   Gate-Array and Behavioral Interface

The `arrayrelatedinfo` statement, which is used in gate-array specification, is allowed only in symbolic, mask-layout, and stranger views. This can be used to declare the background array:

(`arrayrelatedinfo basearray` (`socket` *info*))

or the individual cells:

(`arrayrelatedinfo arraysite` (`plug` *info*))

or macros of cells:

(`arrayrelatedinfo arraymacro` (`plug` *info*))

These statements define a grid that can be connected in a rigid manner, specified by the plugs and sockets. Sockets define permissible connection options and plugs make these connections to give precise gate-array interface. Consult the EDIF document for more information.

The final interface section constructs, which will not be described in detail, are `timing` and `simulate`. The `timing` statement gives port delays for various transitions, and gives stability requirements for the signal values. The `simulate` statement lists test data and expected results.

D.4 View Maps

To relate different views, a viewmap section can exist in each cell, which associates ports from different interface sections or instances from different contents sections. Port mapping is done with:

(portmap *ports*)

where the list of *ports* is of the form:

(qualify *viewname portname*)

Thus to equate port C of the mask-layout view with port D of the schematic view, the map would look like this:

```
(viewmap
    (portmap
        (qualify real-geometry C)
        (qualify more-abstract D)
    )
)
```

Note that the *viewname* is the declared name given to the view.

To relate instances of a cell in different views, the same format applies except that a many-to-one mapping is allowed. For example,

```
(instancemap
    (qualify real-geometry pullup pulldown)
    (qualify more-abstract inverter)
)
```

will map both the pullup and the pulldown in the mask-layout view to the inverter in the schematic view.

D.5 Technology

The technology section provides a background of information for the description of a library. Defaults can be set for other statements in the library, such as the figuregroup. Also, the real units of distance, time, power, and so on can be established. The technology section has the following format:

(technology *name*
    defines renames

```
      figuregroupdefaults
      numberdefinitions gridmaps
      simulation
)
```

where *name* is an identifier for this technology. A set of `define` statements can be used to declare default figuregroups for various signal types and `rename` statements can be used to establish name bindings in the library. The `figuregroupdefault` statement takes a name and a list of `pathtype`, `width`, `color`, `fillpattern`, and `borderpattern` constructs to establish the defaults for subsequent `figuregroup` statements in the library. The `numberdefinition` statement is important because it sets the scale of all EDIF units as follows:

```
(numberdefinition SI
      (scale distance edif real)
      (scale time edif real)
      (scale capacitance edif real)
      (scale current edif real)
      (scale resistance edif real)
      (scale voltage edif real)
      (scale temperature edif real)
)
```

The name `SI` is a standard that should always appear unless an alternate set of unit values is being declared. Any of the `scale` clauses may be used to declare the number of units in the EDIF file that correspond with real units. Real units for distance are in meters, which means that the clause:

```
(scale distance 1000000 1)
```

causes one million EDIF units to be a meter (or one EDIF unit to be a micron). The real-time unit is the second, capacitance is in farads, current is in amperes, resistance is in ohms, voltage is in volts, and temperature is in degrees celsius.

The `gridmap` clause of the technology section can be used to declare nonuniform scaling in the *x* and *y* axes. For example,

```
(gridmap 3 4)
```

will set the *x* units to be three times the `numberdefinition` distance and the *y* coordinates to be four times that amount. This nonuniform scaling of all coordinates has limited application.

A final use of the technology section is for simulation defaults. As with all other simulation constructs, these will not be discussed here.

APPENDIX: EBES FORMAT

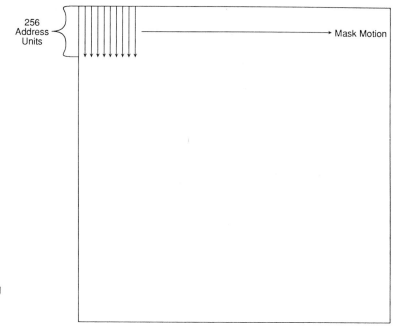

FIGURE E.1   EBES raster motion. Actual mask making sweeps out 256 rows as it advances horizontally.

One of the most popular formats for producing integrated-circuit masks is the **Electron Beam Exposure System (EBES)**, which can specify the very small features needed in high-density chips. Although the actual format varies with each different company's mask-making machine, an original standard, designed at Bell Laboratories, forms the basis on which extensions are made. This basic format is described here.

The electron beam is controlled in a digital raster fashion, such that a line of 256 points (called **address units**) can be written at a time. By convention, this sweep is run vertically and the mask is moved horizontally to make the sweep cover the top 256 rows of the chip (see Fig. E.1). It thus takes multiple passes across the width of the chip to write a complete pattern. When multiple copies of an IC die are being produced on a wafer, a single sweep of 256 rows is made on every die position before advancing to the next 256 rows. This means that the pattern file, which is organized in 256 row **stripes**, needs to be read only once.

E.1   File Structure

The EBES file is organized about these 256 row stripes. The beginning of the file contains a START DRAWING command followed by the stripes and an END DRAWING command. Each stripe consists of a START STRIPE command, a series of **figure** commands, and an END STRIPE command. The figures can be either RECTANGLEs, PAR-ALLELOGRAMs, or TRAPEZOIDs.

EBES files are binary, with 16 bits per word. Commands can take any number of words but must be aggregated into 1024-word blocks. If a command is near the end of a block and would span into the next block, then an END OF BLOCK command must appear followed by pad data to the block end. All blocks must end with the END OF BLOCK command except for the last block, which ends with the END DRAWING command.

E.2   Control Commands

The START DRAWING command is the first in an EBES file and contains 16 words. The first word has a 2 in the high byte and a code in the low byte that gives the size of an address unit. The address-unit size will be 1 micron if the code is 0, one-half micron if the code is 3, and one-quarter micron if the code is 6. The second and third words of the START DRAWING command are the x and y size of the entire pattern. Words 4, 5, and 6 contain the EBES file creation date: Word 4 is the month, word 5 is the day, and word 6 is the year. For these fields, the high byte is the first digit and the low byte is the second digit in ASCII. Words 7 through 13 contain the name of this file by having a name length in word 7 (always the value 6) and by having the next six words (8 to 13) contain an ASCII string with the format "XXXXXXXX.XX". Finally, words 14 through 16 describe the current mask layer by having word 14 contain a 2 and words 15 and 16 contain four ASCII characters.

After the START DRAWING command comes the first START STRIPE command. This command is one word with the value 7 in the low byte and the stripe number (from 1 to 255) in the high byte. Note that the word and byte sizes form restrictive limits on the overall pattern size. For larger dies to be made, multiple pattern files must be abutted.

After the START STRIPE come the figures. When all figures in a stripe have been listed, an END STRIPE command appears. The END STRIPE is a single word with the value 8. At the end of a block is an END OF BLOCK command, which is a single word with the value 9. At the end of the EBES file is an END DRAWING command, which is a single word with the value 4.

E.3   Rectangles

The RECTANGLE is one of the figures allowed in a stripe. It contains four words that describe the width, height, and corner position within the stripe. The first word contains the value 16 in the low 6 bits and contains the rectangle height (minus one) in the high 10 bits. Since the rectangle must fit in a 256-tall stripe, this height can have only the values 0 to 255. The second word of the rectangle command is the width, the third word is the starting x position, and the fourth word is the starting y position. Again, the y position can range only from 0 to 255 since it must be within the stripe.

E.4   Parallelograms

The PARALLELOGRAM command is similar to the RECTANGLE command except that it has two more words to give a skew distance and extra precision bits (see Fig. E.2). Word 1 has the value 17 in the low 6 bits and a value that is one less than the height in the top 10 bits. Words 2, 3 and 4 define the width, x position, and y position as in the RECTANGLE command, except words 2 and 3 are multiplied by 16. Word 5 contains the x offset between the bottom and the top (also multiplied by 16). The sixth word contains extra precision bits for the width (the low 5 bits), the x location (the next 5 bits), and the x offset (the high 6 bits). These extra precision bits are high bits for

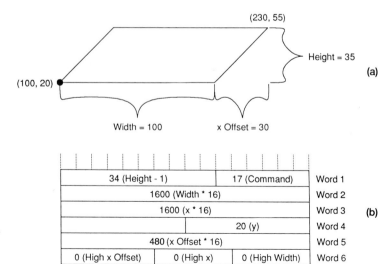

FIGURE E.2   EBES parallelogram example: (a) Figure (b) Record.

the three horizontal-position values, and define 21- or 22-bit fields for each. The reason for this extra precision is not to extend the addressing range but to allow fractional coordinates: These three values are in 1/16 address units. Thus there is an implied decimal point 4 bits from the right on all these numbers.

E.5   Trapezoids

There are three types of TRAPEZOIDs, as illustrated by Fig. E.3. Type 1 has a vertical right edge, type 2 has a vertical left edge, and type 3 has no vertical edges. The type 1 and 2 TRAPEZOIDs can be

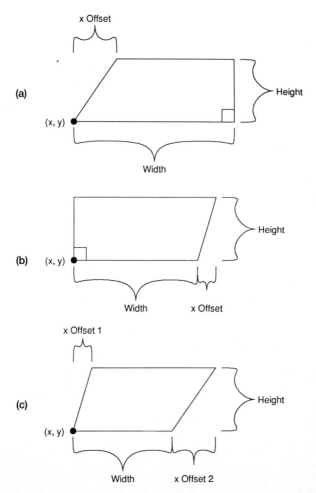

FIGURE E.3 EBES trapezoid types: (a) Type 1 has vertical edge on right (b) Type 2 has vertical edge on left (c) Type 3 has no vertical edges.

described in exactly the same format as is used for the PARALLEL-OGRAM because they have the same fields: x, y, width, height, and x offset. Type 1 TRAPEZOIDs have the value 18 in the low 6 bits of word 1, and type 2 TRAPEZOIDs have the value 19 in this field.

Type 3 TRAPEZOIDs require a seven-word description because of the extra x offset field. Word 1 has a 20 in the low 6 bits and the height minus one in the top 10 bits. Word 2 is the width, word 3 is the x location, word 4 is the y location, word 5 is the first x offset, word 6 is the second x offset, and word 7 contains additional precision bits for the width (low 5 bits), the x location (next 5 bits), and the first x offset (top 6 bits). The extra precision bits for the second x offset are in word 4 (the top 6 bits), which is otherwise used only for its low 8 bits to contain the y location.

APPENDIX:
ELECTRIC
TUTORIAL*

F.1   Introduction

Electric is a computer-aided design system that is tailored for VLSI circuits. It supports MOS, bipolar, and printed-circuit-board layout in addition to many abstract design environments such as schematics, artwork, and architectural specifications. A growing set of tools is available including design-rule checkers, simulators, compacters, routers, generators (such as PLA), and more. The most valuable aspect of Electric is its layout-constraint system, which enables top-down design by enforcing consistency of connections.

Electric views all circuits as collections of **nodes** and **arcs**, woven into a network. The nodes are electrical components such as transistors, contacts, and logic gates. Arcs are simply wires that connect the components. Although layout may appear to be connected when two components touch, the wire must still be used to make the connection. This wire is important because it indicates the connectivity to Electric.

Besides creating meaningful electrical networks, Electric can also place **constraints** on arcs to control geometric changes. For example, an arc that is constrained to be **rigid** does not change its length. This constraint means that, if a node on one end moves, both the arc and the other node move by an equal distance. These constraints propagate through the circuit and hold the layout together, even across hierarchical levels of a design.

This tutorial introduces the concepts and commands necessary to use Electric. It begins with essential features and builds on them to explain all aspects of the system. As with any computer system tutorial, the reader is encouraged to have a machine handy and to try out each operation.

Readers are cautioned that Electric is an everchanging system; thus, by now, has made some aspects of this tutorial incorrect. This document reflects version 3.26 of Electric. Use `-show environment` to find out what version you are running.

F.2   The Working Environment

Electric is very flexible and can run on any number of physical environments. The ideal situation consists of two display areas, a keyboard, and a mouse. This tutorial presumes a three-button mouse, although there can be four-button tablet pucks, one-button pens, or even systems with no pointing device at all. In the unfortunate case in which there is no pointing device, arrow keys on the keyboard position the cursor. However, since Electric can rearrange key and button functions on the fly, hardware configuration is no obstacle to effective use.

The Electric display consists of two distinct areas: the **editing** window and the **status** window. Although these areas can be drawn on the same physical screen, they may also be on two different displays. The status window is simply a video terminal that contains only text. It is divided into a **fixed** part in the top few lines and a **scrolling** part below that. The fixed part has useful pieces of status information that will be described throughout the tutorial. The scrolling part shows all conversation with the system.

The editing window is a full graphics display, usually in color, for viewing circuitry. Again, Electric is flexible and can work on a black-and-white editing window by using stipple patterns instead of colored areas. In extreme circumstances, Electric can use a video terminal as its editing window, writing ''T'' where the transistor belongs and generally limping along with cursor control.

One common configuration worth describing in detail is the main-frame computer (for example, a VAX) with attached **frame buffer**. Here, the user has a video terminal that is used to communicate with the computer. Sitting next to the video terminal is the frame buffer (for example, an AED, a Raster Technologies, or a Tektronix) with a tablet or mouse. The user runs Electric from the video terminal, which then becomes the status window. Electric activates the frame buffer for use as the editing window. It may be necessary to tell the system about the frame buffer so that it knows which one is near your terminal. It may also be necessary to initialize the frame buffer so that Electric can control it. Site-specific guidelines, available from your local Electric guru, tell you exactly what to do.

Another common machine configuration is the **workstation** (for example, Sun or Apollo) that is able to display multiple windows. Here, you invoke Electric from one window, which then becomes the status display. A new window is created for editing, all on the same screen. Some workstations have separate color displays, in which case these are used in a manner similar to the mainframe, as discussed previously. On some workstations, the mouse cursor can roam any-where, even crossing screen boundaries where they exist. In such situations, the user is reminded that Electric accepts commands only when the cursor is in the editing window. As usual, consult local guidelines for the exact procedure to use.

F.3   Basic Editing Concepts

This section introduces the first few commands to Electric and gets you started in circuit design. The lesson that follows is available in the online tutorial, so you could let the system teach this part. In fact, when Electric begins, it suggests that you type —help if you need assistance. Typing this command (do not forget the "-" before the word "help") further informs you that the tutorial is available with the command —com tutor, and from there you get to here.

F.3.1   Introductory Example

Before you can do any layout creation, there must be a **cell** in which to place layout. The command editcell creates a new cell (or calls up an old one) and should be invoked in this way:

```
-editcell wuzzy
```

which creates the cell called "wuzzy." The contents of the editing window remains blank, but the status display changes to show this new cell name at the top.

Now it is possible to create some circuitry. To do this, first select a component from the menu. Since the initial design environment is "nmos," select a metal-polysilicon contact. This contact can be found in the forth entry from the upper-left corner (if the menu runs across the top) or in the forth entry up on the lower-right corner (if the menu runs down the side). To select a menu entry, press any button while the cursor is in the box surrounding that entry. You know that you have the correct component because the status terminal prints:

```
-getproto node Metal-Polysilicon-Con
```

If you get the wrong component, try again. Note that this command, associated with the menu entry, can also be typed in directly.

Once you have selected the proper component, it is highlighted in the menu and its name appears as the "NODE:" at the top of the status display. Now you can place it in the layout by pointing to its destination and pressing the button with the create command (middle button on mice, blue button on tablets). The status display echos the command: —create and the component appears under the cursor.

Now create a second component of a different type. To select this other component, press the space bar. This key is attached to the command —getproto next-proto, which advances the component selection to the metal-diffusion contact. Use the `create` button to place this component elsewhere in the cell.

The screen now has two objects, one of which is **highlighted** with a white box. To highlight the other, place the cursor over it and press the button with the `find` command (left button on mice, white button on tablets). The highlight switches to the other contact.

For a wire to be run between two components, both components must be highlighted. A special option of the `find` command allows you to select a second component. For this example, place the cursor over the unhighlighted component and press the ''o'' key (use the green button on tablets). Both contacts are now highlighted and an enclosing box covers them.

Actual creation of the connecting wire is also done with the `create` command (middle or blue button). Because two components are selected, Electric knows to connect them. Now, unless the two components are in a line, the connecting wire makes a bend. This path is actually two wires and a ''pin'' component at the bend point. Since the connection runs along two sides of the rectangle enclosing the components, there are two ways that the wires may be placed. The choice is determined by the location of the cursor when the `create` command is issued.

To change the way the wires run, you can issue the `undo` command (the ''u'' key or the big ''X'' in the menu) to remove the wires. Then reselect the other component (''o'' key or green button) and make sure the cursor is in the opposite corner when you press the button with the `create` command.

Use the button with the `find` command (left or white) to select all five objects on the screen. There should be two contact nodes, a metal pin node, and two metal arcs.

Any node can be moved, so try moving a contact. To do this, `find` a contact, move the cursor to a new location, and press the button with the `move` command (right on mice, yellow on tablets). Notice that the arcs stretch and move to maintain the connections. What has actually happened is that the programmable constraint system has followed instructions stored on the arcs that react to node changes. The default arc is **Manhattan** and **slidable**, so the letters ''MS'' are shown when the arc is highlighted. Pick an arc and type the `arc rigid` command (the '' | '' key). The letters change to ''R'' on the arc and the wire no longer stretches when components move. Now move the contact on that arc and notice that the pin node always stays rigidly attached. Find the other arc and type the `arc not manhattan` command (the ''/'' key). Now observe the effects of an unconstrained arc as its neighboring nodes move. These arc constraints can be reversed with `arc manhattan` (the ''+'' key) and `arc not rigid` (the ''~'' key).

Electric supports hierarchy by allowing a node to be an instance of another cell. To see hierarchy in action, it is best first to split the display into two windows. Type:

```
–window split
```

The current cell ("wuzzy") appears in both windows, but only one window is highlighted with a green border. Now type:

```
–editcell higher
```

which creates a new cell in the highlighted window. Try creating a few simple nodes in this window (place a contact or two). Then place an instance of "wuzzy" by typing:

```
–getproto wuzzy
```

followed by the `create` button. The box that appears is a node in the same sense as are the contacts and pins: it can be moved, wired, and so on. In addition, since the box contains subcomponents, you can see its contents by pressing the menu entry with the open eye (the `node expand` command). The closed eye (`node not expand`) returns the currently highlighted cell to outline form. If the objects in the cell no longer fit in the display window, use the menu entry with the four outward-pointing arrows at the outside edge (`window all–displayed`).

Before you can attach wires to the "wuzzy" node, there must be connection sites, or **ports**, declared in the cell definition. **Primitive** nodes such as contacts and transistors already have their ports established, but you must explicitly create ports in cells. Move the cursor to the "wuzzy" window and select a contact node. Then type:

```
–port export toe
```

This takes the connection site on the contact node and **exports** it to the outside world. You now see it on the unexpanded "wuzzy" node in the "higher" cell. You can connect wires to that node in just the same way as you wired the contacts before. Keep playing until it becomes obvious.

By now you should have a reasonable feel for Electric. The rest of this tutorial provides increasingly greater detail for the commands to the system.

F.3.2   Typing Commands

The sample editing session illustrates an important notion about Electric. There is a distinction between typing full commands and using abbreviated methods. When you hit the space bar or the "o" key, you have used an abbreviation for a full command that is both executed and printed on the status display. All full commands begin with the "-" key, which tells Electric to read a command, rather than to take an immediate action. Once you have typed the "-", Electric helps you with the command and its parameters by filling out any partially typed words. For example, if you type "-ed" and a space, Electric completes the `editcell` command by writing the letters "itcell". If, at any time, you are unsure about what to type, hit the "?" key; Electric will explain what it wants and will present a list of options. Of course, the backspace key and the ^U key delete the last character and the entire line, just the same as in the operating system.

Throughout the rest of this tutorial, commands such as `create` are printed without their introductory "-". If an abbreviated method is available (a button, single key, or menu entry) then you are told. Otherwise, do not forget to use the "-" before typing a command.

F.3.3 "Find"

The first command to be described in detail is the `find` command, which controls highlighting in the editing window. In general, only one object is highlighted at a time. However, a special option of the `find` command allows a second object to be highlighted also. Besides highlighting objects, Electric can also highlight an arbitrary rectangular area. This notion of a **highlighted area**, as opposed to a **highlighted object**, is useful in some commands. In addition, when two objects are highlighted, the rectangle that encloses them is considered to be a highlighted area.

Highlighted objects have a white box drawn around them. In some cases, the object extends beyond the white box, but the box encloses the essential part of the object. For example, MOS transistors are highlighted where the two materials cross, even though the materials extend on all four sides. Also, CMOS diffusion arcs have implants that surround them, but the highlight covers only the diffusion part.

Besides the basic white box, there may be other things drawn in white when an object is highlighted. If the `find` command uses the `port` option, then highlighted nodes also have highlighted ports. The port that is highlighted is the one closest to the cursor. If the port is a single point, you see a white "+" at the port. If the port is an area, a white line or rectangle indicates its extent.

When the `extra-info` option is used in the `find` command, even more highlighting takes place. Nodes highlighted in this way have dashed lines drawn down the center of all connected arcs. Arcs highlighted with this option have their constraint characteristics printed in white. For example, a highlighted Manhattan arc displays the letter "M" in its center.

The `find` command highlights the object that is closest to the cursor. When more than one object is under the cursor, multiple `find` commands cycle through the direct hits. Many options may be included as parameters: `another` indicates that this is the second object being highlighted; `port` requests a port to be highlighted on the node; `extra-info` requests that additional information be displayed; and `exclusively` requests that only objects of the current node or arc type be highlighted. As examples of these options, the left (or white) button is the command `find port extra-info` and the "o" key (or green button) is the command `find another port extra-info`.

Area highlighting is done with the commands `find area-move` and `find area-size`. The former moves the highlighted area rectangle so that its lower-left corner is at the cursor, and the latter stretches the upper-right corner to the cursor. These commands are available as the "," and "." keys, respectively.

The currently highlighted objects or area can be saved and restored with a number of options. The command `find down-stack` pushes the highlight onto a stack, and `find up-stack` restores the top of the stack as the current highlight. Besides simply being stacked, the highlight can be named for subsequent retrieval with `find save HIGHNAME`. To show a named highlight, use `find name HIGHNAME`. Retrieval of saved or stacked highlights may switch to the second object by using `find another up-stack` or `find another name HIGHNAME`.

The `find clear` command removes all highlighting on the screen. Other `find` options are discussed where appropriate in later sections.

F.3.4   "Help"

The `help` command is very useful. When Electric starts, it suggests that you use this command, and the basic `help` command explains all its options. In general, any other command can be given as a parameter and you then get a manual entry for that command. For example, `help find` will describe the previous command. Special topics can also be mentioned. The command `help apropos` takes another parameter and searches for relevant help entries. You can also type `help *` to see all the commands in Electric, organized by category.

A news facility exists to show users any changes that have been made to the system. Typing `help news` shows any recent additions to the systemwide news file.

F.4   Circuit Creation

This section describes the commands that you use to change the circuit. Two basic commands—`create` and `move`—have already been discussed, but there is more to know about these commands and others.

F.4.1   "Create"

The `create` command (middle or blue button) places nodes and arcs in the current cell. The type of node that is created can be found at the top of the status display after the heading "NODE:". You can change this selection with the `getproto` command, or by using `find` to select an existing node on the screen. When you issue the `create` command, the new node appears such that its lower-left corner is at the cursor. Slight adjustments may be made if the cursor is not exactly on a grid unit, but the size of this adjustment can be controlled with the `grid` command.

The `create` command is also used to run an arc between two nodes. This happens when you have highlighted two nodes simultaneously (with the `find` and `find another` commands). Arc creation first attempts to run a single wire between the two nodes. Generally, this can happen only if the ports are lined up horizontally or vertically. Failing single wire placement, arc creation attempts to make the connection with two wires and an intermediate pin. This usually works, except in situations in which the two ports are directional and face each other in a difficult orientation. For example, if two transistors are stacked vertically and you wish to connect two side ports, then three wires are created to go out, down, and back in. After arc creation, only the original object (selected with `find`) is highlighted unless the command `create remain-highlighted` is issued, in which case the original highlighting of two objects is retained.

In addition to creating an arc between two nodes, you can also select arcs as the starting or ending point of an arc creation. If it is sensible, the `create` command actually uses one of the nodes on an end of the highlighted arc. However, if the perpendicular connection falls inside the arc, the arc is split and a new pin appears to make a "T" connection.

Another way to insert a node within an existing arc is to use `create insert` (the "I" key). This command requires that an arc be highlighted and a current node be selected. A node of the current type is then inserted in the highlighted arc at the position of the cursor.

Another option to the `create` command produces one arc and one node. The `create angle` command places a pin node near the cursor and connects it to the currently highlighted node. The new pin node is then left highlighted. For example, to draw the path of a serpentine wire, select a starting node and then type "d" (`create angle 90`) every time you have moved the cursor to the next bend along the path. The parameter to `create angle` is the number of degrees about the starting node along which possible arcs may appear. The value 90 means that the arc must be in 90-degree increments about the starting node (always vertical or horizontal). The value 45 allows eight possible arc angles, and the value 0 leaves the arc totally unrestricted in its final angle.

F.4.2   "Erase"

The next useful editing command is `erase` (the "e" key or the menu entry with the pencil eraser). This deletes the currently highlighted node or arc. If there is a highlighted area rather than a highlighted object, all objects in the area are erased. Note, however, that an arc cannot remain if its node is gone. Therefore, when a node is erased, all connecting arcs are also deleted. Also, if an erased node has an exported port, then the port disappears and so do all arcs connected to the port on instances of the current cell.

If there are exactly two arcs connected to a node and you use `erase pass-through` (the "E" key) on that node, then a new arc is created to replace the two that were previously attached.

F.4.3   "Move"

The `move` command moves objects. The currently selected node is moved so that its lower-left corner is at the location of the cursor. As in the `create` command, the grid may affect the actual cursor position. The `move angle` option, like its counterpart in the `create` command, moves the node along only the radial lines coming out of the node at the specified angle increments. Thus the `move angle 90` command (the "l" key) always moves either horizontally or vertically toward the cursor, but not both.

The `move` command can also specify relative motion. The commands `move up` ("w" key), `move down` ("z" key), `move left` ("a" key), and `move right` ("s" key) move one unit in the specified direction. The four keys sit in a diamond on the keyboard and their direction is intuitively correct. When they are capitalized, the motion is eight times greater. Part of this setup includes three other keys: "f" for full, "h" for half, and "q" for quarter. Thus, after you type "f," the "W" key moves an object up 8. After a "q" however, the amounts are 0.25 for "w" and 2 for "W". These relative motion commands can actually take any value if they are issued in long form.

Arcs may also move, but not if they cause node motion. Thus the only permissible move command when an arc is highlighted is one that wiggles the arc within its ports. Both ports must have nonzero area and the arc must not have constraints that restrict its motion.

F.4.4 "Size"

Once a node has been placed, its size and orientation can be altered with the three commands size, rotate, and mirror. The size command takes two values: the new x and y size of the node. This command also works for arcs and takes one value: the new width. Note that the size in question is the size of the highlighted area. Thus a default nMOS transistor has a highlighted area (where diffusion and polysilicon cross) of 2 × 2, even though the component is 6 × 6 if you include the four overlap regions that stick out. A CMOS diffusion arc is highlighted on the 2-wide diffusion area, even though the complete arc has implant regions that are much larger.

There are two options to the size command that allow cursor position to control the scaling. Both options find the corner that is closest to the cursor, and pull that corner to the cursor. In the size center—fixed option, the center of the node is held at a fixed position and the object expands uniformly to reach the cursor. Thus the close corner stretches to the cursor and the diagonally opposite corner moves away by the same amount. Arcs pull the close edge to the cursor and push the far edge away. In the size corner—fixed command (the "b" key), the close corner moves to the cursor but the opposite corner remains fixed. The center thus travels one-half of the distance of the close corner. Cursor location is snapped to the grid with these options, but the size center—fixed command allows half-grid locations to account for the fact that it doubles cursor distance.

F.4.5 "Rotate" and "Mirror"

The other two node modification commands are rotate and mirror. The rotate command (the menu entry with the counterclockwise curving arrow) rotates the current node by 90 degrees from its current orientation. If the command is given a numeric parameter, then the node is oriented at exactly that many degrees, regardless of its current rotation. The mirror command takes a single parameter: horizontal or vertical. The node is flipped about its horizontal center or its vertical center, accordingly.

Be aware that mirroring is not the same as rotating, even though both may produce the same visual results. Mirroring causes the node to be **transposed**, and transposed nodes rotate in the opposite direction. A second mirroring removes the transposition of the node and restores normal rotation. To see the orientation of an object, use the show object command (the "i" key).

F.4.6   "Copynode"

Once you have created, scaled, rotated, and mirrored a node, it may be desirable to have many other identical nodes. Two choices face the designer: change the defaults so that newly created nodes begin this way (see the defnode command), or make a duplicate of the existing node. To make a duplicate, use copynode (the "t" key), which creates a new node at the current cursor location, identical to the highlighted node.

F.4.7   "Array"

If one copy is not enough, you may want an array of copies. The array command takes the currently highlighted node and replicates it many times. The first two parameters to this command are the x replication and the y replication, where a value of 1 means no additional nodes in this direction. If either replication value has the letter "F" after it, the array elements are flipped (mirrored) in alternate rows or columns. If only one replication factor is given, it is used in both directions.

Two additional parameters may appear after the replication values: these values are the node-overlap amounts. The default is to abut each array entry with no overlap. Positive overlap values cause the array entries to move in, and negative overlap values spread the array. Note that the x and y overlap values are always the third and fourth parameters to this command so, in order to specify them, you must give both replication factors.

A fifth and final parameter to the array command is the optional keyword no-names. By default, array entries are given names that index each entry. The original node is named "0-0" and the entry to its right is named "1-0". These names are simply visual tags that have no bearing on the node contents (see the node command for name manipulation). If you do not want these names to be placed, use the no-names option. This option must come in the fifth parameter position, so the first four must be specified. For example, to create an array that is 3 × 3, flipping the middle column vertically, abutting entries exactly, and using no names, type array 3f 3 0 0 no-names.

F.4.8 "Spread"

When a large amount of circuitry has been placed too close together or too far apart, Electric's constraint system can help. All that is necessary is to make all arcs in an area rigid and then to move one node. Of course, you may have to move more than one node if the one you pick is not connected to everything else you want to move. In addition to making arcs within the moved area rigid, you must make sure that arcs connecting across the area boundary are nonrigid. Finally, setting arc rigidity should be done temporarily so that it does not spoil an existing constraint setup. All these operations are handled for you by the spread command.

With the spread command, the highlighted node is a focal point about which objects move. The command takes two parameters: a direction (up, down, left, or right) and an amount. An infinite line is passed through the node center and everything above, below, to the left of, or to the right of the line is moved by the specified amount. Negative spread distances compact the circuit. If the spread distance is not given, it defaults to the minimum design-rule spacing for the highlighted node.

F.4.9 "Replace"

The final circuit-modification command described here is replace, which removes the currently highlighted node or arc and reinstalls a new one of a different type. The new type is named in the first parameter. When you are replacing nodes, the existing arcs must be able to reconnect properly and the ports must match. When you are replacing arcs, the two nodes must be able to accept the new type of arc.

Two options for this command that may appear after the new type are this-cell and universally. When the keyword this-cell is used, all nodes or arcs in the current cell of the same type as the currently highlighted one are replaced. The universally option does the operation on every cell in the library.

F.4.10 "Undo"

To finish up this discussion of circuit change commands, it is appropriate to mention that no change is permanent. In fact, any of the functions described here can be undone with the undo command (the "u" key or the menu entry with the large "X"). This command is sometimes misleading because it does not undo the last *command*; rather it undoes the last *change*. Some commands, such as those that set defaults or rearrange display windows, do not actually change the circuit and so are not affected by undo. However, all the commands discussed in this section do affect the database, and can be undone.

The Electric database tracks the creation, deletion, and layout modification of all nodes, arcs, ports, and cells. When a change command is issued, that change and any constrained side effects are stored as a **batch** in the **history list**. The command undo removes the top change batch from the history list and reverses it (if intermediate changes were made by other tools, more than one batch would be reversed up to the last change made by the user). The reversed change batch is then left on the history list so that a second undo redoes and restores the database. To see the state of this list, use show history.

A twist to the undo command is its parameter, which indicates the number of change batches to undo. The command undo 3 removes the top three change batches (from the three previous change commands), undoes them, and leaves one change batch in their place. A subsequent undo therefore restores all three changes. However, an undo 2 after the undo 3 is an illegal request because it suggests that the top batch of changes can be redone while the previous batch of changes (four change commands ago) can be undone. This could result in an inconsistent database if the redone changes are dependent on the previous but now undone change. To undo the command four back after you have undone the last three, you must first do an undo to restore the last three, then an undo 2 to undo all four changes. When in doubt, examine the history list.

F.5 Hierarchy

As the introductory example showed, hierarchy is certainly supported in Electric. Any collection of nodes and arcs can be a cell, and instances of cells can be placed in other cells. This section describes the commands available for manipulating cells and their contents.

F.5.1 "Editcell"

The editcell command is used to display the contents of a cell in the window. If the name given as a parameter to the command is new, a new cell with that name is created. If the name is that of an old cell, its contents are displayed. Cell names may not contain spaces, tabs, unprintable characters, or a colon. Uppercase and lowercase characters are not distinguished: The cell "UPPER" is the same as the cell "Upper."

Changes may be issued for only the current cell, not for any others that are visible as cell instances. However, if a cell instance is highlighted, you may issue the `editcell` command with no parameters to switch editing to the highlighted cell. The window scaling is preserved so that the cell definition is shown where the instance appeared. Additionally, the path down the hierarchy is remembered for subsequent `outcell` commands.

F.5.2 "Outcell"

The `outcell` command pops you to the next higher cell in the hierarchy. If a parameterless `editcell` was used to arrive at the current window, then `outcell` places you at the previous window display, with the former scale and position. The `outcell` command can take a numeric parameter to specify how many levels of hierarchy to rise. Without this parameter, the command rises one level. If the hierarchy was not descended with parameterless `editcell` commands, `outcell` attempts to figure out the next higher cell in the hierarchy and uses that. The command fails if no descent path is saved and none can be clearly determined from the hierarchy.

F.5.3 "Killcell"

What is created must be destroyed, and cells are no exception. The `killcell` command takes a cell name and deletes it. There cannot be any instances of this cell, or the command fails. As a side effect of failure, the command lists all other cells that have instances of this cell, so you can see the extent of the cell's use.

F.5.4 "Getproto"

To place an instance of a cell, one simply uses the `create` command. However, the cell must be the currently selected type, as shown at the top of the status display. Although most of the primitive nodes can be selected from the menu, there are no cells in the menu so the nodes must be selected explicitly with the `getproto` command. Typically, the command is issued with a node or arc name as the first parameter. Thus `getproto Upper` sets cell "Upper" to be the currently selected node. If the node name is ambiguous and might be confused with an arc name, insert the keyword `node` before the node name. This ambiguity distinction can also be made with arcs that might be mistaken for nodes. For example, if you have a cell called "metal" but you want to set the current arc type to Metal, you must use `getproto arc metal`. When selecting primitive nodes and arcs that are in a technology other than the current one, you can use the technology name as a prefix to the node or arc name, separated by a colon. For example, the command `getproto logic:meter` gets the meter primitive from the Logic technology (a useful command when graphically specifying simulation).

There are three options to `getproto` that may be used instead of an explicit node or arc name. The command `getproto next-proto` (the space key) advances the current node to the next primitive in the list. If a nonprimitive is selected, the first primitive is chosen. If two objects are highlighted, such that the next `create` command will create an arc, then this `getproto next-proto` command advances the current arc. The opposite of this option is `getproto prev-proto` (the " ↑ " key), which backs up in the same manner. Either of these options can take the `node` or `arc` keywords to force correct action.

The final option to `getproto` is `this-proto` (the "T" key) that sets the current node to be the currently edited cell. Obviously, you cannot do a `create` in the same cell after this action because that would imply self-referencing in the hierarchy. However, the command is useful when you are editing two cells in different windows, and you wish to place an instance of one cell inside the other. By placing the cursor in the lower-level cell and issuing this command, you are then able to move the cursor to the upper-level cell and do a `create`.

F.5.5   "Port"

All nodes in Electric have connection sites, called **ports**, that indicate where wires may be attached. The primitives have predefined ports, and these form the basis for ports on cells. The user simply selects an existing port on a node, and **exports** it, which makes it available on all instances of the current cell. The `port` command handles all aspects of port exporting, unexporting, changing, and display.

The basic export command is `port export` (the "x" key), which takes the currently highlighted port on the currently highlighted node and exports it, giving it the name specified as a parameter to the command. All port names on a cell must be unique; if a nonunique name is given, it is modified (by adding ".1", ".2", and so on) to be unique. Like cell names, port names may not contain spaces, tabs, or unprintable characters, and no distinction is made between uppercase and lowercase characters. If there is no port highlighted on the currently highlighted node, Electric attempts to determine the correct port to use from a number of factors, including cell-edge proximity and naming conventions. If it cannot guess, it asks you to select graphically by displaying all the possible ports on the node. You can force this graphic selection by using the keyword `specify` after the port name. Also, you can give an actual port name on the highlighted node with the keyword `use`. For example, to export the "trans-poly-left" port of the currently highlighted transistor, calling it port "in" of the current cell, type `port export in use trans-poly-left`.

Two types of characteristics can be associated with a port: behavior and location. The behavior is a simple list of keywords that may appear at the end of the port command: `input`, `output`, `bidirectional`, `power`, `ground`, `clock`, `clock1`, `clock2`, `clock3`, `clock4`, `clock5`, or `clock6`. These behavior characteristics are stored with the port and used primarily by simulators. The location characteristics are `top`, `bottom`, `left`, and `right`, which may be combined to specify eight offsets for the port label. For example, to export the currently highlighted port, name it "corner," mark it as a power connection, and display the port label to the upper-left, type `port export corner power top left`.

Once a port has been exported, the characteristics can be modified with the `port change` option. This command is similar to `port export` except that it operates on an existing port and so does not take a port name. If necessary, you can provide `specify` and `use` keywords to identify the particular port on the highlighted node. You can also use `find export PORTNAME` to highlight directly the node with the exported port PORTNAME. The `port change` command accepts any behavior characteristic, and any location characteristic. One additional location characteristic, `middle`, is available for resetting the port name to the center. To change the port name, use the `rename` command.

Another possible change is to unexport a port. The `port unexport` command removes the port from the current cell. It can take the keywords `specify` and `use` to identify which port on the node to unexport. It can also take the `all` option to unexport all exported ports on the highlighted node. Arcs will be deleted if they connect to unexported ports on instances of the current cell.

It is sometimes desirable to keep an exported port but to transfer it to another node. If a cell is in use higher in the hierarchy, unexporting and then reexporting loses all the existing connections. Instead, the `port move` command can be used. This command is completely graphical in its specification: Two nodes and their ports must be highlighted with `find port` and `find another port`. The exported port is moved from the first node to the second (another) node.

There are special cases of the `port` command that are primarily used in array-based layout. If a cell is replicated many times and the instances are wired together, there are ports on the edge of the array that do not get wired. These ports define the connections into the current cell for the next level of hierarchy. What you would like to do is to export all unwired ports on all cell instances. To do this, use `port re-export-all`, which generates unique port names as it exports all unwired ports on cell instances. To do this same function, but in only the highlighted area, use `port highlighted-re-export`. Note that ports on primitive nodes are not exported with these commands. Use the "stitching" router tools to help with the internal array wiring.

The final `port` option relates to the display of port names. Although the location characteristic of the `port export` and `port change` commands determines where a port name appears, this option controls how they appear in the current window. The `port labels` command may take five options. `Port labels none` indicates that no ports are displayed. `Port labels fit` (the default) indicates that the port name is displayed, but that the letters are truncated if there is not enough room in the area of the node. If there is not even room for one letter, the port name does not appear, but instead a cross is drawn at the port. `Port labels cross` forces crosses to be used always. `Port labels long` forces full text names to appear, regardless of display space. Finally, `port labels short` forces port names to be displayed, but only the first alphabetic characters. Thus the ports "Power-left" and "Power.1" are both written as "Power."

F.5.6   "Library"

Having examined cells and ports on cells, let us step back to consider cells within libraries. A **library** is a collection of cells that form a consistent hierarchy. To enforce this consistency, Electric stores entire libraries in one disk file that is read or written at one time. It is possible, however, to have multiple libraries in Electric at once. Only one library is the current one, as shown at the top of the status display. You can see a list of cells in the current library with the `show cells` command. This command also lists cells in an alternate library if that library's name is mentioned as a parameter.

The `library read` command brings a new library into Electric from disk. The first parameter to this command is the disk file name, which may be qualified with operating-system-specific directory information. Thus the command `library read /usr/rubin/chip` reads the requested file "/usr/rubin/chip," and names the library "chip." Use `show libraries` to see the current libraries and their associated disk file names.

Additional parameters to `library read` specify the format of the disk file. Electric can store complete databases in a binary format or in a less compact text format. The text format is slower to read and larger to store, but can be transferred between machines more reliably and can be edited if necessary. Electric can also read CIF files, although these files do not contain connectivity and other useful information.

Although `binary` is the default, you can specify the keywords `text` or `cif` after the disk file name. Check to see what other formats are available, as more will surely be added. You can also provide the keyword `make-current` to request that this library be the current one, once read.

Switching libraries is done with `library use`, which takes an existing library name that should be made current. If there is a current cell in that library, it is displayed. The current cell of a library is the one currently being edited when the library is saved.

Writing libraries to disk is done with `library write`, which takes a library name and an optional format. The formats include `binary` (the default), `text`, `cif`, and `plot`. The plot format is not readable by Electric, but rather is used by the "splot" utility to produce hardcopy images (you can obtain more information on "splot" simply by running it with no parameters). Note that the `cif` and `plot` outputs write only the current cell and any hierarchy below it, whereas the `binary` and `text` options save the entire library.

When you exit Electric, any modified but unsaved libraries do not have their changes recorded. Electric double-checks with you before exiting. To relieve the problem of unsaved libraries, the `library save` command (the menu entry with an arrow pointing to a filing cabinet) writes all modified libraries in binary format.

The final library option is `library kill`, which deletes the library given as a parameter. This removes only the memory representation, not the disk file. Note that library changes are too vast to be tracked by the database-change mechanism and so are not undoable.

F.5.7   "Copycell"

In general, libraries are completely separate collections of cells that do not relate. For example, two cells in different libraries can have the same name without being related. The only way to combine libraries is to copy cells from one library to another. The `copycell` command takes two parameters: a source and a destination. The source cell name can be in an alternate library if the library name precedes the cell name, separated by a colon. The destination cell is always in the current library. For example, to copy cell "OutPad" from library "Pads" into the current library, type `copycell Pads:OutPad OutPad` which keeps the same name in the current library. The `copycell` command can also be used to copy within the current library; simply omit the library qualification.

When you are copying cells across libraries, it may be necessary to copy subcells also. All necessary subcells are copied unless there already exists a cell by that name in the current library. If a subcell already exists, it must be correct in its size and ports, or else the `copycell` fails.

Another option to `copycell` is `skeletonize`, which copies only the important parts of the source cell (the exported port nodes and a few others that define the cell boundary). For example, the command `copycell adder addframe skeletonize` will create a cell called "addframe" with the skeletal representation of "adder."

F.5.8   "Packagecell" and "Yankcell"

The final two hierarchy-manipulation commands are able to create and delete levels of the hierarchy. The `packagecell` command takes a new cell name as a parameter and creates this cell with a copy of all objects in the highlighted area. Every node touching that area is included in the new cell. All arcs between nodes in the area are also included. The highlighted circuitry is not affected.

The opposite command from `packagecell` is `yankcell`. This command takes the currently highlighted cell instance and replaces it with the cell's contents. All arcs that were connected to the cell are reconnected to the correct parts of the cell. If the option `yankcell keep-ports` is used, any exported ports inside the cell become ports of the current cell.

The combination of `packagecell` and `yankcell` can be used to move an area of circuitry. The area is first packaged and then erased. The new cell can then be placed appropriately and yanked. Of course, you should be sure that the packaged cell does contain the necessary circuitry: Remember that arcs crossing the highlighted boundary are not packaged but are erased.

F.6   Constraints

Electric has the ability to specify graphically layout actions that occur as the circuit is modified. This is done by placing **constraints** on the arcs that react to node changes. Although there are multiple constraint systems in Electric, the default system is the most useful and best supported. This is the **hierarchical-layout** constraint system, which has four constraints.

The first constraint in Electric's hierarchical-layout system is the **rigid** constraint. When an arc is made rigid, it cannot change length. If a node on either end is moved, the other node and the arc move by the same amount. Besides keeping a constant length, rigid arcs attach in a fixed way to their nodes. This means that, if the node rotates or mirrors, the arc spins about, so the overall configuration does not change. Without this rigidity constraint, arcs simply stretch and rotate to keep their connectivity.

The second constraint, which is used only if an arc is not rigid, is the **Manhattan** constraint. Manhattan wires always run horizontally or vertically, and this constraint forces them to remain that way. Thus, if a vertical Manhattan arc connects two nodes, and the bottom node moves left, then the arc and the top node also move by the same amount. If that bottom node moves down, the arc simply stretches without affecting the other node. If the bottom node moves down and to the left, the arc both moves and stretches. Rotation of nodes causes no change to Manhattan arcs unless the arc is connected to an off-center port, in which case a slight translation and stretch may occur.

The third constraint, also considered only for nonrigid arcs, is **slidability**. When an arc is slidable, it may move about within its port. To understand this fully, you should know exactly where the arc **endpoint** is located. Most arcs are defined to extend past the endpoint by one-half of their width. This means that the arc endpoint is centered in one side of the arc rectangle. If the arc is 2 wide, then the endpoint is 1 in from the edge of its rectangle. All arc endpoints must be in the area of the port to which they connect. If the port is a single point, then there is no question of where the arc may lie. If, however, the port has a larger area, as in the case of contacts, then the arc can actually connect in any number of locations.

Slidable arcs may adjust themselves within the port area rather than move. For example, if a node's motion is such that the arc can slide without moving, then no change occurs to the arc or to the other node. Without the slidable constraint, the arc moves to stay connected at the same spot in the port. Slidability propagation works both ways, because if an arc moves but can slide within the other node's port, then that node does not move. Note that slidability occurs only for complete motions and not for parts of a motion. If the node moves by 10 and can slide by 1, then it pushes the arc by the full 10 and no sliding occurs. In this case, only motions of 1 or less slide.

The last of Electric's hierarchical-layout constraints is the only one that is not actually programmable by the user. This is the constraint that all arcs must stay in their ports, even across hierarchical levels of design. When a node moves, and has an exported port, all the ports on instances of that cell also change. The hierarchical-layout constraint system therefore adjusts all arcs connected to those instances, following their constraints. If those constraints change nodes with exported ports in the higher-level cell, then the changes propagate up another level of hierarchy.

This bottom-up propagation of changes guarantees a correctly connected hierarchy, and allows top-down design. Users can create mostly empty cells containing exported ports on unconnected nodes.

They can then do high-level design with these skeleton cells. Later, when the cells are actually designed, the constraint system will maintain the higher levels of hierarchy automatically.

The hierarchical-propagation aspect of the constraint system leaves open the possibility of an overconstrained situation. For example, if two different cells are connected with two horizontal Manhattan wires, and one connection point moves down, then it is not be possible to keep both wires Manhattan. Electric jogs an arc, converting it into three arcs that zig-zag, to retain the connection.

F.6.1 "Arc"

The `arc` command is used to set constraints and other properties on the currently highlighted arc. If an area is highlighted, all arcs in that area are affected. The keywords `rigid`, `manhattan`, and `slide` set the constraint properties, and the keyword `not` before them removes the properties. Many versions of this command are bound to keys: `arc manhattan` is "+", `arc not manhattan` is "/", `arc rigid` is "|", and `arc not rigid` is "~". An additional constraint option is `temp-rigid`, which sets the arcs to be temporarily rigid or not rigid for the next change only.

In addition to the constraining properties, there are a set of characteristics that can be placed on arcs with this command. **Directional** arcs have arrows that show a flow direction. The `arc directional` and `arc not directional` commands control the display of this arrow. The arrow always runs from the tail of the arc to the head of the arc unless you issue the `arc reverse` command.

Arcs in the Logic technology may be **negated**, which causes them to have a bubble drawn where they attach to schematic elements. The commands `arc negated`, `arc not negated`, and `arc reverse` control the bubble placement. Since there can be only one negating bubble on an arc, users who want two must use two arcs, connected by a pin. Negated arcs make no sense in layout technologies and are automatically reset.

Arcs with nonzero width have their geometry extended beyond their endpoints by one-half of their width. The command `arc not ends-extend` changes this, making the arc terminate exactly at the endpoint. The `arc ends-extend` command restores normal geometry.

Individual ends of an arc may be controlled with the commands `arc skip-head` and `arc skip-tail`, and with the negated versions `arc not skip-head` and `arc not skip-tail`. When an end has been skipped, the property associated with it does not occur. This means that directional arcs have their arrowhead omitted (when the head is skipped); negated arcs do not have the bubble drawn (when the tail is skipped); and arcs with their ends not extended have that end extended.

Another property of an arc is its name. This is simply a string that is displayed with the arc as an identifier to the designer. The command `arc name` takes a parameter that is then displayed on the arc. This name can be used in the `find arc` command to select the object directly; however, there are no restrictions on arc-name uniqueness so beware of duplication. The `arc not name` command deletes the name.

The final `arc` options are used in circular geometry. `Arc center` states that the currently highlighted arc is to curve between its endpoints such that the center of curvature is at the location of the cursor. Two curvatures are possible (clockwise and counterclockwise), so use `arc reverse` to make a distinction. You can also use `arc curve` to request an arc the curvature of which passes through the location of the cursor. The command `arc not center` removes the curvature.

F.6.2 "Defarc"

Whereas the `arc` command changes properties on existing arcs, the `defarc` command changes default properties for arcs that are subsequently created. The first parameter to `defarc` is an arc type (for example, Metal), which indicates that this command sets defaults for that kind of arc. If this first parameter is an "*", the command applies to all arcs. Following the arc type come the usual options: `manhattan`, `rigid`, `slide`, `directional`, `negated`, and `ends-extend`. These options can all have a `not` in front of them. There is also a way to remove these properties when they have been applied to all arcs: type `defarc * default` so that individual arc defaults apply.

One other arc default option is the default width. The command `defarc TYPE width WID` sets the default width of TYPE arcs to be WID. Note that newly created arcs may be wider than their default if they electrically connect to an existing arc that is wider.

F.6.3 "Constraint"

As it was mentioned before, there are multiple constraint solvers in Electric, one of them being the hierarchical-layout system. Another is a **linear-inequality** constraint system, which allows you to place x or y inequalities on arcs. For example, the constraint "$x \geq 8$" forces the arc to remain 8 or more long in the x axis.

To switch constraint solvers, use the `constraint use` command. The two possible solvers are `layout` (the current one) and `linear` (the alternate), although you can use `show solvers` to see any others.

Once in linear constraints, the `arc properties` command sets inequality clauses on arcs. Additional directives may be given to the constraint systems with `constraint tell`. For example, the linear-inequality system can be told to re-solve all constraints in a cell with the command `constraint tell linear solve`. Many other options exist and should be checked with `help`.

F.7   The Display

This set of commands affects the editing display. There are commands to control the menus, the editing windows, colors, and more.

F.7.1   "Redraw"

The `redraw` command (the "ˆL" key) simply erases all the displays and draws their contents again. Although this should not be necessary, it is useful if you suspect that the display is wrong, or if the image gets somehow corrupted. In those environments in which multiple devices are known to the system, you can specify another device name as a parameter to this command. Electric then draws everything on that device, effectively providing hardcopy. However, the `library` command has a hardcopy facility that is much better, so the former option finds little use. The command `redraw *` not only redraws the displays, but also reinitializes the display device (potentially useful for certain failure modes of displays).

F.7.2   "Menu"

There are two kinds of menus in Electric: **fixed** and **pop-up**. The fixed menus appear as an array of menu squares along one side of the display. Pop-up menus are drawn near the cursor only when selection is desired. The `menu` command controls the size of these menus, and the placement of fixed menus. The `bind` command is used to set entries.

The fixed menu can be displayed or hidden. Each issuance of the `menu` command with no parameters reverses this state. To force display of the fixed menu, use `menu on`. To force the fixed menu to disappear, use `menu off`. The fixed menu can also be placed along any of the four sides of the display with `menu top`, `menu bottom`, `menu left`, and `menu right`. Finally, the size of the fixed menu can be set with `menu size COLUMNS ROWS`. The placement and size options can be combined into one `menu` command.

Pop-up menus are very different from fixed menus because there can be many pop-ups and they are only displayed when you want to select from them. So that the system can distinguish and invoke pop-up menus, each one is actually a new, user-defined command to Electric. For example, if a new pop-up menu called "elucidate" is created, then typing the command `elucidate` displays the menu. Each entry of the pop-up menu has a command in it, so selection of that entry executes the command. Since menu entries contain commands, they may contain another pop-up menu command, which is then displayed in a hierarchical manner.

Initially, there are no pop-up menus in Electric. To create a pop-up menu, use `menu popup`, which takes a name as its parameter. This name must not conflict with any other command name. Following the pop-up menu name come the keyword `size` and a number of entries to provide in the menu. This size option must be used when you are creating the menu, but it may also be used later to change the number of entries. For example, the `elucidate` command can be created as a pop-up menu with 10 entries by typing `menu popup elucidate size 10`.

The other option to pop-up menus, available only after they have been created, is to set a header string. For example, to place the string "Further information" at the top of the "elucidate" menu, type `menu popup elucidate header "Further information"`.

When a pop-up menu appears, the cursor changes shape to indicate that you must select from the menu. As the cursor roams across the menu, each entry is highlighted to show which command will be executed when you push a button. Pop-up menu entries can be parameterizable such that, once the cursor is over them, the user can type additional information into the menu before pushing the button. By default, newly created pop-up menus have all their entries set this way so that actual commands, typed into each entry, become permanent entries. To see this work, try creating a pop-up menu, invoke it, type random Electric commands into different entries, push a button, and then invoke the menu again. The entries will have been changed. For further explanation of how this works, see the `bind` command.

F.7.3 "Window"

The editing display can be divided into a number of windows, each of which can show a different cell, magnification, and location. Control of window layout, panning, and zooming is all accomplished with the `window` command.

The initial layout consists of one window that fills the editing display. The command `window split` divides this in half, either vertically or horizontally. The aspect ratio of the editing window and its contents are used to determine how the window is split. You can also provide the keywords `vertical` or `horizontal` to override this computation. Once the original is split, only one window is the current window, highlighted with a green border. The current window changes, however, whenever the cursor moves to another window and a command is issued. The `window split` command can be used repeatedly to subdivide windows into halves, quarters, and so on. Once the initial split has been made, there is no longer a choice between horizontal and vertical splits: Each division follows an alternating sequence to achieve a clean binary tree of windows. Notice that each window has a separate line at the top of the status display, showing its location, contents, and other relevant information. The initial window is listed in the status display as "entire"; as division continues, the names become cryptic encodings of window location. For example, "bot-l-t" is the window in the upper half of the lower-left quadrant.

To undo the division of windows, use `window kill`, which deletes the current window and merges it with its neighbor. This command can be issued only to a window at the bottom of the subdivision tree. To back out completely from a mess of multiple windows, type `window 1-window`, which returns to only one window.

Besides using cursor location to select windows, you can invoke the `window use` command to set the current window. A window name, such as "left" or "top-r", must follow. This option is useful only in macros in which the subsequent command does not override the window selection with its cursor location.

The scale of the window contents can be controlled with a number of options. The command `window in-zoom` (the "v" key and the menu entry with the four inward-pointing arrows) zooms in, magnifying the contents of the display. The default amount of zooming is 2, which doubles all sizes and therefore shows one-quarter of the area. However, you can provide any value of zooming as a parameter to this command. The command `window out-zoom` (the "V" key and the menu entry with the four outward-pointing arrows close to the center) does the opposite—it shrinks the display.

The most useful scale change option is `window all-displayed` (the menu entry with the four outward-pointing arrows at the edge), which makes the current cell fill the window. Yet another option is `window highlight-displayed` (the menu entry with the four arrows emanating from a box), which makes the highlighted objects fill the display.

A less used scale option is `window match`, which takes another window name as a parameter and forces the two windows to have the same display scale. This is helpful when you are trying to make comparisons.

Besides scaling, you can also pan the window contents, shifting it about on the display. The `window up`, `window down`, `window left`, and `window right` commands (the menu entries with the large arrows) all shift the window contents appropriately. The default amount is one-half of the window size, but any amount can be specified as a parameter. If a number is given, it indicates the number of windows to shift. If the number ends with the letter "L", it is the number of lambda units to shift (these are simply grid units as will be explained later).

Two less used shift options are `window cursor-centered` (the "c" key) and `window center-highlight`. The former makes the circuitry at the current cursor location be in the center of the window. The latter makes the highlighted objects be centered in the window.

The final window command is `window peek` (the "p" key), which causes everything highlighted to be completely drawn, down through all hierarchical levels. Any cells currently being drawn with bounding boxes are expanded only in the highlighted area. This peek operation is a one-shot display that reverts to unexpanded form if shifting or scaling is done.

F.7.4   "Node"

Although not specifically for display control, the `node` command does have its primary option concerned with the expansion of cell instances. This command, like the `arc` command, affects existing objects, in this case nodes. The `node expand` command (the menu entry with the open eye) and the `node not expand` command (the menu entry with the closed eye) alter the visibility of cell contents. By default, newly created cell instances are not expanded, but rather are drawn as outlines with the cell name inside. When the cell instances are expanded one level (`node expand 1`), the contents are drawn. If the contents have unexpanded cells, then a second `node expand 1` command draws the contents of these lower-level cells. If the number of levels of expansion is not given, an infinite value is assumed, which causes expansion all the way down to the leaf cells. When unexpanding is performed, the order is reversed such that the lowest visible cell instances are the first to be converted back to outline form.

The difference between node expansion and window peeking is that expansion is remembered with every cell instance. Once a node is expanded, it always has its contents drawn whenever it appears on the display. This means that there may be an unexpected amount of contents drawing in some circumstances. For example, suppose the current cell has 100 instances of a subcell, and that subcell has four instances of a lower cell. Selecting one of the 100 instances and expanding one level shows the four subinstances. Expanding again displays the entire contents down to the lower cell. Now, if you expand any other of the 100 instances, you also see the entire contents. This is because the subcell has had its four subinstances expanded, so whenever its contents are drawn, full geometry appears. If, at the start of this example, you had selected one of the 100 instances and peeked inside, you would instantly have seen the entire contents, but that would not have affected the expansion information on that node or on the four subnodes.

Another `node` command option is `name`, which works the same way as arc naming. Any node can be named with `node name` or unnamed with `node not name`. The name has no bearing on the node, but can be used in the `find node` command. Remember that the `array` command sets node names purely for helpful display of the elements.

F.7.5   "Defnode"

To continue this discussion of node properties, it is useful to cover the `defnode` command, which sets defaults for uncreated nodes. Although newly created cell instances are not expanded, the command `defnode CELL expanded` changes that. The specified CELL may be a cell name or "*" to indicate expansion of all subsequently created cells. The reverse of this command is `defnode CELL not expanded`.

Another default node creation option, available only for primitives, is size. The command `defnode PRIMITIVE size X Y` changes the initial size for newly created nodes of type PRIMITIVE. Also, default rotation and transposition can be set for all nodes (cells or primitives) with the `defnode * placement ANGLE`, where ANGLE is a rotation in degrees and may have the letter "T" after it to indicate transposition. If ANGLE is omitted, the default placement, shown in the status display, advances by 90 degrees.

F.7.6   "Visiblelayers"

Returning to display control, there is the `visiblelayers` command, which determines the layers of a circuit actually drawn. Each layer has a unique letter or letters associated with it. When you type the parameterless `visiblelayers` command, you are informed of the currently visible layers (all of them) and their letters. If you give a parameter, it is a collection of layer letters to display. For example, if you find that the metal layer has the letters "mb" and the diffusion layer has the letters "dg", then the command `visiblelayers gb` redraws the window with only these two layers. On the other hand, the command `visiblelayers *-md` does just the opposite, drawing all but the metal and diffusion layers. This is because "*" is short

for "all layers" and subtraction works in this case. Clearly, the command
visiblelayers * restores all layers to the display.

F.7.7 "Color"

The color command affects the actual appearance of the layers drawn
on the screen. On color displays, the layers are composed of red,
green, and blue factors. On black-and-white displays, a 16×8 bit
pattern is used. Some color displays also use the bit pattern to achieve
better layer distinction. In the most extreme conditions, when video
terminals are used for editing, a single character represents each layer.
The color command examines and sets all of these values.

The option color entry takes a layer specification as its first
parameter. This layer is one of the layer letters, as used in the visiblelayers command. Layers that overlap to form a new color indicate
that this combination of layers also has a color entry that can be
manipulated. Use all layer letters at once to indicate the combination
color. Layer specification can also be an integer that indexes the color
table. Using integer layers, you can reference special parts of the
display, such as the highlight color (2), the grid color (1), or the
background color (0). Type help map to see how the color map is
organized.

If the color entry command has only one parameter, that entry's
value is printed. If three more parameters follow it, they set the new
values for red, green, and blue. The values range from 0 (dark) to
255 (bright). For example, to make the highlight color be bright yellow,
type color entry 2 255 255 0. To set the color of transistors (where
polysilicon and diffusion cross) to purple, type color entry pd 255
0 255. On video terminals, the character used for metal may be set
to "W" with color entry m W.

Electric allows you to mix up to 10 entries interactively with the
color mix option. The layers listed as parameters to this command
are displayed along with a mixing palette. When you are done mixing,
the colors change.

Layer patterns can be examined or set with the color pattern
option. The first parameter selects a layer and an optional second
parameter changes the pattern. This second parameter consists of up
to eight numbers, separated by slashes. Each number is a 16-bit value
for a row of the pattern. If fewer than eight numbers are given, the
list is cycled until eight numbers are obtained. For example, to set
the metal layer to a simple checkerboard pattern, type color pattern
m 0x5555/0xaaaa. Note that all layers have individual patterns, so
you cannot set the pattern for layer combinations. Also, the special
parts of the display, such as the highlight, have no pattern.

When you have mired yourself deep in color changes and want to back out, the `color default` command restores original colors. It does not affect patterns, however.

Color and pattern settings can be saved on disk and restored later. The command `color write` takes a file-name parameter and saves all the colors, patterns, and video letters for the current technology. The `color read` command reads the named file. For both commands, the default extension of the disk file is ".map".

F.7.8   "Grid"

The final window control command is `grid`, which controls the display of grids and the alignment of cursor locations. The parameterless `grid` command (the "g" key and the menu entry with the mesh) toggles the display of a grid in the current window. If the command `grid size` is issued, the grid is displayed with mesh points spaced according to the parameter. When only one parameter is specified, a square mesh is displayed. If two parameters are given, they are the x and y grid spacing. Large grid meshes attempt to align themselves with objects in the current cell, so turning them off and on may shift their location.

The `grid alignment` command sets the way in which cursor position is adjusted for other commands. Initially, the alignment is 1, as shown in the status display. The cursor is therefore snapped to the nearest whole lambda unit (see the `lambda` command for an explanation of this value). The command `grid alignment 0.5` causes the cursor to stop on half-lambda units. If the command `grid size 0.5` is issued, then the cursor is snapped to grid mesh points also.

The `grid edges` command is a special case of grid alignment that causes newly created objects to lie with their edges on a specified grid. Initially, there is no edge alignment so, for example, a 3-wide metal wire that runs between two 4-wide contacts is centered, with its edges on half-lambda boundaries. The command `grid edges 1` forces that wire to move to one side, so that its edge is on a whole-lambda unit. The edge-alignment factor is shown in the status display after the grid alignment, separated by a slash. When the edge alignment is zero, it does not appear. Note that it is not always possible to align edges properly, so you should always check your geometry if you insist on this attribute.

F.8   Technologies

When Electric begins, the status display shows that the current technology is nMOS. There are, however, many other technologies from which to choose. A **technology** is an environment of design, consisting of a set of primitive nodes, arcs, and all the associated information necessary to produce layout. This information includes design rules, simulation attributes, and more. The command `show primitives` lists the nodes in the current technology.

Electric has different variations of technologies as well as vastly different technologies. Thus there are many CMOS variations including "mocmos" for MOSIS design rules and "rcmos" with round geometry. In addition to simple MOS technologies, there is "logic" for schematic capture, "art" for purely graphic artwork, "pcb" for circuit-board layout, and others. To see which technologies are available, use `show technologies`.

One particularly interesting technology is the Generic technology, which is a grab bag of miscellaneous facilities. The **universal** arc in the Generic technology is able to make a connection between any two components, even if they are in different technologies. This is useful when mixing technologies while still maintaining proper connectivity for simulation. The **invisible** arc attaches any two components, but makes no electrical connection. It is useful for constraining otherwise unrelated components. The **unrouted** arc makes arbitrary electrical connections, like the universal arc, but routers know to replace them with real geometry.

F.8.1 "Technology"

The `technology` command controls Electric's use of technologies. The option `technology use` takes a technology name as a parameter and switches to that environment of design. The primitives in the menu change and the color map is updated for the new technology. The current cell is not affected, because you can mix technologies freely, placing components together regardless of whether it makes sense.

Some technologies can receive instructions to act in specific ways with the `technology tell` command. For example, the MOSIS CMOS technology can switch from *n*-well to *p*-well with the command `technology tell mocmos p-well`.

Besides technology switching and control, this command can also be used to translate circuits between technologies. The command `technology convert` takes a new technology name as the first parameter and a new cell name as the second. The new cell is created in the new technology. This command works in only limited environments, such as between similar MOS technologies (for design-rule conversion) and when going from MOS to schematics. More advanced conversion may someday be available, such as silicon compilation that converts from schematics to MOS.

F.8.2   The Logic Technology

Electric is primarily tailored for IC layout technologies. Nevertheless, the Logic technology allows schematic capture to be done. Although you can enter this technology with the command `technology use logic`, a more complete conversion occurs if you use the command file "logic.mac". Since this file is usually kept in the system library, you need only issue the command `commandfile logic`. The commands in this file switch to the Logic technology and create some useful macros for schematic editing.

Two new keys, defined in "logic.mac", are useful in schematic capture. The "n" key negates the currently highlighted arc, placing a bubble on one end. The "r" key reverses the arc so that the bubble is on the other end. Besides creating new key commands, the "d" key is modified so that it allows drawing of Manhattan or 45-degree wires (its original binding is `create angle 90` but now the key is `create angle 45`).

Digital logic is done with the And, Or, Xor, and Buffer nodes. Negated arcs turn these into NAND, NOR, Inverter, and many other nodes by allowing any combination of negated inputs or outputs. A Flip-Flop node can be parameterized with the new command `setff`, which takes a parameter of the form TYPECLOCK. The TYPE can be `rs`, `jk`, `d`, or `t`; and the CLOCK can be `ms` (master-slave), `n` (negative), or `p` (positive). For example, to set the currently highlighted Flip-Flop to be JK master-slave, type `setff jkms`.

Analog schematic components can all be parameterized with types and values. The Resistor, Capacitor, and Inductor nodes can be given values with the `setresistor`, `setcapacitor`, and `setinductor` commands. The Transistor node can be styled as *n*-channel MOS (`settrans nmos`), *p*-channel MOS (`settrans pmos`), depletion MOS (`settrans dmos`), bipolar junction (`settrans npn` and `settrans pnp`), junction FET (`settrans njfet` and `settrans pjfet`), or metal semiconductor (`settrans emes` and `settrans dmes`). MOS transistors can take length and width specifications (for example, `settrans dmos4/5`) and other transistors can take area specifications (for example, `settrans pjfet42`). Diodes can also be given area with `setdiodearea`.

General-purpose schematic components exist for both digital and analog applications. The Offpage node can be marked with the `setoffpage` command. A Black-Box node can have any function described with the `setbb` command. Although there are Power and Ground nodes, the Source node can do both functions and, in addition, can be set to be either a voltage source (`setsource v`) or a current source (`setsource c`). Both the Source and Meter nodes can be further customized to describe graphically simulation inputs and outputs to SPICE.

F.8.3   The Artwork Technology

Another unusual technology is Artwork, which provides general-purpose sketching facilities. Like the Logic technology, the command file "art.mac" properly sets up the Electric environment. This technology has nodes for many typical graphic objects such as rectangles, triangles, circles, and arrowheads. There are Message nodes that contain text in nine different scales. Polygonal and Spline nodes allow arbitrary shapes to be defined. Of course, nodes from all other technologies can be used as special electronic symbols when artwork is generated. Conversely, these artwork nodes can be used to embellish designs done in all other technologies.

There are four Message nodes that display their text in different ways. The primitive called Centered-Message displays its text centered in the node area, Left-Message puts the lower-left text corner at the lower-left of the node, Right-Message places the lower-right text corner at the lower-right of the node, and Message centers just like Centered-Message, but forces the text to remain in the node area by shrinking scale and clipping letters.

The `node message` command, used exclusively for these four primitives, is able to manipulate text and font scale. The command `node message set LINE` places a string in the currently highlighted message node. Multiline messages (stacked vertically) can be built with `node message add-line LINE` and any line can be removed with `node message delete-line NUMBER`, which addresses the 1-based list of lines. Finally, the text scale may be set with `node message font SCALE`, where SCALE is `vtiny`, `tiny`, `vsmall`, `small`, `medium` (the default), `large`, `vlarge`, `huge`, or `vhuge`. Not all scales are supported on all displays, but the information is preserved, and is distinguished when the text is printed on devices with larger scale ranges.

The artwork command file provides some simple macros for text manipulation. The command `setletters LINE` places a single string into the message node. The command `setfont SCALE` is shorthand for the text-scale command. Also useful is `askfont`, which displays the font of the currently highlighted message node.

Users can sketch their own shapes when they use the Opened-Polygon, Closed-Polygon, Filled-Polygon, and Spline nodes. The first three take a set of points and connect them with lines, potentially closing off the polygon and filling its interior. The Spline node uses the definition points as control points for an arbitrary curve. Refer to the `node trace` command, described later in the section Miscellaneous, for information about manipulating points on these nodes. The artwork command file also has the `draw` command that lets you set a large collection of freehand-sketched points (best used with the Opened-Polygon node).

The final feature of the Artwork technology is its ability to set the color of any of its nodes. The command `setcolor` takes a color parameter that can be `black` (the default), `white`, `gray`, `lgray` (light gray), `dgray` (dark gray), `red`, `lred`, `dred`, `green`, `lgreen`, `dgreen`, `blue`, `lblue`, `dblue`, `cyan`, `magenta`, `yellow`, `orange`, `purple`, or `brown`. The `askcolor` command displays the color of the currently highlighted Artwork primitive.

F.9   Tools

There are many different tools available in Electric for doing both synthesis and analysis of circuitry. Synthesis tools include routers, compacters, cell generators, and so on. Analysis tools include design-rule checkers and many simulators.

In an attempt to uniformly deal with the tools, Electric treats every center of database activity as a tool. This means that database I/O is a separate tool and that the user interface is a tool. Tools have characteristics, such as whether they function incrementally, whether they do analysis, and whether they are currently "on." Type `show aids` to see a list of tools and their characteristics.

F.9.1   "Onaid" and "Offaid"

To control which tools are functioning, the `onaid` and `offaid` commands set state. Each takes a tool name as a parameter.

Electric controls its tools in a round-robin fashion, cycling through the list continuously and stopping at each "on" tool for a **turn**. When the user interface gets its turn, it accepts one command. If the command turns on another tool, then that tool gets a turn before the next user command is accepted.

Some tools are incremental and tend to remain "on." While on, tools are informed of all database activity. This way, the design-rule checker can evaluate all changes made by the user. Also, the user interface can properly display all changes made by other tools. When incremental tools are turned off, they lose track of database changes. However, when they are turned back on, the database informs them of general activity that occurred while they were off. This allows incremental tools to operate in a batch style. If, when you turn a tool on, you do not want this "catch-up" activity, use the `no-catch-up` option, for example: `onaid drc no-catch-up`.

Some tools are strictly batch-oriented and generally remain "off." Turning them on requests a complete activity, for example running a simulation. The entire action occurs during the turn, after which the tool turns itself off.

F.9.2 "Tellaid"

Tools can be given specific instructions with the `tellaid` command. A tool name is the first parameter, followed by a tool-specific set of options. Use `help` to see each tool's options; for example, `help compaction` explains what may follow `tellaid compaction`.

Incremental tools typically accept instructions to perform larger operations. For example, the design-rule checker examines only the most recent change, but it can be told to check the entire cell with `tellaid drc check`.

Another common use of `tellaid` options is to indicate which of the many suboperations of the tool you desire. For example, the router has many different interfaces, each of which can be selected with `tellaid`. Thus `tellaid routing river-route`, issued prior to the router being turned on, causes river routing to be performed.

F.9.3 The Network Maintainer

There is one tool that deserves special attention because it exists as a helper to all other tools. The network maintainer is always on, watching all database activity and updating connectivity information. Many other tools such as the design-rule checkers and simulators depend on the information created by this tool. Therefore, you should not turn it off (besides, it does not consume much time).

The network maintainer can label and highlight on a network basis. For example, if you highlight an arc and type `tellaid network highlight` (the " = " key), then the tool highlights all arcs connected to the selected one. You can label a network by highlighting any arc on it and typing `tellaid network name` NETNAME. Information about named nets is then available with `tellaid network show` NETNAME, where NETNAME can be `all` to see all named nets. The `highlight` option can also take a named net as a parameter. Finally, you can delete a network name with `tellaid network delete` NETNAME.

One problem with the network maintainer is that it may get confused in pathological situations. This causes network naming to act strangely and also causes failures in other tools. You can check network connectivity by examining the net-number information printed by `show object long` on arcs. If you suspect bad network information, there are two commands you can use to renumber. The command `tellaid network re-number` CELL regenerates correct network information for the named cell (which defaults to the current cell) and the command `tellaid network total-re-number` redoes the entire library.

F.9.4   Simulation

Another tool worthy of elaboration is the simulation tool, which is able to produce input specifications for a number of different simulators. All the simulators work on the current cell and require that all named points be exported. It is also necessary to export power and ground ports. Switch-level simulators such as "esim," "rsim," "rnl," and "mossim" can all be selected by saying `tellaid simulation SIMULATOR` followed by `onaid simulation`. There are also functional simulators— "mars" and "cadat"—which work the same way. Note that the "esim," "rsim," and "rnl" interfaces go beyond specification: Electric actually starts the simulator and allows you to communicate with it. Type `help` to see your options when connected to the simulator.

For circuit simulation, there is SPICE. The interface to this simulator is completely graphical and somewhat complicated. Therefore, SPICE users are encouraged to invoke the command file "spice.mac". This command file defines some of the commands described in this section. Users unfamiliar with SPICE will find the following examples confusing.

All input values to SPICE are controlled with Source nodes from the Logic technology. Similarly, output values are specified with Meter nodes. For example, a Source node, connected to the power and ground ports of the circuit, can be set to be a 5-volt supply by typing `setsource v "DC 5"`. A Source node with piecewise linear information can connect an input port to ground and be parameterized with `setsource v "PWL(0NS 0 5NS 0 6NS 5)"`. Notice that actual SPICE cards are given in the `setsource` command.

Voltage meters are hooked up between the two points that they measure. A meter that reads from 0 to 5 volts may connect an output port to ground and be parameterized with `setmeter "(0,5)"`. You can specify a current meter as part of a source node by typing `setsource vm` rather than `setsource v`.

There are other parameters to SPICE that are also given graphically. For transient analysis, a Source node is placed unconnected and parameterized with the TRAN card when you type, for example, `setsource t ".5NS 20NS 0NS .1NS"`. For DC analysis, a Source node is connected to the loop point and parameterized with `setsource vd "0V 5V .1V"`.

Some nongraphical information can also be given with special commands. The `setlevel` command selects SPICE level 1, 2, or 3. Model cards for any level may be given with `setmodel1`, `setmodel2`, and `setmodel3`. Although some technologies already have model cards for some levels, you can override or augment. For example, to set a two-line model card for level 2 of simulation in the Logic technology, type `setmodel2 logic 0 "First model card"` and `setmodel2 logic 1 "Second model card"`.

When the simulation setup is complete, a SPICE input deck may be produced with the `deck` command. Then, the command `onaid simulation` actually runs SPICE, producing an output listing on the status display. After a successful run, a new cell is created from Artwork primitives that plots the metered signals.

To run SPICE remotely, take the file "sim.spi", which is generated by the `deck` command, and obtain an output file. This output file can be converted to a waveform cell with the command `tellaid simulation spice parse–output FILE` or, for Spice3 output, `tellaid simulation spice parse–spice3–output FILE`.

F.10   Controlling the User Interface

There is a set of commands devoted exclusively to controlling the user interface. These commands rearrange other commands and generally control the system. This section outlines simple control commands; the next section explains the more advanced activity of macro writing.

F.10.1   "Quit"

Certainly the most powerful control command is `quit`, which exits Electric. If any libraries have been changed but not saved, you are stopped and asked if you really want to ignore the changes. To prevent this foolish question, answer it as part of the command: `quit yes`.

F.10.2   "Rename"

This is a simple command with a powerful capability to change any name in Electric. The format is `rename OLDNAME NEWNAME`, where the existence of OLDNAME in the database indicates the kind of object being renamed. Where duplication exists, you are asked which object you want to rename. The renamable objects are: libraries, cells, primitive nodes, arcs, ports in the current cell, technologies, commands, macros, pop-up menus, and database variables.

F.10.3   "Bind"

The `bind` command controls the attachment of commands to all buttons, single keys, and menus. Attachment is done with `bind set`. To attach a command to a mouse button, use `bind set button WHICH COMMAND`, where `WHICH` is a button name (`left`, `middle`, or `right` on mice; `white`, `yellow`, `green`, or `blue` on tablets). The `COMMAND` can be any other Electric command, including macros, pop-up menus, or even another `bind` command. For example, if there is a pop-up menu called "elucidate," then the command `bind set button middle elucidate` causes that menu to appear whenever the middle button is pushed.

Single keystrokes can be bound with `bind set key LETTER COMMAND`, where `LETTER` is any single key. To bind the space bar, you have to quote it (" ") and to bind the quote key, use two quotes in a row ("""). Note that the "-" key is special: it opens up full command evaluation because it is bound to the command `tellaid user`. You can rebind this function to any other key, and then that key introduces full commands. Keys can be unbound by setting the command to the string "-". For example, the command `bind set key " " -` unbinds the space bar. As another example, you can change the full command key from "-" to ":" by typing `bind set key : tellaid user` followed by `bind set key - -`. You will then have to use ":" before full commands such as `bind`.

Fixed menu entries are bound with the command `bind set menu ROW COLUMN COMMAND`, where the `ROW` and `COLUMN` are zero-based values that start at the lower-left. Menu entries are normally displayed by writing the command name in the menu square. As a special case, the `getproto` command displays the object that is named as a parameter. In addition, any command bound to a menu entry can be given a cell or primitive to display in the menu square. The option `glyph NODE` inserted after the `bind set menu` uses `NODE` in the menu. For example, if you want to set menu entry (2, 3) to the command `technology use logic`, and you want a logic gate displayed in the menu entry, type `bind set menu glyph logic:or 2 3 technology use logic`. Now that's a command!

Pop-up menu entries come in two styles: those that are merely selected and those that allow the user to type into the menu entry. The format for binding them is the same except for the object type being bound: `popup` or `input-popup`. For example, to set the third entry of the "elucidate" pop-up menu to the command `show environment`, type `bind set popup elucidate 3 show environment`. Users who issue the `elucidate` command see this entry and can select it to execute the `show` command.

Input pop-ups allow additional information to be typed after the command that is bound. For example, the command `bind set input-popup elucidate 1 port export` places the `port export` command (without the port-name parameter) in the "elucidate" pop-up menu, entry 1. The user who types `elucidate` and points to this entry can then type the port name directly into the menu to complete the command. Note that pop-up menus are set up initially to redefine themselves interactively. This is done by issuing a command such as: `bind set input-popup elucidate 0 bind set popup elucidate 0`, which prepares entry 0 of "elucidate" to execute `bind set popup elucidate 0` followed by whatever command is typed by the user. Thus, user input in the menu defines that entry.

You can see the current bindings by using the command `show bindings`. This command takes a parameter: `keys`, `menus`, `popups`, or `buttons` to select the particular bindings to display. The `popups` option can take a pop-up-menu name to give detail about that set of menu bindings. The command `show bindings short` (the "?" key) briefly describes button and key bindings. Use `show bindings all` if you really want to see a lot of information.

The final `bind` option is related to the way bound commands are displayed. Normally, when a key, menu, or button is pushed, the bound command is printed on the status display. This lets you see what is happening and better explains the phenomenon of binding. The command `bind no-verbose` turns this display off and the command `bind verbose` restores printing.

F.10.4   "System"

Sometimes it is not enough to control Electric: You need to issue commands to the operating system as well. The `system` command opens a path to the outside world. On the UNIX system, you get a shell and can type commands until you exit that shell and return to Electric. You can issue a single command by typing that command as a parameter to `system`. The command `system *` (the "!" key) prompts for a single command and executes it. Due to a quirk in the Sun window system, you must move the cursor into the status window to communicate with the operating system and complete the `system` command.

F.10.5   "Terminal"

The `terminal` command is a collection of facilities concerned with the status terminal and the user session. When the status window is variable in size (not a common occurrence), the commands `terminal length` and `terminal width` effect a change. Following these options is the new number of lines or number of characters per line.

The **more** facility in Electric pauses and awaits acknowledgment when the status display is filled with new messages. You can type a space to allow another screenful, carriage-return to allow one more line, or "q" to flush any more messages. The command terminal more off disables this facility and allows messages to be displayed without stopping. The terminal more on command restores the facility.

The terminal command is also used to control the logging and playback of Electric sessions. Normally, each keyboard and mouse action is recorded in the file ".electric__log". If you lose your session (because Electric bombs out?) then this file can be played back. You must first rename the file (or else the new Electric session will overwrite it) and then issue the command terminal session playback FILE. Certain conditions must be met for a session to be played back properly. The Electric environment must be the same; the windows must be the same size; and it helps if the libraries are the same (not written out during the session). This is because the log file contains only keys, buttons, and mouse coordinates, which can be interpreted differently if bindings have changed. Also, although the log file is reset when the database is saved, internal search lists may get rearranged when written and then read back. For example, a sequence of find commands may select different objects on playback due to the altered database condition.

Other terminal session options include end-record to stop logging, begin-record to resume, rewind-record to truncate the log file, and checkpoint-frequency to set or examine the number of actions between flushes of the log file. By default, the file is flushed every 10 key or button pushes.

F.10.6   "Iterate"

The iterate command (the "&" key) repeats the last command. If a numeric parameter is given, the command is repeated that many times.

F.10.7   "Commandfile"

The commandfile command takes a file name and executes all the Electric commands in the file. If the keyword verbose follows the named command file, then each command is printed as it is executed. Commands in a file are full names (for example, show rather than the letter "i"), without the introductory "-". Lines that begin with a semicolon are ignored.

Note that there is an implicit command file that is executed at the start of every session. The file ".cadrc" in the user's home directory, if it exists, is searched for lines that begin with the keyword "electric." Thus to execute the command file "startup" at every session, place the line "electric commandfile startup" in the file ".cadrc".

F.11   Macro Writing

The most advanced use of Electric involves creating new commands
to tailor the user interface precisely. A **macro** is a new, user-defined
command that contains a collection of other Electric commands. Besides
describing the essential commands for defining macros, this section
covers a few other commands that help with user interaction, database
access, and control flow.

F.11.1   "Macbegin" and "Macend"

The `macbegin` command starts the definition of a macro. If the macro
name that follows is new, a new macro is created. Otherwise, an old
macro is redefined. All commands issued after this command until a
`macend` command are remembered as part of the macro. The "[" key
is `macbegin macro`, and the "]" key is `macend`; they can be used to
define conveniently a macro called "macro." In addition, the "%"
key is the `macro` command that executes these instructions.

There are two options to `macbegin` that may follow the macro
name. The `no-execute` option indicates that the commands in the
macro are not to be executed as the macro is being defined. This is
useful when macros are defined in a command file, and their contents
have no immediate meaning. If this option is used interactively, you
are told after each subsequent command that the command is not
being executed.

The other option to `macbegin` is `verbose`, which requests that
each command of the macro be displayed when the macro is invoked.
Normally, the commands are executed silently.

Macros can be parameterized, just like normal commands. The
use of the construct %1 in any of the macro commands means "substitute
the first parameter of the macro at this point." Since macros have no
parameters when they are first being defined, an initial and default
value can be supplied for these parameter constructs by placing the
value in square brackets following the parameter declaration. For
example, here is a macro to set the color of the metal layer, defaulting
to the color blue:

```
macbegin metalcolor
color entry m %1[0] %2[0] %3[255]
macend
```

This command sets the metal color to blue when it is defined. It can then be invoked by typing `metalcolor 0 255 0`, which sets the color to green. The square bracket values also act as defaults during execution, so the parameterless command `metalcolor` sets the color to blue.

The contents of a macro can be displayed with `show macros` followed by the macro's name. You can also see a list of all macros by issuing the command with no parameter.

F.11.2 "Var"

The `var` command is able to create and to manipulate variables, both within the database and locally for use in macros. This command is very powerful—it is able to examine any of Electric's internals—and should be used with caution. Before the command is described, some explanation must be given of the syntax of database variables.

Any time that a "$" appears in an Electric command, it signals a reference to a **database variable**. Following the "$" comes a reference to an object, followed by qualifiers of that object. For example, "~" is the currently highlighted object, so, if a node is highlighted, $~ refers to that node. If the variable reference is ambiguous within its context in the command, it may be surrounded with parenthesis; for example, $(~).

There are many possible object references. Technology objects are invoked with `tech:` followed by a technology name. Thus, the MOSIS CMOS technology is referenced with $tech:mocmos and the use of "~" indicates the current technology ($tech:~). Besides technologies, you can reference primitives in the current technology ($prim:), cells in the current library ($cell:), either primitives or cells ($node:), arcs in the current technology ($arc:), libraries ($lib:), and tools ($aid:). Abbreviation is permitted, so a reference to the Transistor primitive can be $p:transistor.

Qualification of these objects opens up a large can of worms. Following the basic object reference may appear any string of qualifiers, separated by dots. For example, the width of the currently highlighted arc can be obtained with $~.width, and the value of lambda in the current technology can be obtained with $t:~.deflambda. Multiple qualifications can be used to walk arbitrarily through the database. For example, the disk file to which the currently highlighted object will be saved is $~.parent.lib.libfile which takes the currently highlighted object, finds its parent cell, finds that cell's library, and finds that library's disk file. The same information can be obtained with $cell:~.lib.libfile or $lib:~.libfile.

To describe all the possible qualifications of a database object is beyond the scope of this document. A complete list is provided by

Part I of the Electric internals manual. You can also learn a great deal by typing "?" after a "." so that all your options are shown.

Arrays exist in the Electric database, and can be indexed in the standard way. For example, to obtain the name of the third layer in the current technology, use $tech:~.TECH_layer_names[2]. You can refer to the entire array if you leave off the index specification.

As an aid to obtaining useful information, there are some special constructs that describe current attributes. The variables $~x and $~y are the *x* and *y* coordinates of the cursor. When two objects are highlighted with find and find another, the other object is $~o. When a port has been selected with find port, that port is $~p and the port selected with find another port is $~op. If a port has been exported, the exported ports are $~e and $~oe. Finally, the left, right, top, and bottom edges of the currently highlighted area are available with $~hl, $~hr, $~ht, and $~hb.

Database variables are certainly extensive and can be used to describe information anywhere. However, users who write macros need a convenient way to store working data without having to scour the database for a good place to keep these things. To address this need, there is a set of 52 **command-interpreter variables** with single letters "a" through "z" and "A" through "Z". Each is referenced with a "%" in front, rather than the standard "$". Thus two characters refer to these objects in a uniform and concise way. In actuality, these variables are stored on the user-interface aid object, so the variable %b can also be addressed as $aid:user.local_b. By convention, macro packages use the uppercase variables, leaving the lowercase ones for temporary use by the user.

Returning, finally, to the var command, there are a set of options for examining, setting, and manipulating variables. The command var with no parameters displays all the currently set command-interpreter variables (those that begin with "%").

The var examine option takes a single variable specification and prints its value. The parameter to this command does *not* have its leading "$" or "%", since the command knows what to expect. The introductory character is needed only in other commands, when substitution of the value is desired.

The var delete option takes a variable specification and removes it. You can delete neither the established parts of the database nor the objects. Essentially, you can delete only what you have created.

The var set option creates and changes variables. It takes as its first parameter a variable name, and takes a new value in the second parameter. For example, to create a new attribute called "power-consumption" on the currently highlighted object, and set it to 17, type `var set ~.power-consumption 17`. Following the new value can be a set of modifiers to qualify this variable. The modifier `prolog` indicates that the attribute is Prolog code (not very useful if you do not have a Prolog interpreter in Electric). The modifier `redisplay` requests a redisplay of the currently highlighted object (presumably because the variable change affects the object's appearance). The modifier `temporary` requests that this variable not be stored on disk when the library is saved. The modifier `cannot-change` prevents further tampering with this variable. Finally, the modifier `display` indicates that the value of the variable should be displayed on the object. For example, the command `var set ~.node__name Clock display` sets the variable called "node__name" to the string "Clock" and displays that string on the node. This is essentially what is done in the command `node name Clock`.

Besides setting variables, you can do arithmetic by using "+", "-", "*", "/", and "mod" in place of the word "set". Bit arithmetic can be done with "and" and "or." String concatenation can be done with " | ". For example, the "power-consumption" can be increased by 3 with the command `var + ~.power-consumption 3` and the node name can be updated from "Clock" to "Clock time" with the command `var | ~.node__name " time" redisplay`.

The final `var` option is `var pick`, which permits interactive examination and modification of the variables on an object. The command `var pick ~` displays a pop-up menu with the variables on the currently highlighted object. You can change those that you have created simply by typing new values into the menu.

F.11.3 "Echo" and "Terminal Input"

Some macros are structured to communicate with the user. For output, there is the `echo` command, which simply repeats its parameters on the status display. For example, the command `echo This is a $~.proto object` prints the type of the currently highlighted object.

The other need of interactive macros is the ability to take input from the user. The `terminal input` command takes two parameters: a variable letter and a prompt string. For example, the command `terminal input c "Port name: "` prompts with the message "Port name: " and then places the typed response into the command-interpreter variable %c.

F.11.4 "If"

For real control of macro execution, there is the `if` command. The `if` command takes four parameters: a first value, a comparison, a second value, and a command to execute if the comparison succeeds. The two values can be integers or strings, and the comparison is one of the C language conditions, "==", "!=", ">", ">=", "<", or "<=". For example, the command if $\sim$.proto == transistor echo "You got it" prints "You got it" if the currently highlighted object is a transistor. Another example is the command if $\sim$.width > $\sim$.length size $\sim$.length, which checks to see whether the currently highlighted arc is wider than it is long, and reduces the width if so.

F.11.5 "Create to" and "Move to"

Inside a macro, it may be useful to compute precise positions for components so that they can be created or moved according to formula. Both the create and move commands can accommodate this with the to option, which takes a coordinate value rather than the cursor location. For example, the command create to 17 5 places a new node with its lower-left corner at (17, 5).

One problem that occurs in position computation is a result of the fact that variables are expressed in centimicrons but commands such as move to use lambda. For example, here is a macro that moves the currently highlighted node to the same vertical level as the cursor:

```
macbegin movehorizontally
var set a $~x
var / a $tech:~.deflambda
var set b $~.lowy
var / b $tech:~.deflambda
move to %a %b
macend
```

To understand the need for these commands, execute them individually and observe the values they produce.

F.11.6 Advanced "Undo" Options

If many changes are made inside a macro, the constraint system may break down trying to handle all the side effects. For example, two moves of the same object may cause the second move to be ignored. To solve this problem, the undo broadcast-now command can be inserted between changes. This fakes completion of the command, causing constraints to be applied, finishing the batch of changes, and distributing the changes to all tools. The only unfortunate side effect is that execution of the macro appears as two batches in the history list.

The size of the history list is set initially to be somewhat larger than the number of tools. Since each tool can issue a separate batch of changes, there must be enough space to save a user change and all possible responses so that it all can be undone. If, however, you want to change the number of retained batches, use undo save followed by the number to save. This is useful if macros have too many undo broadcast-now statements that flood the history list.

F.12 Miscellaneous

This last section describes an unrelated collection of topics that have not previously been discussed. Most of these topics are fairly advanced but a discussion of them completes the coverage of Electric's facilities.

F.12.1 "Lambda"

The lambda command is used to change the size of the basic unit of design. Following Mead and Conway, the term **lambda** is used to refer to 1 grid unit of layout. However, Electric retains a true value for this unit, measured in centimicrons, so that mixed technologies show correct proportions. The status window displays the current value of lambda, usually 100 or 200 to indicate 1- or 2-micron spacing. You can change the value of lambda with the lambda change command, which takes a new value as its parameter. This does not cause any displayable change, but does update all internal values to the appropriate scale.

An unusual option to the lambda command is lambda no-scale-change, which also takes a new value but does not scale the existing components. For example, a contact that is 4 × 4 when lambda is 100 becomes a 2 × 2 contact if lambda is changed to 200. This makes the contact's appearance change, probably incorrectly. Note that changes to the value of lambda affect the current technology only and the current library if the change option is used. This means that a library containing mixed technologies requires multiple lambda commands: one lambda change to scale the library and the current technology, and other lambda no-scale-change commands in the other technologies to bring them in line without further altering the library.

F.12.2 "Prolog"

You probably do not have a Prolog interpreter as part of your Electric system. If you do, the prolog command connects you with that interpreter, in which you can write arbitrary Prolog code that is able to access the Electric database. Special Prolog predicates exist to examine and set objects in Electric (see Part II of the internals manual for a description of these and other routines). Type ^D to return to Electric from Prolog.

F.12.3   "Debug"

The debug command is not for general use. It does, however, do some interesting operations that people in the know can use to advantage. The most important option is debug check–database, which examines and repairs the Electric data structures. This operation should be done whenever Electric acts strangely. It is too bad that a valid database cannot be guaranteed, but some operations combine to produce uncontrollable results.

The option debug namespace shows the additional variables in Electric. The option debug internal–errors toggles the printing of database error messages (initially on). The option debug erase–bits resets certain values that may be improperly set. The option debug arena shows the structure of allocated memory. Ignore these options and do your work.

F.12.4   Cell Center

The Generic technology has a node called Cell-Center that can be placed in a cell to define its origin. Once the Cell-Center node is placed, instances of the cell will use that origin, rather than the lower-left corner, as the **grab point** for cursor-based references. For example, if you place this node in the upper-right corner of a cell, then the create and move commands place instances such that their upper-right corner is at the cursor. Deleting this node restores the lower-left corner as the grab point.

F.12.5   Cell Views

A cell may be given a **view**, which is a description of its contents. The command defnode CELL view VIEW sets the named cell to be the specified view. Possible views are layout, layout–alt1, layout–alt2 (all for IC layout), schematics (for logic designs), icon (to describe a cell symbolically), and simulation–output. In addition to being labeled, any collection of cells can be linked together so that they are different views of the same thing. For example, if you create cell "mosadder" in CMOS with the layout of a full adder, and you create cell "adderlogic" in the Logic technology with the schematic of this circuit, then you can issue the command defnode adderlogic view schematics associated-cell mosadder, which ties together the cells and gives them views.

Disassociation of cell views is done by typing defnode CELL remove-view-links, which removes all links to other cells. To remove the cell view label, give it the view type unknown.

The technology convert command creates automatic view links between the cell it generates and the original cell. The SPICE simulator links its waveform output to the simulated cell with the "simulation-output" view. To see which cells are associated, use defnode CELL view, which shows the associations with the named cell.

A particularly useful cell association is the view type icon. The icon cell is used for instances of an associated **contents** cell, which contains layout or schematics. For example, you may create a cell called "plus" in the Artwork technology that shows a circle with a plus sign inside. This is the icon that you wish to use for the adder cell. You may then say defnode plus view icon associated-cell adderlogic to tie this icon to the schematic cell with real contents. Now, if you do a getproto adderlogic and a create, the "plus" cell—rather than the "adderlogic" cell—is actually created, because it is the icon symbol that gets used for instances. To force the creation of the named contents cell, use create contents.

The icon cell is correctly tied to its contents in most respects. If you descend into it (by highlighting it and typing editcell), then you actually find yourself editing the associated contents cell. Typing out-cell properly returns you to the location of the icon cell. Also, simulators such as MOSSIM correctly substitute the contents whenever an icon appears. In order for this to work, all ports on the contents cell must appear with the same name on the icon cell.

F.12.6   "Node Trace"

For some primitive nodes, rotation, mirroring, and scaling are not enough. These primitives need to have arbitrary polygonal information to describe their contour. The node trace command controls these polygonal descriptions by adding, deleting, and moving individual points in a node's **trace**.

When you are editing a trace, it is useful to have a "current" point that commands can reference. The find vertex command highlights the closest point to the cursor when it selects a node with trace information. In addition, find line highlights an edge of the trace data. When a point is highlighted, the command node trace add-point adds the cursor location as a new trace point after the currently highlighted point. The command node trace delete-point removes the currently highlighted point from the trace and the command `node trace move-point` moves the currently highlighted trace point to the cursor. When an edge is highlighted instead of a point, the point closest to the start of the trace is used in these operations.

There are quite a few primitive nodes that make use of this trace information. All the **pure-layer** nodes in the IC layout technologies can be described polygonally. The pure-layer nodes are those near the end of each technology that contain only one layer and have the word ''-Node'' at the end of their names. For example, the node called Metal-Node in nMOS looks like a rectangle of the Metal layer until you add trace information. With trace data, this node can take any shape. The Artwork technology has four nodes that use trace information: Opened-Polygon, Closed-Polygon, Filled-Polygon, and Spline. Also available are the MOS transistors, which can take trace information and use it as the serpentine path of the gate (polysilicon). To change the width of serpentine transistors, use `transistorwidth` rather than `size`.

The ''useful.mac'' command file, which is automatically loaded on startup, has macros for editing trace information. The command `polygonedit` (the ''P'' key) rebinds some keys and buttons so that you can conveniently manipulate points. Rather than the left, middle, and right buttons being `find`, `create`, and `move`, they become `find vertex`, `node trace add-point`, and `node trace move-point`. The ''e'' key becomes `node trace delete-point`, which completes the analogy between objects and trace points. You can now manipulate points as easily as you formerly manipulated nodes. Typing ''P'' again exits this mode and restores the key and button functions. One additional feature of this macro is that it replaces the currently highlighted node with the Opened-Polygon of the Artwork technology. This node is the simplest polygon to draw and makes editing steps faster to run. When you exit the mode, the original polygon node is replaced.

Besides adding single points, you can input a continuous freehand sketch of points. The `node trace` command reads a sequence of points while any mouse button is held down, and stores the outline in the command-interpreter variable %T. The command `node trace store-trace` also places this information in the currently highlighted node. The option `wait-for-down` used with either of these commands requests that the system wait for a button to be pushed before starting the trace. This is useful when the command has been bound to a button, such that one is already pushed when the command is invoked.

A side effect of the freehand-sketching options is that the trace information lingers in the command-interpreter variable %T. This can be used by other commands, for example `window trace-displayed`, which zooms and pans the window so that the most recent trace fills the screen. One could also envisage the addition of cursive-script recognition that analyzes the trace and issues appropriate commands.

F.13   Conclusion

By now you have seen every aspect of Electric. This means that you are ready to modify the system code for your own particular needs. Electric is able to accommodate additional tools, technologies, display drivers, and even constraint systems, so you are invited to incorporate your favorite CAD code. There are three internals manuals that explain (1) the data structures, (2) the database routines, and (3) the details of writing technologies. Now you can do anything.

REFERENCES FOR THE ENTIRE BOOK

Abramovici, M.; Levendel, Y.H.; and Menon, P.R., "A Logic Simulation Machine," Proceedings 9th Symposium on Computer Architecture, SIGArch Newsletter, 10:3, 148-157, April 1982. [6]

Ackland, Bryan; Dickenson, Alex; Ensor, Robert; Gabbe, John; Kollaritsch, Paul; London, Tom; Poirier, Charles; Subrahmanyam, P.; and Watanabe, Hiroyuki, "CADRE—A System of Cooperating VLSI Design Experts," Proceedings IEEE International Conference on Computer Design, 99-104, October 1985. [1, 4]

Ackland, Bryan and Weste, Neil, "Realtime Animation Playback on a Frame Store Display System," *Computer Graphics*, 14:3, 182-188, August 1980. [9]

Ackland, Bryan and Weste, Neil, "An Automatic Assembly Tool for Virtual Grid Symbolic Layout," *VLSI '83* (Anceau and Aas, eds), North Holland, Amsterdam, 457-466, August 1983. [4]

Adobe Systems Incorporated, *PostScript Language Tutorial and Cookbook*, Addison-Wesley, Reading, Massachusetts, 1985. [9]

ANSI, *Programmer's Hierarchical Interactive Graphics Standard (PHIGS)*, American National Standards Institute X3H3/84-40, February 1984. [9]

Applicon, *Bravo3 User's Guide*, Applicon Incorporated, Ann Arbor, Michigan, 1986. [10]

Applicon, *IAGL User's Guide*, Applicon Incorporated, Burlington, Massachusetts, June 1983. [8]

Arnold, John E., "The Knowledge-Based Test Assistant's Wave/Signal Editor: An Interface for the Management of Timing Constraints," Proceedings 2nd Conference on Artificial Intelligence Applications, 130-136, December 1985. [3, 8]

Arnold, Michael H. and Ousterhout, John K., "Lyra: A New Approach to Geometric Layout Rule Checking," Proceedings 19th Design Automation Conference, 530-536, June 1982. [5]

Atkinson, William D.; Bond, Karen E.; Tribble, Guy L.; and Wilson, Kent R., "Computing with Feeling," *Computers and Graphics*, 2:2, 97-103, 1977. [9]

Atwood, Thomas M., "An Object-Oriented DBMS for Engineering Design Support Applications," Proceedings Compint Conference, Montreal, 299-307, September 1985. [3]

Ayres, Ronald F., *VLSI Silicon Compilation and the Art of Automatic Microchip Design*, Prentice-Hall, Englewood Cliffs, New Jersey, 1983. [4]

Baird, Henry S., "Fast Algorithms for LSI Artwork Analysis", Proceedings 14th Design Automation Conference, 303-311, June 1977. [5]

Baird, H. S. and Cho, Y. E., "An Artwork Design Verification System," Proceedings 12th Design Automation Conference, 414-420, June 1975. [5]

Baker, Clark M. and Terman, Chris, "Tools for Verifying Integrated Circuit Designs," *Lambda*, 1:3, 22-30, 4th Quarter 1980. [4, 5, 11]

Balraj, T. S. and Foster, M. J., "Miss Manners: A Specialized Silicon Compiler for Synchronizers," Proceedings 4th MIT Conference on Advanced Research in VLSI (Leiserson, ed), 3-20, April 1986. [4]

Baray, Mehmet B. and Su, Stephen Y. H., "A Digital System Modeling Philosophy and Design Language," Proceedings 8th Design Automation Workshop, 1-22, June 1971. [2]

Barrow, Harry G., "VERIFY: A Program for Proving Correctness of Digital Hardware Designs," *Artificial Intelligence*, 24:1-3, 437-491, December 1984. [5, 11]

Barsky, Brian A. and Beatty, John C., "Local Control of Bias and Tension in Beta-splines," *Computer Graphics*, 17:3, 193-218, July 1983. [9]

Barton, E. E. and Buchanan, I., "The Polygon Package," *Computer Aided Design*, 12:1, 3-11, January 1980. [3]

Batali, J. and Hartheimer, A., "The Design Procedure Language Manual," AI Memo 598, Massachusetts Institute of Technology, 1980. [3, 8, 11]

Baumgart, Bruce Guenther, *Geometric Modeling for Computer Vision*, PhD dissertation, Stanford University, August 1974. [3]

Bell, C. Gordon; Grason, John; and Newell, Allen, *Designing Computers and Digital Systems Using PDP 16 Register Transfer Modules*, Digital Press, Maynard, Massachusetts, 1972. [2]

Bell, C. Gordon and Newell, Allen, *Computer Structures: Readings and Examples*, McGraw-Hill, New York, 1971. [2]

Bell, J. L. and Slomson, A. B., *Models and Ultraproducts: An Introduction*, North-Holland and American Elsevier, New York, 1971. [2]

Bentley, Jon Louis; Haken, Dorthea; and Hon, Robert W., "Fast Geometric Algorithms for VLSI Tasks," Proceedings 20th IEEE Compcon, 88-92, February 1980. [5]

Beresford, Roderic, "Comparing Gate Arrays and Standard-Cell ICs," *VLSI Design*, IV:8, 30-36, December 1983. [4]

Bezier, P, *Numerical Control—Mathematics and Applications*, (A. R. Forest, trans), Wiley, London, 1972. [9]

Blank, Tom, "A Survey of Hardware Accelerators Used in Computer-Aided Design," *IEEE Design and Test*, 1:3, 21-39, August 1984. [4]

Bobrow, Daniel G.; Burchfiel, Jerry D.; Murphy, Daniel L.; and Tomlinson, Raymond S., "TENEX: A Paged Time Sharing system for the PDP-10," *CACM*, 15:3 135-143, March 1972. [10]

Borning, Alan, "ThingLab—A Constraint-Oriented Simulation Laboratory," PhD dissertation, Stanford University, July 1979. [3, 8]

Borriello, Gaetano, "WAVES: A Digital Waveform Editor for the Design, Documentation, and Specification of Interfaces," unpublished document. [3, 8]

Borriello, Gaetano; Katz, Randy H.; Bell, Alan G.; and Conway, Lynn, "VLSI System Design by the Numbers," *IEEE Spectrum*, 22:2, 44-50, February 1985. [7]

Breshenham, J. E., "Algorithm for Computer Control of Digital Plotter," *IBM Systems Journal*, 4:1, 25-30, 1965. [9]

Breshenham, J. E., "A Linear Algorithm for Incremental Digital Display of Circular Arcs," *CACM*, 20:2, 100-106, February 1977. [9]

Breuer, Melvin A., "A Class of Min-Cut Placement Algorithms," Proceedings 14th Design Automation Conference, 284-290, June 1977. [4]

Brown, Harold; Tong, Christofer; and Foyster, Gordon, "Palladio: An Exploratory Environment for Circuit Design," *IEEE Computer*, 16:12, 41-56, December 1983. [1, 3, 8]

Bryant, Randal Everitt, *A Switch-Level Simulation Model for Integrated Logic Circuits*, PhD dissertation, Massachusetts Institute of Technology Laboratory for Computer Science, report MIT/LCS/TR-259, March 1981. [6, 11]

Bryant, Randal, "Preface," Proceedings 3rd Caltech Conference on VLSI (Bryant ed), Computer Science Press, v-viii, March 1983. [3]

Buric, Misha R. and Matheson, Thomas G., "Silicon Compilation Environments," Proceedings Custom Integrated Circuits Conference, 208-212, May 1985. [4, 8]

Burstein, Michael; Hong, Se June; and Pelavin, Richard, "Hierarchical VLSI Layout: Simultaneous Placement and Wiring of Gate Arrays," *VLSI '83* (Anceau and Aas, eds), North Holland, Amsterdam, 45-60, August 1983. [4]

CAE Corporation, *CAE 2000 Command Language User's Manual*, August 1984. [8]

Calma Corporation, *GDS II Stream Format*, July 1984. [7]

Calma, *GPL II Programmers Reference Manual*, GE Calma Company, February 1981. [8]

Card, Stuart K.; Moran, Thomas P.; and Newell, Allen, *The Psychology of Human-Computer Interaction*, Lawrence Erlbaum, Hillsdale, New Jersey, 1983. [10]

Catmull, Edwin, "A Hidden Surface Algorithm with Anti-Aliasing," *Computer Graphics*, 12:3, 6-11, August 1978. [9]

Chao, Shiu-Ping; Huang, Yen-Son; and Yam, Lap Man, "A Hierarchical Approach for Layout Versus Circuit Consistency Check," Proceedings 17th Design Automation Conference, 270-276, June 1980. [5]

Chawla, Basant R.; Gummel, Hermann K.; and Kozak, Paul, "MOTIS—An MOS Timing Simulator," *IEEE Transactions on Circuits and Systems*, CAS-22:12, 901-910, December 1975. [6]

Chen, C.F.; Lo, C-Y.; Nham, H.N.; and Subramaniam, Prasad, "The Second Generation MOTIS Mixed-Mode Simulator," Proceedings 21st Design Automation Conference, 10-17, June 1984. [6]

Cheng, Chung-Kuan and Kuh, Ernest S., "Module Placement Based on Resistive Network Optimization," *IEEE Transactions on CAD*, 3:3, 218-225, July 1984 [4]

Cherry, James; Shrobe, Howard; Mayle, Neil; Baker, Clark; Minsky, Henry; Reti, Kalman; and Weste, Neil, "NS: An Integrated Symbolic Design System," *VLSI '85*, (Horbst, ed), 325-334, August 1985. [8]

Chiba, Toshiaki; Takashima, Makoto; and Mitsuhashi, Takashi, "A Mask Artwork Analysis System for Bipolar Integrated Circuits," 21st IEEE Compcon, 175-183, September 1981. [5]

CMC, *Guide to the Integrated Circuit Implementation Services of the Canadian Microelectronics Corporation*, version 2:0, Kingston Ontario, January 1986. [7]

Clark, G. C. and Zippel, R. E., "Schema: An Architecture for Knowledge Based CAD," *ICCAD '85*, 50-52, November 1985. [3]

Clarke, Edmund and Feng, Yulin, "Escher—A Geometrical Layout System for Recursively Defined Circuits," Proceedings 23rd Design Automation Conference, 650-653, June 1986. [8]

Clocksin, W. F. and Mellish, C. S., *Programming In Prolog*, 2nd Edition, Springer-Verlag, Berlin, 1984. [11]

Computervision, *CADDS II/VLSI Integrated Circuit Programming Language User's Guide*, Computervision Corporation Document 001-00045, Bedford, Massachusetts, April 1986. [8]

Conway, Lynn; Bell, Alan; and Newell, Martin E., "MPC79: The Demonstration-Operation of a Prototype Remote-Entry Fast-Turnaround VLSI Implementation System," Proceedings MIT Conference on Advanced Research in Integrated Circuits, January 1980 (also reprinted in *Lambda*, 1:2, 10-19, 2nd Quarter 1980). [7]

Crawford, B. J., "Design Rules Checking for Integrated Circuits Using Graphical Operators," *Computer Graphics*, 9:1, 168-176, 1975. [5]

Curry, James E., "A Tablet Input Facility for an Interactive Graphics System," Proceedings IJCAI '69, 33-40, May 1969. [10]

Davis, A. L. and Drongowski, P. J., "Dataflow Computers: A Tutorial and Survey," University of Utah UUCS-80-109, July 1980. [2]

Davis, Tom, and Clark, Jim, "SILT: A VLSI Design Language," Stanford University Computer Systems Laboratory Technical Report 226, October 1982. [8]

Deas, Alex R. and Nixon, Ian M., "Chromatic Idioms for Automated VLSI Floorplanning," *VLSI '85*, (Horbst, ed), 61-70, August 1985. [4]

De Man, Hugo J.; Bolsens, I.; Meersch, Erik Vanden; and Cleynenbreugel, Johan Van, "DIALOG: An Expert Debugging System for MOSVLSI Design," *IEEE Transactions on CAD*, 4:3, 303-311, July 1985. [5]

Denneau, Monty M., "The Yorktown Simulation Engine," Proceedings 19th Design Automation Conference, 55-59, June 1982. [6]

Dennis, J. B., Fosseen, J. B., and Linderman, J. P., "Data Flow Schemas," Proceedings International Symposium on Theoretical Programming, 187-216, 1972. [2]

Denyer, Peter B.; Murray, Alan F.; and Renshaw, David, "FIRST—Prospect and Retrospect," *VLSI Signal Processing*, IEEE press, New York, 252-263, 1984. [4]

Deutsch, David N., "A 'Dogleg' Channel Router," Proceedings 13th Design Automation Conference, 425-433, June 1976. [4]

Deutsch, L. P. and Bobrow, D. G., "An Efficient Incremental Automatic Garbage Collector," *CACM*, 19:9, 522-526, September 1976. [3]

Digital, *PDP10 Timesharing Handbook*, Digital Press, Maynard, Massachusetts, 1970. [10]

Do, James and Dawson, William M., "Spacer II: A Well-Behaved IC Layout Compactor," *VLSI '85*, (Horbst, ed), 283-291, August 1985. [4]

Dobes, Ivan and Byrd, Ron, "The Automatic Recognition of Silicon Gate Transistor Geometries: An LSI Design Aid Program," Proceedings 13th Design Automation Conference, 327-335, June 1976. [5]

Department of Defense, "Graphic Symbols for Logic Diagrams," MIL-STD-806B, Washington, D.C., February 1962. [2]

Doreau, Michel T. and Koziol, Piotr, "TWIGY: A Topological Algorithm Based Routing System," Proceedings 18th Design Automation Conference, 746-755, June 1981. [4]

Dunlop, A. E., "SLIM—The Translation of Symbolic Layouts into Mask Data," Proceedings 17th Design Automation Conference, 595-602, June 1980. [4]

Ebeling, Carl and Zajicek, Ofer, "Validating VLSI Circuit Layout by Wirelist Comparison," *ICCAD '83*, 172-173, September 1983. [1, 5]

Electronic Design Interface Format Steering Committee, *EDIF—Electronic Design Interchange Format Version 1 0 0*, Texas Instruments, Dallas, Texas, 1985. [7, 11]

Entenman, George and Daniel, Stephen W., "A Fully Automatic Hierarchical Compactor," Proceedings 22nd Design Automation Conference, 69-75, June 1985. [4]

Eustace, R. Alan and Mukhopadhyay, Amar, "A Deterministic Finite Automaton Approach to Design Rule Checking For VLSI," Proceedings 19th Design Automation Conference, 712-717, June 1982. [5]

Factron, "CADDIF Version 2.0 Engineering Specifications," Schlumberger Factron, October 1985. [7]

Fairbairn, D. G. and Rowson, J. A., "ICARUS: An Interactive Integrated Circuit Layout Program," Proceedings 15th Design Automation Conference, 188-192, June 1978. [10]

Feller, A., "Automatic Layout of Low-Cost Quick-Turnaround Random-Logic Custom LSI Devices," Proceedings 13th Design Automation Conference, 79-85, June 1976. [2]

Fishburn, J. P. and Dunlop, A. E., "TILOS: A Posynomial Programming Approach to Transistor Sizing," *ICCAD '85*, 326-328, November 1985. [5]

Foley, J. D. and Van Dam, A., *Fundamentals of Interactive Computer Graphics*, Addison-Wesley, Reading, Massachusetts, 1982. [9]

Freeman, William J. III and Freund, Vincent J. Jr., "A History of Semicustom Design at IBM," *VLSI Systems Design*, Semicustom Design Guide, 14-22, Summer 1986. [4]

Frey, Ernest J., "ESIM: A Functional Level Simulation Tool," *ICCAD '84*, 48-50, November 1984. [6]

Fuchs, Henry; Poulton, John; Paeth, Alan; and Bell, Alan, "Developing Pixel-Planes, A Smart Memory-Based Raster Graphics System," Proceedings MIT Conference on Advanced Research in VLSI (Penfield, ed), 137-146, January 1982. [9]

Gajski, Daniel D., "ARSENIC Silicon Compiler," Proceedings International Symposium on Circuits and Systems, 399-402, June 1985. [4]

Garey, Michael R. and Johnson, David S., *Computers and Intractability, A Guide to the Theory of NP-Completeness*, W.H. Freeman, San Francisco, 1979. [4]

Gerber Corporation, "Gerber Format," Gerber Scientific Instrument Company document number 40101-S00-066A, July 1983. [7]

German, Steven M. and Wang, Yu, "Formal Verification of Parameterized Hardware Designs," Proceedings IEEE International Conference on Computer Design, 549-552, October 1985. [5]

Gibson, Dave and Nance, Scott, "SLIC—Symbolic Layout of Integrated Circuits," Proceedings 13th Design Automation Conference, 434-440, June 1976. [1, 2]

Glasser, Lance A. and Dobberpuhl, Daniel W., *The Design and Analysis of VLSI Circuits*, Addison-Wesley, Reading, Massachusetts, 1985. [2, 6]

Goates, Gary B.; Harris, Thomas R.; Oettel, Richard E.; and Waldron, Harvey

M. III, "Storage/Logic Array Design: Reducing Theory to Practice," *VLSI Design*, III:4, 56-62, 1982. [4]

Gordon, M., "A Very Simple Model of Sequential Behaviour of nMOS," *VLSI '81* (Gray, ed), Academic Press, London, 85-94, August 1981. [5, 11]

Gordon, Mike, "HOL—A Machine Oriented Formulation of Higher Order Logic," University of Cambridge Computer Laboratory, technical report 68, July 1985. [5]

Gosling, James, *Algebraic Constraints*, PhD dissertation, Carnegie-Mellon University, CMU-CS-83-132, May 1983. [3, 8]

Gosling, James, personal communications. [10]

GPSC, "Status Report of the Graphic Standards Planning Committee," *Computer Graphics*, 13:3, August 1979. [9, 11]

Griswold, Thomas W., "Portable Design Rules for Bulk CMOS," *VLSI Design*, III:5, 62-67, September/October 1982. [11]

Gross, A. G.; Raamot, J.; and Watkins, S. B., "Computer Systems for Pattern Generator Control," *Bell Systems Technical Journal*, 49:9, 2011-2029, November 1970. [7]

Grundmann, John W., "Event-Driven MOS Timing Simulator," *ICCAD '83*, 141-142, September 1983. [6]

Guttman, Antonin, "R-Trees: A Dynamic Index Structure for Spatial Searching," *ACM SIGMOD*, 14:2, 47-57, June 1984. [3, 11]

Haber, Ralph Norman, "How We Remember What We See," *Scientific American*, 222:5, 104-112, May 1970. [10]

Hachtel, G. D.; Newton, A. R.; and Sangiovanni-Vincentelli, A. L., "Techniques for Programmable Logic Array Folding," Proceedings 19th Design Automation Conference, 147-155, June 1982. [4]

Hamachi, Gordon T. and Ousterhout, John K., "A Switchbox Router with Obstacle Avoidance," Proceedings 21st Design Automation Conference, 173-179, June 1984. [4]

Harrison, Richard A. and Olson, Daniel J., "Race Analysis of Digital Systems Without Logic Simulation," Proceedings 8th Design Automation Workshop, 82-94, June 1971. [5]

Heller, W. R., "An Algorithm for Chip Planning," Caltech Silicon Structures Project file #2806, 1979. [4]

Heller, William R.; Sorkin, G.; Maling, Klim, "The Planar Package Planner for System Designers," Proceedings 19th Design Automation Conference, 253-260, June 1982. [4]

Hellestrand, G. R.; Tan, C. H.; Yong, F. N.; and Forster, R. L., "Australian Multi-Project Chip Activities, 1982-1986," Joint Microelectronics Research Centre, University of New South Wales, October 1986. [7]

Henderson, Peter, "Functional Geometry," Proceedings ACM Symposium on LISP and Functional Programming, 179-187, August 1982. [8]

Hennion, B. and Senn, P., "A New Algorithm for Third Generation Circuit Simulators: The One-Step Relaxation Method," Proceedings 22nd Design Automation Conference, 137-143, June 1985. [6]

HHB, *CADAT User's Manual*, Revision 5.0, HHB-Softron, Mahwah, New Jersey, June 1985. [11]

Hightower, D. W., "A Solution to Line-Routing Problems in the Continuous

Plane,'' Proceedings 6th Design Automation Workshop, 1-24, June 1969. [4]

Holt, Dan and Sapiro, Steve, ''BOLT—A Block Oriented Design Specification Language,'' Proceedings 18th Design Automation Conference, 276-279, June 1981. [8]

Hon, Robert W., *The Hierarchical Analysis of VLSI Designs*, PhD dissertation, Carnegie-Mellon University Computer Science Department, CMU-CS-83-170, December 1983. [5]

Hon, Robert W. and Sequin, Carlo H., ''A Guide to LSI Implementation,'' 2nd Edition, Xerox Palo Alto Research Center technical memo SSL-79-7, January 1980. [4, 7, 11]

Horowitz, Mark, ''Timing Models for MOS Pass Networks,'' Proceedings International Symposium on Circuits and Systems, 198-201, May 1983. [5]

Hsueh, Min-Yu and Pederson, Donald O., ''Computer-Aided Layout of LSI Circuit Building-Blocks,'' Proceedings International Symposium on Circuits and Systems, 474-477, July 1979. [4, 8]

IBM, *Advanced Statistical Analysis Program (ASTAP)*, IBM Corporation Data Products Division, Publication SH20-1118-0, White Plains, New York. [6]

Insinga, Aron K., ''Behavioral Modeling in a Structural Logic Simulator,'' *ICCAD '84*, 42-44, November 1984. [6]

Jain, Sunil K. and Agrawal, Vishwani D., ''Modeling and Test Generation Algorithms for MOS Circuits,'' *IEEE Transactions on Computers*, C-34:5, 426-433, May 1985. [6]

Jarvis, J. F.; Judice, C. N.; and Ninke, W. H., ''A Survey of Techniques for the Image Display of Continuous Tone Pictures on Bilevel Displays,'' *Computer Graphics and Image Processing*, 5:1, 13-40, March 1976. [9]

Johannsen, D. L., ''Bristle Blocks: A Silicon Compiler,'' Proceedings 16th Design Automation Conference, 310-313, June 1979. [4]

Johnson, Dean P. and Lipman, Jim, ''IC Packaging: An Introduction For the VLSI Designer,'' *VLSI Systems Design, VII:6, 108-116, June 1986.* [7]

Johnson, Stephen C., ''Hierarchical Design Validation Based on Rectangles,'' Proceedings MIT Conference on Advanced Research in VLSI (Penfield, ed), 97-100, January 1982. [5, 8, 11]

Jouppi, Norman P., ''TV: An nMOS Timing Analyzer,'' Proceedings 3rd Caltech Conference on VLSI (Bryant, ed), Computer Science Press, 71-85, March 1983. [5]

Kahrs, Mark, ''Silicon compilation of a very high level signal processing specification language,'' *VLSI Signal Processing*, IEEE press, New York, 228-238, 1984. [4]

Kaplan, David, ''A 'Non-Restrictive' Artwork Verification Program for Printed Circuit Boards,'' Proceedings 19th Design Automation Conference, 551-558, June 1982. [5]

Karplus, Kevin, ''Exclusion Constraints, a new application of Graph Algorithms to VLSI Design,'' Proceedings 4th MIT Conference on Advanced Research in VLSI (Leiserson, ed), 123-139, April 1986. [3, 5]

Kedem, Gershon, ''The Quad-CIF Tree: A Data Structure for Hierarchical

On-Line Algorithms,'' Proceedings 19th Design Automation Conference, 352-357, June 1982. [3]

Keller, John, *Power and Ground Requirements for a High Speed 32 Bit Computer Chip Set*, Masters thesis, University of California at Berkeley, UCB/CSD 86/253, August 1985. [4]

Kernighan, Brian W. and Ritchie, Dennis M., *The C Programming Language*, Prentice-Hall, Englewood Cliffs, New Jersey, 1978. [11]

Kernighan, B. W.; Schweikert, D. G.; and Persky, G., ''An Optimum Channel-Routing Algorithm for Polycell Layouts of Integrated Circuits,'' Proceedings 10th Design Automation Workshop, 50-59, June 1973. [4]

Ketonen, Jussi and Weening, Joseph S., ''EKL—An Interactive Proof Checker User's Reference Manual,'' Stanford University Department of Computer Science, report STAN-CS-84-1006, June 1984. [5]

Kim, Jin H.; McDermott, John; and Siewiorek, Daniel P., ''Exploiting Domain Knowledge in IC Cell Layout,'' *IEEE Design and Test*, 1:3, 52-64, 1984. [4]

Kingsley, C., *Earl: An Integrated Circuit Design Language*, Masters Thesis, California Institute of Technology, June 1982. [8, 11]

Kirkpatrick, S.; Gelatt, C. D. Jr.; and Vecchi, M. P., ''Optimization by Simulated Annealing,'' *Science*, 220:4598, 671-680, May 1983. [4]

Kollaritsch, P. W. and Weste, N. H. E., ''A Rule-Based Symbolic Layout Expert,'' *VLSI Design*, V:8, 62-66, August 1984. [4]

Koppelman, George M. and Wesley, Michael A., ''OYSTER: A Study of Integrated Circuits as Three-Dimensional Structures,'' *IBM Journal of Research and Development*, 27:2, 149-163, March 1983. [1]

Kors, J. L. and Israel, M., ''An Interactive Electrical Graph Extractor,'' Proceedings 21st Design Automation Conference, 624-628, June 1984. [5]

Kowalski, T. J. and Thomas, D. E., ''The VLSI Design Automation Assistant: Prototype System,'' Proceedings 20th Design Automation Conference, 479-483, June 1983. [4]

Kozminski, Krzysztof and Kinnen, Edwin, ''An Algorithm for Finding a Rectangular Dual of a Planar Graph for Use in Area Planning for VLSI Integrated Circuits,'' Proceedings 21st Design Automation Conference, 655-656, June 1984. [4]

Knuth, Donald E., *The Art of Computer Programming, Volume 1/Fundamental Algorithms*, Addison-Wesley, Reading, Massachusetts, 1969. [3]

Knuth, Donald E., *TEX and METAFONT—New Directions in Typesetting*, Digital Press, Bedford, Massachusetts, 1979. [9]

Kroeker, Wallace I., *Integrated Environmental Support for Silicon Compilation of Digital Filters*, Masters Thesis, University of Calgary Computer Science Department, March 1986. [11]

Lanfri, Ann R., ''PHLED45: An Enhanced Version of Caesar Supporting 45 degree Geometries,'' Proceedings 21st Design Automation Conference, 558-564, June 1984. [3]

Lansky, A. L. and Owicki, S. S., ''GEM: A Tool for Concurrency Specification and Verification,'' Proceedings 2nd Annual ACM Symposium on Principles of Distributed Computing, 198-212, August 1983. [11]

Lathrop, Richard H. and Kirk, Robert S., ''An Extensible Object-Oriented

Mixed-Mode Functional Simulation System," Proceedings 22nd Design Automation Conference, 630-636, June 1985. [6]

Lauther, Ulrich, "Channel Routing in a General Cell Environment," *VLSI '85*, (Horbst, ed), 393-403, August 1985. [4]

Lee, C. Y., "An Algorithm for Path Connections and Its Applications," *IRE Transactions on Electronic Computers*, EC-10, 346-365, September 1961. [4]

Leinwand, Sany M., "Integrated Design Environment," unpublished manuscript, April 1984. [3]

Liblong, Breen M., *SHIFT—A Structured Hierarchical Intermediate Form for VLSI Design Tools*, Masters Thesis, University of Calgary Department of Computer Science, September 1984. [7]

Lightner, M.R.; Moceyunas, P.H.; Mueller, H.P.; Vellandi, B.L.; and Vellandi, H.P., "CSIM: The Evolution of a Behavioral Level Simulator from a Functional Simulator: Implementation Issues and Performance Measurements," *ICCAD '85*, 350-352, November 1985. [6]

Lin, Tzu-Mu and Mead, Carver A., "Signal Delay in General RC Networks with Application to Timing Simulation of Digital Integrated Circuits," Proceedings MIT Conference on Advanced Research in VLSI (Penfield, ed), 93-99, January 1984. [5]

Lipton, Richard J.; North, Stephen C.; Sedgewick, Robert; Valdes, Jacobo; and Vijayan, Gopalakrishnan, "ALI: A Procedural Language to Describe VLSI Layouts," Proceedings 19th Design Automation Conference, 467-473, June 1982. [8, 11]

Liu, Erwin S. K., "A Silicon Logic Module Compiler," Project Report, University of Calgary Department of Computer Science, April 1984. [4, 11]

Locanthi, Bart, "Object Oriented Raster Displays," Proceedings 1st Caltech Conference on VLSI (Seitz, ed), 215-225, January 1979. [9]

Lopez, Alexander D. and Law, Hung-Fai S., "A Dense Gate Matrix Layout Method for MOS VLSI," *IEEE Transactions on Electron Devices*, 27:8, 1671-1675, August 1980. [4]

Losleben, P., "Computer Aided Design for VLSI," *Very Large Scale Integration (VLSI)* 5 (Barbe, ed), Springer-Verlag, Berlin, 89-127, 1980. [1]

Losleben, Paul and Thompson, Kathryn, "Topological Analysis for VLSI Circuits," Proceedings 16th Design Automation Conference, 461-473, June 1979. [5]

Luk, W. K., "A Greedy Switch-box Router," Carnegie-Mellon University Department of Computer Science VLSI Document V158, May 1984. [4]

Lyon, Richard F., "Simplified Design Rules for VLSI Layouts," *Lambda*, 2:1, 54-59, 1st Quarter 1981. [5]

Lyon, Richard F., "The Optical Mouse, and an Architectural Methodology for Smart Digital Sensors," Proceedings C-MU Conference on VLSI Systems and Computations (Kung, Sproull, and Steele, eds), 1-19, October 1981. [9]

Lyon, Richard F. and Schediwy, Richard R., "CMOS Static Memory with a New 4-Transistor Memory Cell," Proceedings Stanford Conference on Advanced Research in VLSI, March 1987. [11]

Malachi, Yonatan and Owicki, Susan S., "Temporal Specifications of Self-

Timed Systems," Proceedings C-MU Conference on VLSI Systems and Computations (Kung, Sproull, and Steele, eds), Computer Science Press, 203-212, 1981. [2]

Maley, F. Miller, "Compaction with Automatic Jog Introduction," Chappel Hill Conference on VLSI (Fuchs, ed), 261-283, March 1985. [4]

Mathews, Robert; Newkirk, John; and Eichenberger, Peter, "A Target Language for Silicon Compilers," Proceedings 24th IEEE Computer Society International Conference, 349-353, February 1982. [8]

Mayo, Robert N., "Combining Graphics and Procedures in a VLSI Layout Tool: The Tpack System," University of California at Berkeley Computer Science Division technical report, January 1984. [4, 8]

McCarthy, John; Abrahams, Paul W.; Edwards, Daniel J.; Hart, Timothy P.; and Levin, Michael I., *LISP 1.5 Programmer's Manual*, MIT Press, Cambridge, Massachusetts, 1962. [7]

McCormick, Steven P., "EXCL: A Circuit Extractor for IC Designs," Proceedings 21st Design Automation Conference, 616-623, June 1984. [5]

McCreight, E.M., "Efficient Algorithms for Enumerating Intersecting Intervals and Rectangles," Xerox Palo Alto Research Center, CSL-80-9, 1980. [3, 4]

McWilliams, Thomas M., "Verification of Timing Constraints on Large Digital Systems," Proceedings 17th Design Automation Conference, 139-147, June 1980. [5]

Mead, C. and Conway, L., *Introduction to VLSI Systems*, Addison-Wesley, Reading, Massachusetts, 1980. [1, 3, 5, 6]

Metropolis, Nicholas; Rosenbluth, Arianna W.; Rosenbluth, Marshall N.; Teller, Augusta H.; and Teller, Edward, "Equation of State Calculations by Fast Computing Machines," *Journal of Chemical Physics*, 21:6, 1087-1092, June 1953. [4]

Miczo, Alexander, *Digital Logic Testing and Simulation*, Chapter 2: "Combinational Logic Test," Harper and Row, New York, 1986. [6]

Miller, George A., "The Magical Number Seven, Plus or Minus Two: Some Limits on Our Capacity for Processing Information," *Psychological Review*, 63:2, 81-97, March 1956. [1]

Mitchell, Tom M.; Steinberg, Louis I.; and Shulman, Jeffrey S., "A Knowledge-Based Approach to Design," Proceedings IEEE Workshop on Principles of Knowledge-Based Systems, 27-34, December 1984. [4]

Moore, E. F., "Shortest Path Through a Maze," Harvard University Press, Cambridge, Massachusetts, 285-292, 1959. [4]

Mori, Hajimu, "Interactive Compaction Router for VLSI Layout," Proceedings 21st Design Automation Conference, 137-143, June 1984. [4]

MOSIS, *MOSIS User's Manual*, University of Southern California Information Sciences Institute, 1986. [7]

Mosteller, R. C., "REST—A Leaf Cell Design System," *VLSI '81* (Gray, ed), Academic Press, London, 163-172, August 1981. [1, 4, 8]

Moszkowski, Ben, "A Temporal Logic for Multilevel Reasoning about Hardware," *IEEE Computer*, 10-19, February 1985. [2]

Mukherjee, Amar, *Introduction to nMOS and CMOS VLSI Systems Design*, Prentice-Hall, Englewood Cliffs, New Jersey, 1986. [11]

Nagel, L. W., "Spice2: A Computer Program to Simulate Semiconductor

Circuits,'' University of California at Berkeley, ERL-M520, May 1975. [5, 6, 11]

Nelson, Greg, ''Juno, a constraint-based graphics system,'' *Computer Graphics*, 19:3, 235-243, July 1985. [3, 8]

Newell, Martin E. and Fitzpatrick, Daniel T., ''Exploiting Structure in Integrated Circuit Design Analysis,'' Proceedings MIT Conference on Advanced Research in VLSI (Penfield, ed), 84-92, January 1982. [5, 9]

Newell, Martin E. and Sequin, Carlo H., ''The Inside Story on Self-Intersecting Polygons,'' *Lambda*, 1:2, 20-24, 2nd Quarter 1980. [3, 9]

Newkirk, John and Mathews, Robert, *The VLSI Designer's Library*, Addison-Wesley, Reading, Massachusetts, 1983. [2]

Newman, William M. and Sproull, Robert F., *Principles of Interactive Computer Graphics*, 2nd Edition, McGraw-Hill, New York, 1979. [3, 9, 10]

Newton, Arthur Richard and Sangiovanni-Vincentelli, Alberto L., ''Relaxation-Based Electrical Simulation,'' *IEEE Transactions on CAD*, CAD-3:4, 308-331, October 1984. [6]

Nogatch, John T. and Hedges, Tom, ''Automated Design of CMOS Leaf Cells,'' *VLSI Systems Design*, VI:11, 66-78, November 1985. [4]

Noll, A. Michael, ''Man-Machine Tactile Communication,'' *Society for Information Display Journal*, 1:2, 5-11, July/August 1972. [9]

North, Stephen C., ''Molding Clay: A Manual for the Clay Layout Language,'' Princeton University Department of Electrical Engineering and Computer Science, VLSI Memo #3, July 1893. [8]

Ousterhout, J. K., ''Caesar: An Interactive Editor for VLSI Layouts,'' *VLSI Design*, II:4, 34-38, 1981. [3, 10, 11]

Ousterhout, John K., ''Crystal: A Timing Analyzer for nMOS VLSI Circuits,'' Proceedings 3rd Caltech Conference on VLSI (Bryant, ed), Computer Science Press, 57-69, March 1983. [5]

Ousterhout, John K., ''Corner Stitching: A Data-Structuring Technique for VLSI Layout Tools,'' *IEEE Transactions on CAD*, 3:1, 87-100, January 1984. [3]

Patil, Suhas S., ''An Asynchronous Logic Array,'' Project MAC tech memo TM-62, Massachusetts Institute of Technology, May 1975. [2]

Penfield, Paul Jr. and Rubenstein, Jorge, ''Signal Delay in RC Tree Networks,'' Proceedings 18th Design Automation Conference, 613-617, June 1981. [5]

Pieper, Chris, ''Stimulus Data Interchange Format,'' *VLSI Systems Design*, Part I: VII:7, 76-81, July 1986; Part II: VII:8, 56-60, August 1986. [7]

Piscatelli, R. N. and Tingleff, P., ''A Solution To Closeness Checking of Non-Orthogonal Printed Circuit Board Wiring,'' Proceedings 13th Design Automation Conference, 172-178, June 1976. [5]

Pope, Stephen; Rabaey, Jan; and Brodersen, Robert W., ''Automated Design of Signal Processors Using Macrocells,'' *VLSI Signal Processing*, IEEE press, New York, 239-251, 1984. [4]

Ramsay, Frank R., ''Automation of Design for Uncommitted Logic Arrays,'' Proceedings 17th Design Automation Conference, 100-107, June 1980. [4]

Reddy, D. R. and Rubin, Steven M., ''Representation of Three-Dimensional Objects,'' Carnegie-Mellon University Department of Computer Science, Report CMU-CS-78-113, April 1978. [3]

Ritchie, D. M. and Thompson, K., "The UNIX Time-Sharing System," *Bell Systems Technical Journal*, 57:6, 1905-1929, 1978. [10]

Rivest, Ronald L., "The 'PI' (Placement and Interconnect) System," Proceedings 19th Design Automation Conference, 475-481, June 1982. [4]

Rivest, Ronald L. and Fiduccia, Charles M., "A 'Greedy' Channel Router," Proceedings 19th Design Automation Conference, 418-424, June 1982. [4]

Rodriguez, Jorge E., *A Graph Model for Parallel Computations*, PhD dissertation, Massachusetts Institute of Technology, Report MAC-TR-64, September 1969. [2]

Rogers, David F. and Adams, J. Alan, *Mathematical Elements for Computer Graphics*, McGraw-Hill, New York, 1976. [9]

Rosenberg, Jonathan B. and Weste, Neil H. E., "ABCD—A Better Circuit Description," Microelectronics Center of North Carolina Technical Report 4983-01, February 1983. [8]

Roth, J.P., "Diagnosis of Automata Failures: A Calculus and a Method," *IBM Journal of Research and Development*, 10:4, 278-291, July 1966. [6]

Rowson, James A., *Understanding Hierarchical Design*, PhD dissertation, California Institute of Technology, TR 3710, April 1980. [1]

Rubin, Steven M., "An Integrated Aid for Top-Down Electrical Design," *VLSI '83* (Anceau and Aas, eds), North Holland, Amsterdam, 63-72, August 1983. [11]

Saito, Takao; Uehara, Takao; and Kawato, Nobuaki, "A CAD System For Logic Design Based on Frames and Demons," Proceedings 18th Design Automation Conference, 451-456, June 1981. [8]

Saleh, Resve A.; Kleckner, James E.; and Newton, A. Richard, "Iterated Timing Analysis in Splice1," *ICCAD '83*, 139-140, September 1983. [6]

Sastry, S. and Klein, S., "PLATES: A Metric Free VLSI Layout Language," Proceedings MIT Conference on Advanced Research in VLSI (Penfield, ed), 165-169, January 1982. [8]

Schediwy, Richard R., *A CMOS Cell Architecture and Library*, Masters thesis, University of Calgary Department of Computer Science, 1987. [2, 3]

Scheffer, Louis K., "A Methodology for Improved Verification of VLSI Designs Without Loss of Area," Proceedings 2nd Caltech Conference on VLSI (Seitz, ed), 299-309, January 1981. [1, 5]

Schiele, W., "Design Rule Adaptation of Non-Orthogonal Layouts with Approximate Scaling," *VLSI '85*, (Horbst, ed), 273-282, August 1985. [4]

SCI, *GENESIL System User's Manual*, Silicon Compilers, Incorporated publication 110016, November 1985. [4]

Scott, Walter S. and Ousterhout, John K., "Plowing: Interactive Stretching and Compaction in Magic," Proceedings 21st Design Automation Conference, 166-172, June 1984. [4]

Seattle Silicon, *The Mentor Idea/Concorde User's Manual*, Seattle Silicon Technologies, Incorporated, publication UMC Beta 300 Rev 1, March 1986. [4]

Sechen, Carl and Sangiovanni-Vincentelli, Alberto, "The TimberWolf Placement and Routing Package," Proceedings Custom Integrated Circuit Conference, 522-527, May 1984. [4]

Seiler, Larry, "A Hardware Assisted Design Rule Check Architecture," Proceedings 19th Design Automation Conference, 232-238, June 1982. [5]

Seitz, Charles L., "System Timing," *Introduction to VLSI Systems* (Mead and Conway), Addison-Wesley, Reading, Massachusetts, 1980. [1]

Sequin, Carlo H., "Managing VLSI Complexity: An Outlook," *Proceedings IEEE*, 71:1, 149-166, January 1983. [1]

Shand, Mark A., "Hierarchical VLSI Artwork Analysis," *VLSI '85*, (Horbst, ed), 419-428, August 1985. [5]

Siewiorek, Daniel P.; Bell, C. Gordon; and Newell, Allen, *Computer Structures: Principles and Examples*, McGraw-Hill, New York, 1982. [2]

Simoudis, Evangelos and Fickas, Stephen, "The Application of Knowledge-Based Design Techniques to Circuit Design," *ICCAD '85*, 213-215, November 1985. [4]

Singh N. "MARS: A Multiple Abstraction Rule-Based Simulator," Stanford University Heuristic Programming Project HPP-83-43, December 1983. [11]

Soukup, Jiri, "Circuit Layout," *Proceedings IEEE*, 69:10, 1281-1304, October 1981. [4]

Southard, Jay R., "MacPitts: An Approach to Silicon Compilation," *IEEE Computer*, 74-82, December 1983. [4]

Spickelmier, Rick L. and Newton, A. Richard, "Wombat: A New Netlist Comparison Program," *ICCAD '83*, 170-171, September 1983. [5]

Sproull, Robert F. and Sutherland, Ivan E., *Asynchronous Systems II: Logical Effort and Asynchronous Modules*, to be published. [3]

Stallman, Richard M., "The Extensible, Customizable Self-Documenting Display Editor," Proceedings ACM SIGPLAN SIGOA Symposium on Text Manipulation, Portland Oregon, 147-156, June 1981. [10]

Stallman, R.M. and Sussman, G.J., "Forward Reasoning and Dependency Directed Backtracking in a System for Computer-Aided Circuit Analysis," *Artificial Intelligence*, 9:2, 135-196, October 1977. [8]

Stamos, James W., "A Large Object-Oriented Virtual Memory: Grouping Strategies, Measurements, and Performance," Xerox PARC SCG-82-2, May 1982. [3]

Steele, G. L. Jr., *The Definition and Implementation of a Computer Programming Language Based on Constraints*, PhD dissertation, Massachusetts Institute of Technology, August 1980. [3, 8]

Supowitz, Kenneth J. and Slutz, Eric A., "Placement Algorithms for Custom VLSI," Proceedings 20th Design Automation Conference, 164-170, June 1983. [4]

Sussman, Gerald Jay, "SLICES—At the Boundary between Analysis and Synthesis," AI Memo 433, Massachusetts Institute of Technology, 1977. [8]

Sussman, Gerald Jay and Steele, Guy Lewis, "CONSTRAINTS—A Language for Expressing Almost-Hierarchical Descriptions," *Artificial Intelligence*, 14:1, 1-39, August 1980. [3]

Sutherland, Ivan E., *Sketchpad: A Man-Machine Graphical Communication System*, PhD dissertation, Massachusetts Institute of Technology, January 1963. [3, 8]

Szabo, Kevin S. B.; Leask, James M.; and Elmasry, Mohamed I., "Symbolic Layout for Bipolar and MOS VLSI," *IEEE Transactions on CAD*, to appear, 1987. [2]

Takashima, Makoto; Mitsuhashi, Takashi; Chiba, Toshiaki; and Yoshida, Kenji, "Programs for Verifying Circuit Connectivity of MOS/LSI Mask Artwork," Proceedings 19th Design Automation Conference, 544-550, June 1982. [5]

Teig, Steven; Smith, Randall L.; and Seaton, John, "Timing-Driven Layout of Cell-Based ICs," *VLSI Systems Design*, VII:5, 63-73, May 1986. [4]

Terman, Christopher J. "RSIM—A Logic-level Timing Simulator," Proceedings IEEE International Conference on Computer Design, 437-440, October 1983. [6]

Thornton, Robert W., "The Number Wheel: A Tablet Based Valuator for Interactive Three-Dimensional Positioning," *Computer Graphics*, 13:2, 102-107, August 1979. [10]

Tilbrook, David M., *A Newspaper Pagination System*, Masters Thesis, University of Toronto Department of Computer Science, 1976. [10]

Tompa, Martin, "An Optimal Solution to a Wire-Routing Problem," Proceedings 12th Annual ACM Symposium on Theory of Computing, 161-176, 1980. [4]

Trimberger, Stephen, "Combining Graphics and A Layout Language in a Single Interactive System," Proceedings 18th Design Automation Conference, 234-239, June 1981. [8]

Trimberger, Stephen, "Automated Performance Optimization of Custom Integrated Circuits," *VLSI '83* (Anceau and Aas, eds), North Holland, Amsterdam, 99-108, August 1983. [5]

Turing, A. M., "Computing Machinery and Intelligence," *Mind*, 59:236, 433-460, October 1950. [8]

VanCleemput, W. M., "An Hierarchical Language for the Structural Description of Digital Systems," Proceedings 14th Design Automation Conference, 377-385, June 1977. [2]

Varner, Denise, "Color Avionics," unpublished manuscript. [10]

VLSI Design Staff, "A Perspective On CAE Workstations," *VLSI Design*, IV:4, 52-74, April 1985. [7]

VLSI Systems Design Staff, "1986 Survey of Logic Simulators," *VLSI Systems Design*, VII:2, 32-40, February 1986. [6]

VTI, *VLSI Design System*, VLSI Technologies Inc., 1983. [4]

Wang, Paul K.U., "Approaches to Hardware Acceleration of Circuit Simulation," Proceedings IEEE International Conference on Computer Design, 724-726, October 1985. [6]

Wardle, C. L.; Watson, C. R.; Wilson, C. A.; Mudge, J. C.; and Nelson, B. J., "A Declarative Design Approach for Combining Macrocells by Directed Placement and Constructive Routing," Proceedings 21st Design Automation Conference, 594-601, June 1984. [4]

Watanabe, Hiroyuki, *IC Layout Generation and Compaction Using Mathematical Optimization*, PhD dissertation, University of Rochester Computer Science Department, TR 128, 1984. [4]

Weinberger, A., "Large Scale Integration of MOS Complex Logic: A Layout Method," *IEEE Journal of Solid State Circuits*, 2:4, 182-190, 1967. [4]

Weinreb, Daniel and Moon, David, "Flavors: Message Passing in the Lisp Machine," MIT Artificial Intelligence Lab Memo #602, November 1980. [3]

Weste, Neil, "Virtual Grid Symbolic Layout," Proceedings 18th Design Automation Conference, 225-233, June 1981. [1, 2, 4, 8, 11]

Weste, Neil and Eshraghian, Kamran, *Principles of CMOS VLSI Design*, Addison-Wesley, Reading, Massachusetts, 1985. [2, 4]

Whelan, Daniel S., "A Rectangular Area Filling Display System Architecture," *Computer Graphics*, 16:3, 147-153, July 1982. [9]

Whitney, Telle, "A Hierarchical Design-Rule Checking Algorithm," *Lambda*, 2:1, 40-43, 1st Quarter 1981. [5]

Whitney, Telle, *Hierarchical Composition of VLSI Circuits*, PhD dissertation, California Institute of Technology Computer Science, report 5189:TR:85, 1985. [3]

Wilcox, C. R.; Dageforde, M. L.; and Jirak, G. A., *Mainsail Language Manual*, Version 4.0, Xidak, 1979. [8]

Williams, John D., "STICKS—A graphical compiler for high level LSI design," Proceedings AFIPS Conference 47, 289-295, June 1978. [1, 2, 4, 8, 11]

Wilmore, James A., "Efficient Boolean Operations on IC Masks," Proceedings 18th Design Automation Conference, 571-579, June 1981. [4, 5]

X3H3/83-25r3 Technical Committee, "Graphical Kernel System," *Computer Graphics* special issue, February 1984. [9]

Wing, Omar; Huang, Shuo; and Wang, Rui, "Gate Matrix Layout," *IEEE Transactions on CAD*, 4:3, 220-231, July 1985. [4]

Zippel, Richard, "An Expert System for VLSI Design," Proceedings IEEE International Symposium on Circuits and Systems, 191-193, May 1983. [3, 8]

INDEX